Biochemistry and Molecular Biology of Plant–Pathogen Interactions

PROCEEDINGS OF THE PHYTOCHEMICAL SOCIETY OF EUROPE

88038
M/L

PROCEEDINGS OF THE
PHYTOCHEMICAL SOCIETY OF EUROPE

Biochemistry and Molecular Biology of Plant–Pathogen Interactions

Edited by

C.J. SMITH

School of Biological Sciences,
University College of Swansea, UK

CLARENDON PRESS · OXFORD

1991

Oxford University Press, Walton Street, Oxford OX2 6DP

Oxford New York Toronto
Delhi Bombay Calcutta Madras Karachi
Petaling Jaya Singapore Hong Kong Tokyo
Nairobi Dar es Salaam Cape Town
Melbourne Auckland

and associated companies in
Berlin Ibadan

Oxford is a trade mark of Oxford University Press

Published in the United States
by Oxford University Press, New York

British Library Cataloguing in Publication Data

A catalogue record for this book is available
from the British Library

Library of Congress Cataloging in Publication Data
Biochemistry and molecular biology of plant–pathogen interactions
edited by C.J. Smith
— (Proceedings of the Phytochemical Society of Europe; 32)
Includes bibliographical references and index.
1. Plant–pathogen relationships—Molecular aspects.   I. Smith, C.J. (Christopher John),
1948–   II. Phytochemical Society of Europe.   III. Series.
SB732.7B56 1991
632'.3—dc20   90–14273 CIP
ISBN 0 19 857734 6 ✓

Typeset by
Promenade Graphics Ltd, Cheltenham
Printed and bound by
Biddles Ltd, Guildford and King's Lynn

Preface

In 1982 in his preface to the book *Biochemical Plant Pathology*, Professor James Callow wrote: 'Yet, plant breeding for resistance is presently carried out against a background in which the function of not one single gene for resistance is understood at the molecular level, and we do not yet have a molecular description of the key events in any plant disease comparable to that of certain animal diseases. We need to know for example, what the primary products of resistance genes are, what processes they control in the plant, how they interact with products of the pathogen, and where they are localized in the cell'. Such is the 'lead' time for planning meetings of the Phytochemical Society of Europe that only some four years after Professor Callow wrote these words the Committee of the Society felt sufficiently encouraged by progress made in this area to organize a meeting entitled 'Biochemistry and Molecular Biology of Plant–Pathogen Interactions'. The meeting took place in Norwich in April 1989 and this volume is a record of the papers which were presented there.

With the optimism of those who will no longer be committee members at the scheduled date of the meeting, the organizers selected speakers whose laboratories, in addition to contributing much to the mainstream of their particular research areas, were also involved in investigations the main rewards of which were yet to come. On more than one occasion this high-risk strategy caused a sleepless night, but spirits were sustained by the words of that great optimist Mr Micawber of *David Copperfield* who, when he was anticipating his release from the debtors' prison, said: 'I have no doubt I shall, please Heaven, begin to be beforehand with the world, and to live in a perfectly new manner, if—in short, if anything turns up'.

The choice of such a broad title for the meeting was deliberate, as was the choice of speakers. Viruses, bacteria, fungi, and nematodes were included as representative pathogens and the organizers did not confine themselves to an isolated view of the role of only the pathogen or the host in the interaction but were eager to present accounts of how both elements contributed. The principle that united all the speakers and their areas of research was the focusing of the newer molecular technologies on some of the problems in plant pathology that have held attention for many years now, including some of those highlighted by Professor Callow. For example, in considering the question 'What are the primary products of resistance?', it has not been of overriding concern whether the products of resistance examined were those of tobacco in response to tobacco mosaic virus (Chapter 6), or of *Brassica* in response to *Xanthomonas campestris* (Chapter 10); both provide elegant examples of the approaches that may be employed. Similarly, the question 'What processes do they (the primary products of resistance) control in plants?' is equally well addressed by the study of lipid peroxidation in membranes (Chapter 7) or the control of gene transcription in bean (Chapter 17). Nor has it been central in considering the problem of 'How do such products interact with the products of the pathogen?' whether the interaction is that between alfalfa mosaic virus and its hosts (Chapter 5) or of *Phytophthora infestans* with potato (Chapter 15); both illustrate the achievements made possible by the molecular approach. Throughout the volume there are examples that answer the question 'Where in the cell are they localized?', but what better (and novel) example of defining location than the response of potato to infection by cyst nematodes (Chapter 14)?

The driving force behind the diverse range of systems and responses included here has been the desire to define and understand at the molecular level the factors that determine the outcome, be it resistance or susceptibility, when host and pathogen come together. It is the application of the powerful modern techniques of molecular biology in an attempt to understand the complex biological phenomenon of host–pathogen interaction that forms the common theme of the book. In some cases emphasis has been placed on the host, in others the pathogen, whilst in some systems sufficient is now known to provide a description of events from both perspectives. If the reader feels the balance is wrong, that there are instances where too much attention has been paid to one aspect of an interaction at the expense of some other, then the fault lies not with the authors, but with those who selected the topics and ultimately with me as editor. The authors were asked to provide contributions in specific areas and to provide manuscripts that have both an element of review and some of the more recent results from their own laboratories. That they have made an excellent job of discharging their tasks will be apparent to the reader, and I am grateful to them.

I must also thank my co-conspirator who worked with so much energy and efficiency in making arrangements at Norwich, Dr Phil Mullineaux. In the annals of conference organizing his particular brand of 'carrot-and-stick' approach was truly awesome and the beer festival he arranged skilfully eradicated any doubts that delegates may have had concerning the success of the conference. I must also thank those members of the AFRC Institute of Plant Science Research, John Innes Institute, for their invaluable assistance; their efforts did much to ensure the smooth running of the meeting. In addition I am grateful to Advanced Technologies (Cambridge) Ltd, Agricultural Genetics Co., Ciba-Geigy Agrochemicals, and Imperial Chemical Industries (International Seeds) for financial and other support.

The success of this volume will be due to the authors. If there are errors they are mine. The reader may judge whether something did indeed turn up, but to borrow Mr Micawber's words again: 'In case of anything turning up (of which I am rather confident), I shall be extremely happy if it should be in my power to improve your prospects'. I hope this volume achieves that.

Swansea   C.J.S.
March 1991

Contents

Contributors

G.P. Accotto: Istituto di Fitovirologia Applicata del CNR, Strada delle Cacce 73, 10135 Torino, Italy.

H.J. Atkinson: Centre for Plant Biochemistry and Biotechnology, University of Leeds, Leeds LS2 9JT, UK.

J.A. Bailey: Department of Agricultural Sciences, University of Bristol, AFRC Institute of Arable Crops Research, Long Ashton Research Station, Bristol BS18 9AF, UK.

C.E. Barber: The Sainsbury Laboratory, John Innes Institute, Colney Lane, Norwich NR4 7UH, UK.

J.F. Bol: Department of Biochemistry, Leiden University, Gorlaeus Laboratories, Einsteinweg 5, 2333 CC Leiden, The Netherlands.

D.J. Bowles: Centre for Plant Biochemistry and Biotechnology, University of Leeds, Leeds LS2 9JT, UK.

A.P. Brown: Department of Biological Sciences, University of Durham, Science Laboratories, South Road, Durham DH1 3LE, UK.

A.K. Chatterjee: Department of Plant Pathology, 108 Waters Hall, University of Missouri, Columbia, MO 65211, USA.

Z. Chen: Department of Biological Sciences, Purdue University, West Lafayette, Indiana 47907, USA.

A.D. Choudhary: Plant Biology Division, The Samuel Roberts Noble Foundation, PO Box 2180, Ardmore, Oklahoma 73402, USA.

D. Collinge: The Sainsbury Laboratory, John Innes Institute, Colney Lane, Norwich NR4 7UH, UK.

A. Collmer: Department of Plant Pathology, Cornell University, Ithaca, New York 14853–5908, USA.

J. Conrads-Strauch: The Sainsbury Laboratory, John Innes Institute, Colney Lane, Norwich NR4 7UH, UK.

G. Creissen: Department of Virus Research, John Innes Institute, AFRC Institute of Plant Science Research, Colney Lane, Norwich NR4 7UH, UK.

K.P. Croft: Department of Applied Biology, University of Hull, Hull HU6 7RX, UK.

M.J. Daniels: The Sainsbury Laboratory, John Innes Institute, Colney Lane, Norwich NR4 7UH, UK.

R.A. Dixon: Plant Biology Division, The Samuel Roberts Noble Foundation, PO Box 2180, Ardmore, Oklahoma 73402, USA.

J.M. Dow: The Sainsbury Laboratory, John Innes Institute, Colney Lane, Norwich NR4 7UH, UK.

D. Evans: Department of Biological Sciences, University of Warwick, Coventry CV4 7AL, UK.

C.S. Garrett: Department of Biological Sciences, University of Durham, Science Laboratories, South Road, Durham DH1 3LE, UK.

C.L. Gough: The Sainsbury Laboratory, John Innes Institute, Colney Lane, Norwich NR4 7UH, UK.

S.J. Gurr: Centre for Plant Biochemistry and Biotechnology, University of Leeds, Leeds LS2 9JT, UK.

K. Hahlbrock: Max-Planck-Institut für Züchtungsforschung, Abteilung Biochemie, Carl-von-Linné-Weg 10, D–5000 Köln 30, Germany.

K.E. Hammond-Kosack: The Sainsbury Laboratory, John Innes Institute, Colney Lane, Norwich NR4 7UH, UK.

M.J. Harrison: Plant Biology Division, The Samuel Roberts Noble Foundation, PO Box 2180, Ardmore, Oklahoma 73402, USA.

K. Hinze: Max-Planck-Institut für Züchtungsforschung, Abteilung Biochemie, Carl-von-Linné-Weg 10, D–5000 Köln 30, Germany.

C.L. Holness: Department of Virus Research, John Innes Institute, AFRC Institute of Plant Science Research, Colney Lane, Norwich NR4 7UH, UK.

M.J. Huisman: MOGEN International NV, Einsteinweg 97, 2333 CB Leiden, The Netherlands.

S.M. Jenkins: Plant Biology Laboratory, The Salk Institute for Biological Studies, PO Box 85800, San Diego, California 92138, USA.

J.E. Johnson: Department of Biological Sciences, Purdue University, West Lafayette, Indiana 47907, USA.

E. Kombrink: Max-Planck-Institut für Züchtungsforschung, Abteilung Biochemie, Carl-von-Linné-Weg 10, D–5000 Köln 30, Germany.

C.J. Lamb: Plant Biology Laboratory, The Salk Institute for Biological Studies, PO Box 85800, San Diego, California 92138, USA.

M.A. Lawton: Plant Biology Laboratory, The Salk Institute for Biological Studies, PO Box 85800, San Diego, California 92138, USA.

H.J.M. Linthorst: Department of Biochemistry, Leiden University, Gorlaeus Laboratories, Einsteinweg 5, 2333 CC Leiden, The Netherlands.

G.J. Loake: Department of Biological Sciences, University of Durham, Science Laboratories, South Road, Durham DH1 3LE, UK.

G.P. Lomonossoff: Department of Virus Research, John Innes Institute, AFRC Institute of Plant Science Research, Colney Lane, Norwich NR4 7UH, UK.

J.L. McEvoy: Department of Plant Pathology, 108 Waters Hall, University of Missouri, Columbia, MO 65211, USA.

A.J. Maule: Department of Virus Research, John Innes Institute, AFRC Institute of Plant Science Research, Colney Lane, Norwich NR4 7UH, UK.

R.L.J. Meuwissen: Department of Biochemistry, Leiden University, Gorlaeus Laboratories, Einsteinweg 5, 2333 CC Leiden, The Netherlands.

D.E. Milligan: The Sainsbury Laboratory, John Innes Institute, Colney Lane, Norwich NR4 7UH, UK.

P.M. Mullineaux: Department of Virus Research, John Innes Institute, AFRC Institute of Plant Science Research, Colney Lane, Norwich NR4 7UH, UK.

H. Murata: Department of Plant Pathology, 108 Waters Hall, University of Missouri, Columbia, MO 65211, USA.

A.E. Osbourn: The Sainsbury Laboratory, John Innes Institute, Colney Lane, Norwich NR4 7UH, UK.

R. Parra: The Sainsbury Laboratory, John Innes Institute, Colney Lane, Norwich NR4 7UH, UK.

H.V. Reynolds: Department of Virus Research, John Innes Institute, AFRC Institute of Plant Science Research, Colney Lane, Norwich NR4 7UH, UK.

M. Schröder: Max-Planck-Institut für Züchtungsforschung, Abteilung Biochemie, Carl-von-Linné-Weg 10, D–5000 Köln 30, Germany.

C. Scollan: Centre for Plant Biochemistry and Biotechnology, University of Leeds, LS2 9JT, UK.

M. Shanks: Department of Virus Research, John Innes Institute, AFRC Institute of Plant Science Research, Colney Lane, Norwich NR4 7UH, UK.

C.H. Shaw: Department of Biological Sciences, University of Durham, Science Laboratories, South Road, Durham DH1 3LE, UK.

A.J. Slusarenko: Institut für Pflanzenbiologie, Zollikerstrasse 107, CH–8008 Zürich, Switzerland.

C.J. Smith: Biochemistry Research Group, School of Biological Sciences, University College of Swansea, Singleton Park, Swansea SA2 8PP, UK.

C.V. Stauffacher: Department of Biological Sciences, Purdue University, West Lafayette, Indiana 47907, USA.

B.A. Stermer: Plant Biology Division, The Samuel Roberts Noble Foundation, PO Box 2180, Ardmore, Oklahoma 73402, USA.

J.L. Tang: The Sainsbury Laboratory, John Innes Institute, Colney Lane, Norwich NR4 7UH, UK.

A. Vivian: Science Department, Bristol Polytechnic, Coldharbour Lane, Frenchay, Bristol BS16 1QY, UK.

C.R. Voisey: Department of Applied Biology, University of Hull, Hull HU6 7RX, UK.

K.R. Wood: School of Biological Sciences, University of Birmingham, Edgbaston, Birmingham B15 2TT, UK.

C.J. Woolston: Department of Applied Biology, University of Hull, Hull HU6 7RX, UK.

L. Yu: Plant Biology Division, The Samuel Roberts Noble Foundation, PO Box 2180, Ardmore, Oklahoma 73402, USA.

1 Spread of viruses within plants

ANDY J. MAULE

Department of Virus Research, John Innes Institute, AFRC Institute of Plant Science Research, Colney Lane, Norwich NR4 7UH, UK

Introduction

In this article I shall give an overview of our current understanding of the way viruses move around plants to establish a systemic infection. This topic is currently attracting attention and has been the subject of several reviews (e.g. Atabekov and Dorokhov 1984; Hull 1989). My aim has been to select some key examples which represent our current knowledge of the complex structural and biochemical interaction that occurs between virus and the host plant, and which results in virus spread, and to suggest some experiments which might advance our understanding. I have also taken the opportunity to speculate as to what might be the mechanism(s) which assists virus movement.

The importance of virus movement as an area for study lies in the fact that it is a process which is fundamental to all but a very few plant virus infections. Clearly, a virus which is able to replicate in a host cell but is unable to move to neighbouring uninfected cells will not establish a systemic infection. One implication of this is that virus spread must be a target for the genetic control of virus disease that may be achieved through natural selection, plant breeding or, in future, genetically engineered virus resistance. It is possible that the first of these has resulted in so-called subliminal virus infections (e.g. TMV in cotton or cowpea; Sulzinski and Zaitlin 1982) where single cells become infected but the infection progresses no further, and no symptoms are apparent.

Type of virus spread

In describing virus spread it is convenient to divide the phenomenon into sectors. Whether these divisions are justified and represent real differences in the mechanisms which operate is a matter for debate, but they do provide a way to dismantle the structural complexity of the host plant into manageable proportions.

When a virus is introduced into the leaf epidermal or mesophyll tissues of a susceptible plant by mechanical abrasion or insect vector, it will establish a focus of infection from which it invades neighbouring cells through a process of 'local spread'. The enlarging infection site eventually encounters the vascular tissues through which the virus may be moved to the more distal tissues of the plant. This latter process has been loosely described as 'long distance' spread and has received little attention experimentally, probably because of the complexity of the cell and tissue types involved. However, in terms of rapidity of spread around the plant, long-distance movement is probably of prime importance, in fact, local cell-to-cell spread may serve only to transport the virus to and from the vascular tissues. It could be argued further that cell-to-cell spread loses significance if the virus reaches the meristem because all subsequent cells could be infected through cell division, and local spread would be important only at the earliest times after infection.

During local spread, the virus or its nucleic acid (see later) must cross the cell wall from infected to uninfected cell. The only channels known for this passage are the plasmodesmata, and so I will devote some space to a description of the structure and function of these intercellular elements, and how they may be affected during virus infection.

Plasmodesmata

Plasmodesmata traverse the cell wall and provide continuous symplastic connections throughout the tissue. The physical organization of these channels has largely been deduced from ultrastructural analyses of fixed tissue in the electron microscope. However, a structure has been described (Fig. 1.1; for reviews see: Robards 1976; Gunning and Overall 1983; Esau and Throsch 1985) that appears to be generally accepted. In healthy leaf mesophyll tissue, plasmodesmata are plasmalemma-lined channels approximately 20–30 nm in diameter, encasing an axial membranous element, the desmotubule, which is believed to be derived from the endoplasmic reticulum. The space between the two membranes is about 5 nm and may contain tubules. Movement of soluble material between cells probably occurs through this space and possibly within the tubules.

It is clear, from a consideration of the internal dimensions of the plasmodesma, that large macromolecular structures such as organelles, free ribosomes, etc., cannot pass from one cell to the next. From elegant experiments (e.g. Terry and Robards 1987; Baron-Epal *et al.* 1988) in which the movement between cells of fluorescent molecules of defined size was recorded, we know that plasmodesmata have a molecular exclusion limit, or gating capacity, of only 800–1000 daltons (0.8–1.0 kDa). The following examples illustrate different approaches to the problem. First, trichomes from Abutilon were microinjected with soluble fluors of different size and the migration of the marker to adjacent non-injected cells was

(a)

(b)

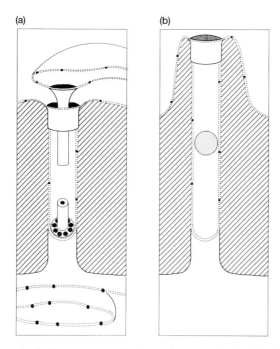

Fig. 1.1. Hypothetical structure of plasmodesmata in healthy (a) and virus-infected (b) tissues. The plasmodesma is a plasma membrane-lined channel which passes through the cell wall (hatched) and contains an axial element, the desmotubule, derived from endoplasmic reticulum. To accommodate virus particles (b) the overall dimension of the plasmodesma is changed and the desmotubule lost.

recorded (Terry and Robards 1987). Second (Baron-Epel *et al.* 1988), suspension-culture cells were allowed to take up fluorescein-labelled esters through the plasma membrane; free fluor which was released by endogenous esterase activity was retained within the cells. Small groups of cells were selected subsequently, and one central cell photobleached with a laser. FRAP (fluorescent recovery after photobleaching) measurements showed that only small molecules could restore fluorescence in the bleached cell. In addition to the movement of water-soluble molecules, these latter experiments showed that lipid-soluble molecules also could migrate between adjacent cells. Experiments to test the movement of larger membrane-associated molecules have not yet been reported. Whilst the experiments described above do not support the idea that plasmodesmata are the routes for virus spread in plants, in the absence of specific secretion and uptake mechanisms, both of which are essential to the systemic invasion of animal tissues with viruses, they are the only option. Therefore, it must be concluded that plasmodesmata are modified to allow the passage of virus particles or virus nucleic acid, both of which have molecular weights in the range 10^6–10^7 i.e. three to four orders of magnitude larger than the normal gating capacity of plasmodesmata in uninfected tissues. At present the detailed structural changes which must take place to achieve this functional modification are not known, but from observations of the ultrastructure of cauliflower mosaic virus (CaMV) particles within plasmodesma it would appear that, in addition to an increase in the overall dimensions of the plasmodesmata the desmotubule is no longer present. Figure 1.2 shows electron micrographs of plasmodesma in both healthy and CaMV-infected turnip leaves. In healthy leaves the structures are visible as channels approximately 20 nm in diameter which traverse the cell wall, either without deviation, or via a reticulate network of channels within the middle lamella. After infection the plasmodesmata show extensive modification. The width of the structure is increased such that it is able to accommodate virus particles, and the plasmodesmal tube is increased in length so that it invades the space of the two neighbouring cells by up to 1μm in each direction, distorting their plasma membrane. The micrographs indicate that the tube extensions are probably associated with new cell wall synthesis.

Several important questions arise out of these observations:

1. How is this modification of plasmodesmata achieved?
2. Do modified plasmodesmata remain open throughout the infection?
3. How does the plant continue to function normally with ungated plasmodesmata?
4. Is the breakdown in the control of intercellular movement the key to symptom expression?

Answers to these questions are not yet available, although some progress has been made; virus movement has been shown to be under the control of a virus-coded protein which has become a focus of current research.

The 30 kDa protein of tobacco mosaic virus controls virus spread

A temperature-sensitive mutant (Ls1) of TMV has been identified which, at the non-permissive temperature, does not establish a systemic infection in host plants, but does replicate normally in host protoplasts (Nishiguchi *et al.* 1978). Hence, the second fundamental requirement for establishment of a successful systemic infection, virus movement, is defective. Analysis of the mutant genome and its protein product showed that the mutation lay within the coding region for the 30 kDa product of TMV (Leonard and Zaitlin 1982; Meshi *et al.* 1987). Restoration of a normal infection could be achieved either by returning the plants to the permissive temperature, or

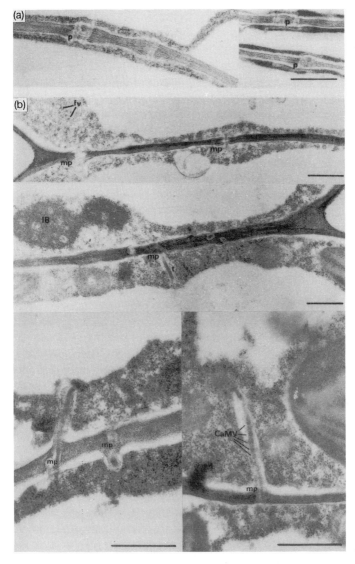

Fig. 1.2. Plasmodesmata in leaf tissue of healthy (a) and CaMV-infected (b) turnip. Modified plasmodesmata are indicated (mp), some show an extended tubular structure (t) and some contain virus particles (CaMV). Virus inclusion body (IB), free virus (fv), plasmodesmata (p). Bar markers represent 500 nm.

by complementation of the defective gene function in transgenic host plants that constitutively express the wild type 30k protein (Deom *et al.* 1987). Although these experiments implicate the 30k protein in virus movement they do not show that the virus product alone is sufficient to bring about the modification in plasmodesmal structure. Evidence for a direct interaction between the 30k protein and plasmodesmata comes from immunogold cytochemical studies (Tomenius *et al.* 1987) which showed that in recently infected tissues the protein was localized within the plasmadesmal channel.

Evidence for determinants of spread from other viruses

The definitive demonstration of the involvement of the TMV 30k protein in controlling spread was followed by the identification of homologues from closely and distantly related viruses, although at present the same range of evidence is not available to support their involvement.

Sequence analysis

The first evidence obtained for other systems came from amino acid sequence analysis. Computer assisted sequence analysis showed some similarity between the '30k' products from a range of sequenced tobamoviruses and the potential product (Pl) of caulimovirus gene I (Hull and Covey 1985; Hull *et al.* 1986; Richins *et al.* 1987). The sequences within the caulimovirus group were well conserved and showed only the same degree of difference from the tobamoviruses as the tobamoviruses showed differences between themselves (Hull *et al.* 1986). The homology between the two groups is concentrated within the central region of both proteins, and broadly corresponds with a region of identified homology between the TMV 30k and the 29 kDa proteins of the tobraviruses (Boccara *et al.* 1986; MacFarlane *et al.* 1989). Tobacco vein mottling virus (Dormier *et al.* 1987) and tomato black ring virus (Meyer *et al.* 1986), representatives of two other virus groups, have been shown also to have potential protein products related to the TMV 30k protein. Sequence homology-based groupings of putative spread-proteins from other viruses (Hull 1989) have been identified after their participation in this role had been indicated by other evidence. This is most notable for the 3A protein of the tricornaviruses, since a role in spread has been indicated for the alfalfa mosaic virus (AlMV) product from its subcellular location in the cell wall (see later).

Mutational analysis

Stronger, independent evidence for a spread function comes from mutational analysis, particularly when it is supported by evidence for competent replication in single cells. Hence, comparative experiments between plants and protoplasts have identified viral genomic components which code for spread-proteins such as RNA3 for AlMV (Nassuth *et al*. 1981; Huisman *et al*. 1986) and brome mosaic virus (BMV) (Kiberstis *et al*. 1981), middle component RNA (MRNA) for cowpea mosaic virus (CPMV; Goldbach *et al*. 1980), RNA2 for tomato blackring virus (TBRV; Robinson *et al*. 1980), RNA2 for red clover necrotic mosaic virus (RCNMV; Osman and Buck 1987), and DNA B for the bipartite geminiviruses, African cassava mosaic virus (ACMV; Townsend *et al*. 1986) and tomato golden mosaic virus (TGMV; Rogers *et al*. 1986). In the latter (i.e. TGMV), transgenic plants containing DNA A integrated in the host genome were substituted for protoplasts.

More detailed mutational evidence has been published only for the geminiviruses. For the bipartite geminiviruses there is good evidence (Etessami *et al*. 1988; Hayes *et al*. 1988) that the two protein products from DNA B are involved in controlling spread. The functional differences between these products is not understood. A different situation exists for the monopartite geminiviruses which have no DNA B component. Mutation of the coat protein gene (V2) of maize streak virus (MSV; Lazarowitz *et al*. 1989; Boulton *et al*. 1989), wheat dwarf virus (WDV; Woolston *et al*. 1989), and beet curly top virus (BCTV; Briddon *et al*. 1989) prevents complete systemic infection of host plants but replication in protoplasts (WDV; BCTV) or at the site of inoculation (MSV) can still occur. For MSV and BCTV, these effects are associated with diminished accumulation of the single-stranded DNA, the DNA form normally found in virus particles, and may represent a requirement in this virus subgroup for encapsidation for effective movement. Interestingly, encapsidation is also of importance for the long distance movement of some RNA viruses (see later).

Mutational analyses of the virion-sense genes of MSV (Lazarowitz *et al*. 1989; Boulton *et al*. 1989) have implicated a second gene product (V1 product; 10.9 kDa) in the control of spread. However, the technique of 'agroinfection' which provides an *Agrobacterium*-mediated delivery of the viral genome directly to the meristematic region of the host, makes it difficult to distinguish between cell-to-cell spread and propagative spread through cell division.

Verification that the 29 kDa protein from tobacco rattle virus (TRV) is a functional homologue of the TMV 30k protein, has been obtained by demonstrating TMV complementation of frameshift mutations in the cistron coding for the 29 kDa protein from TRV (V. Ziegler-Graf and D. Baulcombe, personal communication).

For CaMV, an alternative approach to genetic modification has supported the evidence from sequence homology comparisons for the role of Pl. *In vitro* and *in vivo* recombination has been used to map the genetic determinants for symptom variation between two CaMV isolates, Cabb B-JI and Bari (Stratford and Covey 1989). These isolates varied with respect to the speed of establishment of a systemic infection, and both the degree of chlorosis and stunting induced in infected tissues. The rate of systemic infection mapped to the most well defined locus, at the NH_2-end of Pl, where the difference between the two isolates was limited to just eight out of 139 amino acids, six of which represented familial alterations.

Localization of the spread proteins

As already stated, the TMV 30k product is associated with plasmodesmata in tissues recently infected. Using immunogold cytochemistry, similar locations were identified for the '48k' products of RCMV (Shanks *et al.* 1989) and CPMV (van Lent *et al.* 1990) and for Pl of CaMV (Linstead *et al.* 1988), results which are consistent with these products having a role in cell-to-cell spread. The location of the RNA 3A product of AlMV in inoculated leaves was around the edge of the expanding infection site, where it was restricted to the middle lamella region of the cell wall, but was not specifically associated with plasmodesmata (Stussi-Garaud *et al.* 1987).

Implication in spread by default

Since all transmissible viruses require some form of spread as part of their life cycle it has become generally believed that they will all code for a 'spread protein'. Hence, a cistron coding for a protein to which no function has been assigned, particularly if it occurs on the 5' side of a gene for a structural protein and on a separate genomic segment from the replication proteins or subgenomic RNA, may well draw speculation as to the role of its protein product in virus movement. With time, an increase in the number of examples which break the pattern set by TMV, AlMV, etc., will discourage such speculation. Already, the occurrence of the tobravirus putative spread determinant on the large genomic RNA and the implication of multiple gene products for the geminiviruses indicate that no universal organizational pattern for the expression of these proteins exists.

Function of the spread proteins

Clues as to how the cell-to-cell spread proteins effect the movement of viruses through the tissue may be obtained in various ways: their locations indicate a site of action, their patterns of accumulation in relation to virus replication and location may provide a measure of their longevity; examination of the proteins themselves could provide information on post-translational modification, structure and possibly 'activity'; and

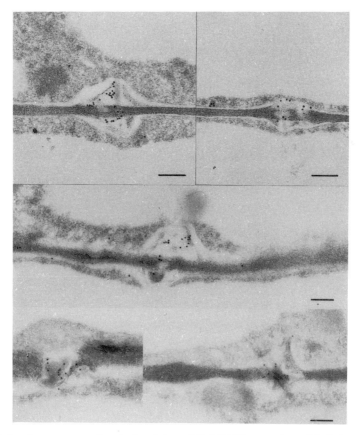

Fig. 1.3. Immunocytochemical staining of CaMV P1 around modified plasmodesmata. Bar markers represent 200 nm.

comparisons of their amino acid sequence with protein databases could identify potential biochemical functions.

Detection and location

As described above the identification of plasmodesmata and/or cell walls as subcellular locations has, in some cases, been considered as a key piece of evidence for the role of putative spread proteins. The location of CaMV P1 in tissue pieces at the edge of chlorotic lesions in inoculated leaves at approximately three weeks post-inoculation (Fig. 1.3) suggests that the product does not constitute a structural component of the plasmodesmal channel, but occurs in the surrounding modified cell wall region, i.e. at an extracellular location (Linstead *et al.* 1988). This is an unexpected position since the primary amino acid sequence does not show a hydrophobic NH_2-terminus characteristic of a signal peptide for protein secretion. It is possible that Pl is relocated by subsequent modification after an initial interaction with plasmodesmata. With TMV (Tomenius *et al.* 1987) and the comoviruses, RCMV (Shanks *et al.* 1989) and CPMV (van Lent *et al.* 1990), the location was within the channel. It is not clear whether this reflects a different function or just a difference in sampling time because in the latter cases tissue was taken at earlier times after infection.

That there is a temporal effect on accumulation of the transport proteins has become apparent from several studies and recent work (Blum *et al.* 1989) with TMV-infected protoplasts would suggest that the effect is achieved through a rapid turnover, rather than specific regulation of expression of the 30k protein. So, although accumulation of TMV 30k protein is transient in infected protoplasts (Ooshika *et al.* 1984; Watanabe *et al.* 1986; Blum *et al.* 1989) and in plants early after infection, (Joshi *et al.* 1983; Tomenius *et al.* 1987) expression in protoplasts treated with host transcriptional inhibitors was continuous throughout the replication cycle.

The AlMV 3A protein was located in cell walls only at the outer edge of the expanding lesion and it disappeared once all the surrounding cells were infected (Stussi-Garaud *et al.* 1987). However in extracts of cell wall taken from infected leaves it appeared to remain throughout the infection (Godefroy-Colburn *et al.* 1986). This apparent contradiction was reconciled on the basis that in the earlier work the averaged accumulation within the leaf was being measured (Stussi-Garaud *et al.* 1987). The 3A protein was also detected transiently in a crude membrane fraction from infected tobacco leaves (Berna *et al.* 1986).

A time course of CaMV Pl accumulation in a series of newly-formed systemically-infected leaves of turnip showed that Pl could be detected sequentially, in a high speed pellet fraction, a fraction enriched for viroplasms, and a cell wall fraction. The protein was present only transiently in the former two (Albrecht *et al.* 1988). A problem with this approach is that it is not possible to relate the observations to the development of virus infection. We have taken an alternative approach (Maule *et al.* 1989) and have assessed accumulation of CaMV products in all the leaves of a systemically infected plant. Western blots of the same high speed pellet fraction as above, probed with an antiserum to a β-gal-Pl fusion protein, indicated that Pl was most abundant in tissues that were incompletely infected, or in tissues of the meristematic region, where invasion of new tissues was presumably necessary. The distribution of the three Pl-specific products 46, 42, and 38 kDa detected in these and previous experiments (Harker *et al.* 1987), also varied with leaf age. The youngest leaves contained predominantly the 46 kDa product, the largest systemically-infected leaves the smaller two products and the inoculated leaves equal proportions of the 46 and 38 kDa products. The possibility that these products may serve separate functions was raised.

In these experiments we were only able to detect CaMV Pl in the high speed pellet fraction where it was presumably attached to some particulate material. Recently, using an antiserum raised to Pl expressed in insect cells via a baculovirus vector (see later), we have also detected Pl products corresponding to those detected previously, both in viroplasm preparations and crude extracts. A comparison of the concentration of Pl in the different fractions suggests that Pl is selectively purified in the high speed pellet fraction. The components of this fraction are not completely characterized but it does contain CaMV DNA replication complexes (Thomas *et al.* 1985) and polyribosomes. Since experiments with immunogold cytochemistry have already identified the cell wall as one site of Pl accumulation, then either the total amount committed to cell walls is small, or the fractionation procedure co-incidentally purified cell wall-derived components in this fraction. Multiple products were also detected for AlMV 3A protein in the cell wall fraction (Godefroy-Colburn *et al.* 1986).

Two reports link putative spread-proteins with nuclei; the TMV 30k protein was located in a nuclear fraction from infected protoplasts (Watanabe *et al.* 1986), but not whole plants (Tomenius *et al.* 1987), and immunogold localization of the cucumber mosaic virus (CMV) 3A protein showed nucleoli to be the site of accumulation in infected tobacco leaves (MacKenzie and Tremaine 1988); the significance of these observations is not understood.

Characterization of the spread proteins

None of the spread proteins has yet been purified in its native state, from either infected protoplasts or intact tissue, so the alternative approach of characterizing the proteins expressed in heterologous cell systems, has been pursued. The only studies carried out of native proteins relate to interpretation of mobility after SDS-PAGE, and an observation that the TMV 30k protein in protoplasts is phosphorylated (D. Zimmerman, personal communication).

As noted before the movement proteins of A1MV and CaMV occur as multiple forms. The CaMV (Gordon *et al.* 1988), CPMV (Wellink *et al.* 1986), and RCMV (Shanks *et al.* 1989) transport proteins migrate on SDS-PAGE gels with larger M_rs than their theoretical values. However when the proteins are obtained through *in vitro* translation either of *in vitro*-generated transcripts, or viral RNA, they show corresponding migrations which argues against a shift in molecular weight attributable to post-transnational processing.

From a detailed analysis of the distribution of Pl of the tissues of CaMV-infected turnips (Maule *et al*. 1989) it was apparent that Pl was not an abundant protein, making further study of the structural and biochemical characteristics of the native protein very difficult. In an attempt to learn more about the nature of CaMV Pl we have obtained high-level and productive expression of the protein in insect cells through the use of baculovirus vector. This expression system was selected because it has been reported to allow faithful post-translational processing of the expressed protein (for reviews see Luckow and Summers 1988; Cameron *et al*. 1989). This expression was achieved using a cloned fragment of CaMV DNA encompassing gene I that was mutagenized to introduce a Bam Hl site at a position either 8 or 18nt upstream of the ATG of gene I. A Bam Hl fragment starting at either of these sites and including all of gene I and the 5' third of CaMV gene II, was introduced into the baculovirus transfer vector pAcRP18 (Matsuura *et al*. 1986). This vector contains the EcoR1-I fragment, from the baculovirus *Autographa californica* nuclear polyhedrosis virus (AcNPV), from which a portion of the polyhedrin coding region had been removed. Transfection of an insect cell line (Sf21) with the transfer vector and wild type AcNPV DNA allowed isolation of seven AcNPV recombinants, each with a polyhedrin -ve phenotype. Of these seven, two (one of each of the two Bam Hl modifications) expressed Pl. The level of expression for both recombinants was high, perhaps only two or three times less than polyhedrin itself, but was different for the two recombinants; that with 18 non-coding upstream nucleotides expressed at the higher level. Pl-specific products were identified using the antiserum to the β-gal-Pl fusion protein. The major product corresponded with the largest Pl product from infected plants. Smaller products were always present but their concentration was greatly reduced in cells analysed immediately after harvesting. Analysis of Pl produced in insect cells showed that it was highly sensitive to the presence of divalent cations *in vivo*, although less sensitive in a semi-purified state. We have not yet established whether the smaller Pl products had the same identity as those detected in infected plants.

By taking advantage of the fact that it was highly insoluble, Pl purification was achieved after removal of contaminant material with high concentrations of urea. During subcellular fractionation the insoluble Pl co-sedimented with nuclei initially, and was obtained in a substantially pure form from nuclear sonicates. In tests of solubility it was shown that the protein was insoluble under all but completely-denaturing conditions i.e. 6 M guanidine HCl or 1 per cent sodium dodecyl sulphate. The cause for this extreme insolubility is not known, but it is probably a reflection of a property of the protein itself rather than the expression system since Pl from cells expressing at a lower level, showed the same solubility characteristics. Analysis of the linear sequence of amino acids in Pl does not indicate the presence of a trans-membrane domain or other protein component which might require a strongly hydrophobic environment. An assessment of the three dimensional distribution of charged and uncharged amino acids is not possible at this time.

In similar work (C. Stussi-Garaud, personal communication), the AlMV 3A protein has been expressed in yeast. Here too the protein could only be solubilized using denaturing conditions i.e. 6 M urea. In addition attempts to extract the TMV 30k protein from cell walls of transgenic tobacco plants were most successful when guanidine HCl (6 M) and dithiothreitol were used in the extraction buffer (K. Schubert, C. M. Deom, and R. N. Beachy, personal communication). Hence, it would appear that low solubility may be a general characteristic of this class of proteins.

In future some clues as to the structure of CaMV Pl may be obtained from an analysis of the insoluble Pl product. Studies with immunogold cytochemistry using the electron microscope have identified the insoluble Pl in Sf21 cells as aggregates with a fibrillar appearance in longitudinal section, and tubular appearance in cross section (Fig. 1.4), and which occur predominantly but not exclusively in the cytoplasm. It will be interesting to see if purified Pl protein refolded after complete denaturation, aggregates in the same manner.

Investigations of post-translational phosphorylation and glycosylation have shown that Pl in insect cells is not phosphorylated. Cells infected with the recombinant AcNPV and with large amounts of ^{32}P-ortho-phosphate did not incorporate radioactivity into Pl but did so into other proteins. This is in contrast to the observation noted earlier that the TMV 30k protein is phosphorylated in tobacco protoplasts. Preliminary data based upon chemical deglycosylation indicate that Pl is not glycosylated, but this aspect needs to be investigated more thoroughly. No information is available yet for the TMV 30k protein. Experiments which are directly comparable for the two systems need to be carried out before drawing conclusions from these observations, but a difference in the post-translational processing for the different proteins is indicated.

Comparative sequence analysis

Attempts have been made both within and between virus groups to identify common structural features on the linear amino acid sequence of the movement proteins. Comparisons that were made between the 30k proteins of the tobamoviruses (Saito *et al.* 1988) identified two regions (I and II) that were particularly well conserved, suggesting that these regions were 'directly important to the functioning of the 30k protein'.

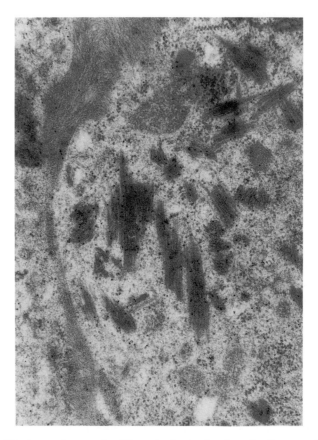

Fig. 1.4. Electron micrograph of the cytoplasm of an insect cell expressing CaMV P1. P1 has been immunogold-labelled and is visible as fibrillar aggregates with a tubular appearance in cross-section.

The region I contained a short hydrophobic sequence flanked by Gly and Asp, and which it was suggested, could be similar to related sequences found in nucleotide-binding protein. Sequence similarity across regions I and II was particularly strong with analogous proteins in the tobraviruses but not with the spread proteins of the caulimoviruses (Saito *et al.* 1988; Hull 1989), although as noted before there is homology between TMV 30k and caulimovirus Pl proteins, mainly between the central region of Pl and the area between regions I and II of TMV (Hull 1989).

Attempts have also been made to relate sequence similarities with potential biochemical activities or functions by using the spread proteins as probe sequences for the protein databases. Sequence homology between CaMV Pl and the ATP-binding site of some kinases, or the Mg^{2+}-ATP-binding sites of other proteins (Martinez-Izquierdo *et al.* 1987), and that between the TMV 30k protein and kinases (Saito *et al.* 1989), has been noted. It has also been suggested that CaMV Pl could be a ribonucleotide reductase (Martinez-Izquierdo *et al.* 1987). There are homologies also between the TMV 30k or CaMV Pl proteins and a variety of proteins including, a yeast cob intron 4 protein of unknown function (Zimmern 1983; Hull *et al.* 1986; Martinez-Izquierdo *et al.* 1987), plastocyanin (Hull *et al.* 1986) and two regions within the polyprotein of the picornavirus, hepatitis A. In this last case, the homologies occur at the C-terminus of the 2C protein (function unknown) and within the 3A protein (Vpg precursor.) Whether any of these sequence relationships really relate to biochemical activities or functions must remain a matter for speculation in the absence of any additional experimental evidence. Many of the similarities identified are not strong and because of the limitation of the current computing software, must relate only to the linear sequence of amino acids and not the three dimensional protein conformation. There is a risk that analyses like these which are both rapid and relatively straightforward for the user, can be misleading, and yet they can also generate a direction to the further study of a protein. For example, experiments to test the capacity of the TMV 30k or CaMV Pl protein to bind ATP or Mg^{2+}-ATP will be carried out in future, and probably, the proteins will also be tested for ribonucleoside reductase activity. It is interesting that in the comparisons between CaMV Pl and the protein database, the yeast cob intron 4 protein and the hepatitis A protein both achieved higher scores than the TMV 30k protein. Does Pl have a function related to that of both of these proteins, or do these proteins have a function related to that necessary to achieve virus spread, or neither?

'The spread function'

It is fair to say that there is no definitive evidence in favour of any specific function for the spread proteins. Atabekov and Dorokov (1984) suggested two possible mechanisms for virus spread. Firstly, interference with the normal gating processes of the plasmodesmata and secondly, a suppression of those host responses that normally prevent virus movement. Association between the spread proteins from several virus groups, and plasmodesmata or cell wall components, suggests the first option, whereas the location of spread proteins in nuclei (Watanabe *et al.* 1986; MacKenzie and Tremaine 1988) was considered to be compatible with the latter hypothesis. It may also be that the process of virus movement is a complex phenomenon involving interplay of these and other options.

If the modification of plasmodesmata which allows virus movement were simply a result of direct interaction between the movement protein and some component of that structure then transgenic plants expressing TMV 30k protein (Deom *et al.* 1987) should show the same modification in the absence of infection. Experiments using fluorescent molecules of varying sizes have been carried out in these plants and have demonstrated an increase in the gating capacity of plasmodesmata (Wolf *et al.* 1989) and an association between the 30k protein and the plasmodesmata in the absence of TMV infection (D. Atkins, R. Hull, K. Roberts, and R. N. Beachy, personal communication). Similarly, tobacco plants transgenic for the AlMV 3A cistron accumulate the 3A protein in cell walls as do infected plants (Loesch-Fries *et al.* 1988). Neither of these transgenic plants show any serious deleterious effects from constitutive spread protein expression, indicating that the consequence of their 'activity' cannot be typical virus-induced symptoms. However, until more extensive experiments are carried out it cannot be categorically stated that plasmodesmata are modified to the same extent as they are in infected plants.

Evidence possibly implicating some other virus factor in controlling spread comes from observations of the ability of the 30k transgenics to complement the spread of other viruses that are restricted in their distribution in normal tobacco. From the range of viruses tested only potato leaf roll virus spread beyond its normal tissue limits (C. Holt and R. N. Beachy, personal communication), whereas evidence from double infection experiments suggests that complementation can occur more widely (see later). Similarly, transgenic tobacco plants constitutively expressing the AlMV 3A protein, did not complement the movement of any from a wide range of hetcrologous viruses tested (Loesch-Fries *et al.* 1988).

The ultrastructural evidence available indicates that the observed modifications of plasmodesmata are the sum of several structural changes, including for example, loss of the desmotubule, increased overall dimension of the outer tube and new cell wall synthesis. This cascade of events, particularly noticeable in tissues infected with CaMV (Fig. 1.2), would indicate that the movement proteins take effect through an interaction with a host component. Two mechanisms can be envisaged to achieve this. Firstly, a classical signal-transduction mechanism where the virus protein binds to a host receptor and triggers the response, or secondly, the biochemical activity of the virus protein could modify the host component, as could occur for example with a protein kinase activity. There is insufficient evidence to favour with justification either mechanism.

How the infectious agent is 'propelled' through the modified plasmodesmata is also unknown. The cytosolic connection between cells may be the medium through which spread occurs, although the possibility that spread may occur through the agency of fluid membranes, does not seem to have been addressed.

A possibility for future understanding of plasmodesmata and the mechanism by which they become modified lies in a comparison between them and the gap junctions found in animal cells. Several pieces of evidence suggest that these two structures may be functionally analogous. Both have a similar basic gating capacity (about 800–1000 Da) and utilize the Ca^{2+} agonist IP_3 to mediate their function (Tucker 1988), and furthermore a 27 kDa protein from gap junctions has been shown recently to be antigenically related to a protein of similar size from plasmodesmata (Meiners and Schindler 1987): what might be the effect of expressing a plant virus movement protein in an animal tissue containing gap junctions?

Virus and host specificity

Comparisons between the amino acid sequence of the putative spread proteins from a range of different viruses with different host ranges (Hull 1989) do not reveal any structural protein features which can be correlated with host range. Yet the existence of subliminal infections indicates that there must be some host limitation in the specificity of their function.

If the majority of viruses code for spread proteins that operate by causing similar structural changes of plasmodesmata then trans-complementation between closely-and distantly-related viruses should be a relatively common phenomenon, both experimentally and in nature. In this event, viruses which normally only established a subliminal, or tissue localized, infection would 'break out' to give a systemic infection. This phenomenon has been demonstrated experimentally with several virus combinations (reviewed in Hull 1989). From this broad range of examples no clear indication of functional relationships becomes apparent and in this respect the situation parallels that regarding sequence comparisons.

The problem here is the number of variables for comparison. In the examples listed (Hull 1989) there is a range of monocot and dicot host plants infected with representatives of several virus groups which may or may not be limited to particular tissues of the plant. All were tested under a variety of experimental conditions. Perhaps, if the phenomenon of trans-complementation stands these diverse tests, then it really is leading us to conclude that the process of virus spread does depend upon common principles amongst all viruses, but that the degree to which the plasmodesmatal modifications can be utilized depends upon both the timing and location of the challenge virus in relation to the replication and spread of the helper virus.

The infectious agent that moves

In the preceding discussion, I have not addressed the question of the nature of the infectious agent which moves from cell to cell and systemically around the plant. The options must be limited to just three, virus particles, virus nucleic acid, or virus-specific nucleoprotein (vNP), although it may be that more than one agent may be effective at any one time.

The strongest ultrastructural evidence comes from observations of virus particles within plasmodesmata in a wide range of host/virus interactions, including viruses with icosahedral or rod-shaped morphology (e.g. Fig. 1.2; Shanks *et al.* 1989; Weintraub *et al.* 1976). The contrary view is obtained from studies of the immunogold localization of the movement proteins from TMV (Tomenius *et al.* 1987) and AlMV (Stussi-Garaud *et al.* 1987). In these cases, the viral protein accumulated in cell wall components **prior** to the accumulation of virus particles in the adjacent cells, possibly arguing against the particle as the movement agent. There are disadvantages with all of these observations. Firstly, it is not possible to deduce whether the particular plasmodesma under examination is in fact functional, blockage at one end could trap particulate material within the channel. Secondly, it may require very few particles in any one cell for the infection to spread on to its neighbour.

There is evidence that spread of the infection can occur in the absence of coat protein and, hence, virus capsids. TMV genomes with mutations in the coat protein region replicate in individual cells and will spread from cell to cell within the inoculated leaf, although they do show a defect in long distance systemic spread (Siegel *et al.* 1962; Dawson *et al.* 1988). Similarly the 'NM' infections of host plants brought about by RNA 1 of the tobraviruses (the coat protein gene of these viruses occurs within RNA 2) are characterized by a spreading infection in the inoculated leaf, but not systemically throughout the plant (Sanger 1968). In complementation experiments with TMV, RCMV B-RNA was able to spread from cell to cell (Malyshenko *et al.* 1988). Coat protein mutants of the bipartite geminiviruses, ACMV (Stanley and Townsend 1986; Etessami *et al.* 1988) and TGMV (Gardiner *et al.* 1988; Brough *et al.* 1988), spread systemically throughout the plant, although the exact form of the spreading agent (ds DNA, ss DNA, or vNP) is not known. Viroids which do not encapsidate also cause systemic disease in their host plants (Diener 1987), however, the stability of viroids is greater than ss viral RNA since the genomic RNA shows extensive self complementarity and is extensively double-stranded.

My personal bias in this discussion is that it is virus particles which normally are the agents that move from cell to cell and systemically through the host. Clearly, there are circumstances (e.g. in the absence of functional coat protein) when another strategy must be utilized, and this alternative may also operate in parallel with movement of virus particles. It is unlikely, however, that in these circumstances movement occurs as free nucleic acid. Since free viral nucleic acid is probably unstable *in vivo* (e.g. de Varennes and Maule 1985), a type of vNP would seem more likely. Furthermore, virus particles are the normal environment of the genomic nucleic acid. For CaMV, viral DNA replication and encapsidation appear to be co-ordinate processes such that the replicating DNA and template RNA remain insensitive to nucleases *in vivo* and *in vitro* (Thomas *et al.* 1985; Marsh and Guilfoyle 1987; Fuetterer and Hohn 1987).

Conclusions

It seems to me, reading back through the body of this review, that I have emphasized what we do not know rather than the advances that have been made in the last few years. Several important milestones have been reached, particularly, the demonstration of complementation of conditional mutants in transgenic plants that confirms the role of the TMV 30k protein in controlling virus spread, the demonstration of trans-complementation between virus groups with different host ranges, the subcellular localization of the putative spread proteins in cell wall components, and the demonstration that the TMV 30k protein alone is able to influence the functional characteristics of plasmodesmata. However, despite extensive sequence analysis and numerous reports of the detection and preliminary characterization of the spread proteins, we are still a long way from understanding how they function.

Evidence from the complemented movement of TRV RNA and RCMV B-RNA shows that the TMV 30k protein will control the transport of non-encapsidated RNA and the ultrastructural evidence indicates that virus particles move as well. The possibility that additional virus-coded products are involved in the overall process of systemic spread seems likely. The bipartite geminiviruses have two products involved and the monopartite geminiviruses may require coat protein in addition to a putative spread protein. The necessity for coat protein for long distance spread of some RNA viruses has been established. Experiments to test whether long distance spread depends **only** on coat protein and can be independent of the 30k homologues have not been carried out.

I have discussed the movement of viruses generally with reference to TMV about which we have most information. This indirectly implies that all viruses exploit a common mechanism for their intercellular and systemic movement. Without doubt this is an oversimplification of a complex process and more complete information for other viruses is required before a clearer picture can emerge. Based upon the diversity of observations obtained for different viruses, it has been suggested (Hull 1989) that virus transport may utilize several different mechanisms.

For the future, we must look to the techniques of protein expression, plant transformation and mutational analysis to reveal how the spread proteins interact with host components in either the nucleus or the cell wall, to stimulate modifications of plasmodesmata. In particular this information will be of interest both to virologists and cell biologists interested in the structure and functioning of plasmadesmata, and will allow an assessment of the feasibility of using cell-to-cell spread as a target for genetic manipulation of non-conventional resistance to virus infection. A complete appreciation of the way in which viruses move through the plant will require an understanding of the gating action of intercellular connections, not only between mesophyll cells but also between the cell types which make up the vascular system, and an understanding of the principles of phleom transport and the interrelationship of cell types at the end of vascular lineages and at their origin in the meristematic region.

The resolution of the details of cell movement, as a phenomenon necessary for the successful infection of plants by most viruses, will only be achieved through the interaction and collaboration of scientists of diverse disciplines and represents one of the most significant current challenges in plant virology.

Acknowledgements

I would like to thank Prof. J. W. Davies, Prof. R. Hull, Dr. J. Stanley, and Margaret Boulton for their critical appraisal of this review during its preparation, Dr Just Vlak for his assistance with the baculovirus expression system and I. Wilson and J. W. M. van Lent for assistance with the electron microscopy. I am also grateful to Drs R. Hull, C. Stussi-Garaud, J. Stanley, S. Covey, and Margaret Boulton for access to reprints of manuscripts submitted for publication and to others cited in the text for allowing me to quote unpublished data.

References

Albrecht, H., Geldreich, A., Menissier de Murcia, J., Kircherr, D., Mesnard, J-M., and Lebeurier, G. (1988). Cauliflower mosaic virus gene I product detected in a cell-wall-enriched fraction. *Virology* **163**, 503–8.

Atabekov, J.G. and Dorokhov, Y.L. (1984). Plant virus-specific transport function and resistance of plants to viruses. *Advances in Virus Research* **29**, 313–64.

Baron-Epel, O., Hernandez, D., Jiang, L-W., Meiners, S., and Schindler, M. (1988). Dynamic continuity of cytoplasmic and membrane compartments between plant cells. *Journal of Cell Biology* **106**, 715–21.

Berna, A., Briand, J-P., Stussi-Garaud, C., and Godefroy-Colburn, T. (1986)

Kinetics of accumulation of the three non-structural proteins of alfalfa mosaic virus in tobacco plants. *Journal of General Virology* **67**, 1135–47.

Blum, H., Gross, H.J., and Beier, H. (1989). The expression of the TMV-specific 30 kDa protein in tobacco protoplasts is strongly and selectively enhanced by actinomycin. *Virology* **169**, 51–61.

Boccara, M., Hamilton, W.D.O., and Baulcombe, D.C. (1986). The organisation and interviral homologies of genes at the 3' end of the tobacco rattle virus RNA1. *EMBO Journal* **5**, 223–9.

Boulton, M.I., Steinkellner, H., Donson, J., Markham, P.G., King, D.I., and Davies, J.W. (1989). Mutational analysis of the virion-sense genes of maize streak virus. *Journal of General Virology* **70**, 2309–23.

Briddon, R.W., Watts, J., Markham, P.G., and Stanley, J. (1989). Agroinoculation of beet curly top virus demonstrates that the coat protein is essential for infectivity. *Virology* **172**, 628–33.

Brough, C.L., Hayes, R.J., Morgan, A.J., Coutts, R.H.A., and Buck K.W. (1988). Effects of mutagenesis *in vitro* on the ability of cloned tomato golden mosaic DNA to infect *Nicotiana benthamiana* plants. *Journal of General Virology* **69**, 503–14.

Cameron, I.R., Possee, R.D., and Bishop, D.L. (1989). Insect cell culture technology in baculovirus expression systems. *Trends in Biotechnology* **7**, 66–71.

Dawson, W.O., Bubrick, P., and Grantham, G.L. (1988). Modifications of tobacco mosaic virus coat protein gene affecting replication, movement and symptomatology. *Phytopathology* **78**, 783–9.

Deom, C.M., Oliver, M.J., and Beachy, R.N. (1987). The 30-kilodalton gene product of tobacco mosaic virus potentiates virus movement. *Science* **237**, 389–94.

de Varennes, A. and Maule, A.J. (1985). Independent replication of cowpea mosaic virus bottom component RNA: *In vivo* instability of the viral RNAs. *Virology* **144**, 495–501.

Diener, T.O. (ed.) (1987). *The viroids*. Plenum Press, New York.

Dormier, L.L., Shaw, J.G., and Rhoads, R.E. (1987). Potyviral proteins share amino acid sequence homology with picorna-, como- and caulimoviral proteins. *Virology* **158**, 20–7.

Esau, K. and Thorsch, J. (1985). Sieve plate pores and plasmodesmata, the communication channels of the symplast: ultrastructural aspects and developmental relations. *American Journal of Botany* **72**, 1641–53.

Etessami, P., Callis, R., Ellwood, S., and Stanley, J. (1988). Delimitation of essential genes of cassava latent virus DNA 2. *Nucleic Acids Research* **16**, 4811–29.

Fuetterer, J. and Hohn, T. (1987). Involvement of nucleocapsids in reverse transcription: a general phenomenon. *Trends in Biochemical Science* **12**, 92–5.

Gardiner, W.E., Sunter, G., Brand, L., Elmer, J.S., Rogers, S.G., and Bisaro, D.M. (1988). Genetic analysis of tomato golden mosaic virus: the coat protein is not required for systemic spread or symptom development. *EMBO Journal* **7**, 899–904.

Godefroy-Colburn, T., Gagey, M-J., Berna, A., and Stussi-Garaud, C. (1986). A non-structural protein of alfalfa mosaic virus in the walls of infected tobacco cells. *Journal of General Virology* **67**, 2233–9.

Goldbach, R., Rezelman, G., and van Kammen, A. (1980). Independent replication and expression of B-component RNA of cowpea mosaic virus. *Nature* (London) **286**, 297–300.

Gordon, K., Pfeiffer, P., Futterer, J., and Hohn, T. (1988). *In vitro* expression of cauliflower mosaic virus genes. *EMBO Journal* **7**, 309–17.

Gunning, B.E.S. and Overall, R.L. (1983). Plasmodesmata and cell-to-cell transport in plants. *Bioscience* **33**, 260–5.

Harker, C.L., Mullineaux, P.M., Bryant, J.A., and Maule, A.J. (1987). Detection of CaMV gene I and gene VI protein products *in vivo* using antisera raised to COOH-terminal β-galactosidase fusion proteins. *Plant Molecular Biology* **8**, 275–87.

Hayes, R.J., Petty, I.T.D., Coutts, R.H.A., and Buck, K.W. (1988). Gene amplification and expression in plants by a replicating geminivirus vector. *Nature* (London) **334**, 179–82.

Huisman, M.J., Sarachu, A.N., Ablas, F., Broxterman, H.J.G., van Vloten-Doting, L., and Bol, J.F. (1986). Alfalfa mosaic virus temperature-sensitive mutants. III Mutants with a putative defect in cell-to-cell transport. *Virology* **154**, 401–4.

Hull, R. (1989). The movement of viruses in plants. *Annual Review of Phytopathology* **27**, 213–40.

Hull, R. and Covey, S.N. (1985). Cauliflower mosaic virus: pathways to infection. *BioEssays* **3**, 160–3.

Hull, R., Sadler, J., and Longstaff, M. (1986). The sequence of carnation etched ring virus DNA: comparison with cauliflower mosaic virus and retroviruses. *EMBO Journal* **5**, 3083–90.

Joshi, S., Pleij, C.W.A., Haenni, A.L., Chapeville, F., and Bosch, L. (1983). Properties of the tobacco mosaic virus intermediate length RNA-2 and its translation. *Virology* **127**, 100–11.

Kiberstis, P.A., Loesch-Fries, L.S., and Hall, T.C. (1981). Viral protein synthesis in barley protoplasts inoculated with native and fractionated brome mosaic virus RNA. *Virology* **112**, 804–8.

Lazarowitz, S.G., Pinder, A.J., Damsteegt, V.D., and Rogers, S.G. (1989). Maize streak virus genes essential for systemic spread and symptom development. *EMBO Journal* **8**, 1023–32.

Leonard, D.A. and Zaitlin, M. (1982). A temperature-sensitive strain of tobacco mosaic virus defective in cell-to-cell movement generates an altered virus-coded protein. *Virology* **117**, 416–24.

Linstead, P.J., Hills, G.J., Plaskitt, K.A., Wilson, I.G., Harker, C.L., and Maule, A.J. (1988). The subcellular location of the gene I product of cauliflower mosaic virus is consistent with a function associated with virus spread. *Journal of General Virology* **69**, 1809–18.

Loesch-Fries, L.S., Merlo, D., Halk, E., Krahn, K., Nelson, S., and Jarvis, N. (1988). Functions of alfalfa mosaic virus RNA3. *5th International Congress of Plant Pathology*, Kyoto, Japan. Abstract 1–6–4, p. 39.

Luckow, V.A. and Summers, M.D. (1988). Trends in the development of baculovirus expression vectors. *Biotechnology* **6**, 47–55.

MacFarlane, S.A., Taylor, S.C., King, D.I., Hughes, G. and Davies, J.W. (1989). Pea early browning virus RNA1 encodes four polypeptides including a putative zinc-finger protein *Nucleic Acids Research* **17**, 2245–60.

MacKenzie, D.J. and Tremaine, J.H. (1988). Ultrastructural location of non-structural protein 3A of cucumber mosaic virus in infected tissue using monoclonal antibodies to a cloned chimeric fusion protein. *Journal of General Virology* **69**, 2387–95.

Malyshenko, S.I., Lapchic, L.G., Kondakova, O.A., Kuznetzova, L.L., Taliansky, M.E., and Atabekov, J.G. (1988). Red clover mottle comovirus B-RNA spreads between cells in tobamovirus-infected tissues. *Journal of General Virology* **69**, 407–12.

Marsh, L.E. and Guilfoyle, T.J. (1987). Cauliflower mosaic virus replication intermediates are encapsidated into virion-like particles. *Virology* **161**, 129–37.

Martinez-Izquierdo, J.A., Futterer, J., and Hohn, T. (1987). Protein encoded by ORF 1 of cauliflower mosaic virus is part of the viral inclusion body. *Virology* **160**, 527–30.

Matsuura, Y., Possee, R.D., and Bishop, D.H.L. (1986). Expression of the S-coded genes of lymphocytic choriomeningitis arenavirus using a baculovirus vector. *Journal of General Virology* **67**, 1515–29.

Maule, A.J., Harker, C.L., and Wilson, I.G. (1989). The pattern of accumulation of cauliflower mosaic virus-specific products in infected turnips. *Virology* **169**, 436–46.

Meiners, S. and Schindler, M. (1987). Immunological evidence for gap junction polypeptide in plant cells. *Journal of Biological Chemistry* **262**, 951–3.

Meshi, T., Watanabe, Y., Saito, T., Sugimoto, A., Maede, T., and Okada, Y. (1987). Function of the 30 kDa protein of tobacco mosaic virus: involvement in cell-to-cell movement. *EMBO Journal* **6**, 2557–63.

Meyer, M., Hemmer, O., Mayo, M.A., and Fritsch, C. (1986). The nucleotide sequence of tomato black ring virus RNA-2. *Journal of General Virology* **67**, 1257–71.

Nassuth, A., Alblas, F., and Bol, J.F. (1981). Localization of genetic information involved in the replication of alfalfa mosaic virus. *Journal of General Virology* **53**, 207–14.

Nishiguchi, M., Motoyoshi, F., and Oshima, M. (1978). Behaviour of a temperature-sensitive strain of tobacco mosaic virus in tomato leaves and protoplasts. *Journal of General Virology* **39**, 53–61.

Ooshika, I., Watanabe, Y., Meshi, T., Okada, Y., Igano, K., Inouye, K., and Yoshida, N. (1984). Identification of the 30k protein of TMV by immunoprecipitation with antibodies directed against a synthetic peptide. *Virology* **132**, 71–8.

Osman, T.A.M. and Buck, K.W. (1987). Replication of red clover necrotic mosaic virus RNA in cowpea protoplasts: RNA 1 replicates independently of RNA 2. *Journal of General Virology* **68**, 289–96.

Richins, R.D., Schlothof, H.B., and Shepherd, R.J. (1987). Sequence of figwort mosaic virus DNA (caulimovirus group). *Nucleic Acids Research* **15**, 8451–66.

Robards, A.W. (1976). Plasmodesmata in higher plants. In *Intercellular communication in plants: studies on plasmodesmata*. (ed. B. E. S. Gunning and A. W. Robards, pp. 15–57. Spinger-Verlag, Berlin.

Robinson, D.J., Barker, H., Harrison, B.D., and Mayo, M.A. (1980). Replication of RNA-1 of tomato black ring virus independently of RNA-2. *Journal of General Virology* **51**, 317–26.

Rogers, S.G., Bisaro, D.M., Horsch, R.B., Fraley, R.T., Hoffman, N.L., Brand, L., Elmer, J., and Lloyd A.M. (1986). Tomato golden mosaic virus A component DNA replicates autonomously in transgenic plants. *Cell* **45**, 593–600.

Saito, T., Imai, Y., Meshi, T., and Okada, Y. (1988). Interviral homologies of the 30k proteins of tobamoviruses. *Virology* **176**, 653–6.

Sanger, H.L. (1968). Characteristics of tobacco rattle virus: evidence that its two particles are functionally defective and mutually complementing. *Molecular and General Genetics* **101**, 346–67.

Shanks, M., Tomenius, K., Clapham, D., Huskison, N., Barker, P., Wilson, I.G., Maule, A.J., and Lomonossoff, G.P. (1989). Identification and subcellular localisation of a putative cell-to-cell transport protein from red clover mottle virus *Virology* **173**, 400–7.

Siegel, A., Zaitlin, M., and Sehgal, O.M. (1962). The isolation of defective tobacco mosaic virus strains. *Proceedings of the National Academy of Sciences USA* **48**, 1845–51.

Stanley, J. and Townsend, R. (1986). Infectious mutants of cassava latent virus generated *in vivo* from intact recombinant clones containing single copies of the genome. *Nucleic Acids Research* **14**, 5981–98.

Stratford, R. and Covey, S.N. (1989). Segregation of cauliflower mosaic virus symptom genetic determinants. *Virology* **172**, 451–9.

Stussi-Garaud, C., Garaud, J., Berna, A., and Godefroy-Colburn, T. (1987). *In situ* location of alfalfa mosaic virus non-structural protein in plant cell walls: correlation with virus transport. *Journal of General Virology* **68**, 1779–84.

Sulzinski, M.A. and Zaitlin, M. (1982). Tobacco mosaic replication in resistant and susceptible plants: in some resistant species virus is confined to a small number of initially infected cells. *Virology* **121**, 12–19.

Terry, B.R. and Robards, A.W. (1987). Hydrodynamic radius alone governs the mobility of molecules through plasmodesmata. *Planta* **171**, 145–57.

Thomas, C.M., Hull, R., Bryant, J.A., and Maule, A.J. (1985). Isolation of a fraction from cauliflower mosaic virus infected protoplasts which is active in the synthesis of (+) and (−) strand viral DNA and reverse transcription of primed RNA templates. *Nucleic Acids Research* **13**, 4557–76.

Tomenius, K., Clapham, D., and Meshi, T. (1987). Localization by immunogold cytochemistry of the virus-coded 30k protein in plasmodesmata of leaves infected with tobacco mosaic virus. *Virology* **160**, 363–71.

Townsend, R., Watts, J., and Stanley, J. (1986). Synthesis of viral DNA forms in *Nicotiana plumbaginifolia* protoplasts inoculated with cassava latent virus (CLV): evidence for the independent replication of one component of the CLV genome. *Nucleic Acids Research* **14**, 1253–65.

Tucker, E.B. (1988). Inositol bisphosphate and inositol trisphosphate inhibit cell-to-cell passage of carboxyfluorescein in staminal hairs of *Setcreasea pur-purea*. *Planta* **174**, 358–63.

van Lent, J., Wellink, J., and Goldbach, R. (1990). Evidence for the involvement of the 58k and 48k proteins in the intercellular movement of cowpea protoplasts. *Journal of General Virology* **71**, 219–23.

Watanabe, Y., Ooshika, I., Meshi, T., and Okada, Y. (1986). Subcellular localization of the 30k protein in TMV-inoculated protoplasts. *Virology* **152**, 414–20.

Weintraub, M., Ragetli, H.W.J., and Leung, E. (1976). Elongated virus particles in plasmodesmata. *Journal of Ultrastructural Research* **56**, 351–64.

Wellink, J., Rezelman, G., Goldbach, R., and Beyreuther, K. (1986). Determination of the proteolytic processing sites in the polyprotein encoded by the bottom component RNA of cowpea mosaic virus. *Journal of Virology* **59**, 50–8.

Wolf, S., Deom, C.M., Beachy, R.M., and Lucas, W.J. (1989). Movement protein of tobacco mosaic virus modifies plasmodesmatal size exclusion limit. *Science* **246**, 377–9.

Woolston, C.J., Reynolds, H.V., Stacey, N.J., and Mullineaux, P.M. (1989). Replication of wheat dwarf virus DNA in protoplasts and analysis of coat protein mutants in protoplasts and plants. *Nucleic Acids Research* **17**, 6029–41.

Zimmern, D. (1983). Homologous proteins encoded by yeast motochondrial introns and by a group of RNA viruses from plants. *Journal of Molecular Biology* **171**, 345–52.

2 Organization and expression of the genome of cucumber mosaic virus

K. ROGER WOOD

School of Biological Sciences, University of Birmingham, Birmingham B15 2TT, UK

Introduction

Cucumber mosaic cucumovirus (CMV) is the type member of the cucumovirus group, a group which also includes tomato aspermy (TAV) and peanut stunt (PSV) cucumoviruses. In the following sections, I shall review the structure, organization and expression of the viral genome, and present evidence, where available, for the function of genome-encoded proteins. This will include sequence comparison with genomes of other viruses for which there is more information concerning protein function.

Genome structure

CMV is one of the tricornaviridae, with a single stranded RNA genome of messenger sense [ss (+)] RNA divided into three species, RNAs 1, 2 and 3 (Fig. 2.1). All three are essential for systemic infection and production of progeny virions (Peden and Symons 1973; Lot *et al.* 1974); the presence of capsid protein is not essential. The nucleotide sequence encoding the capsid protein lies within the 3'-terminal section RNA 3. However, this protein is translated from a subgenomic species, RNA 4, which is produced *in vivo* and which represents the 3'-terminal sequence of RNA 3. All four RNAs are encapsidated in 180 copies of capsid protein (CP), providing icosahedral particles about 30 nm. in diameter containing either RNA 1, RNA 2, or RNA 3 and RNA 4 (Johnson and Argos 1985).

On the basis of several criteria, including nucleic acid hybridization experiments (Gonda and Symons 1978; Piazzolla *et al.* 1979; Owen and Palukaitis 1988; Rizzo and Palukaitis 1988, 1989), those CMV strains investigated to date can be assigned to one of two subgroups. The complete nucleotide sequence of the genome of the Q strain, a member of subgroup II, and the sequences of RNAs 1 and 2 of Fny, which is assigned to subgroup I, have been reported (Table 2.1). The nucleotide sequences of RNAs 1 (Rezaian *et al.* 1985), 2 (Rezaian *et al.* 1984), and 3 (Davies and

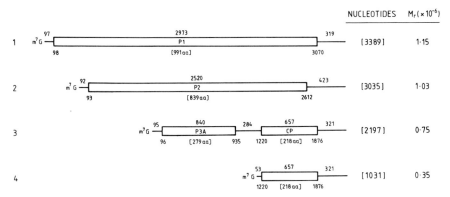

Fig. 2.1. The three genomic RNAs and subgenomic RNA 4 of CMV. Open reading frames are indicated by open boxes. Sizes of RNAs and proteins are those of the Q strain. (Rezaian *et al.* 1984, 1985; Davies and Symons 1988.)

Symons 1988) of CMV Q reveal a striking sequence homology between the 3'-termini of the four RNA species (Figs. 2.2 and 2.3). A region of virtually complete homology extends 138 nucleotides from the 3'-terminus, with regions of complete or partial homology extending 304 nucleotides from the 3'-terminus of RNA 3. There is also a 96 per cent homology between the 3'-terminal 180 nucleotides of RNAs 1 and 2 of CMV Fny (Rizzo and Palukaitis 1989), and a significant homology between the terminal 210 nucleotides (at least) of the RNAs of CMV P (Symons 1985). A similar degree of homology was not found, however, between the 3'-terminal 165 nucleotides of CMV T RNAs. This 3'-terminal sequence forms an extensive base-paired stem-loop region both in the absence (Fig. 2.4a) or presence (Fig. 2.4b) of magnesium (Ahlquist *et al.* 1981a; Rizzo and Palukaitis 1988). This transfer RNA-like structure, which has been found to be present in all CMV RNAs for which nucleotide sequence data is available (cf. also Hayakawa *et al.* 1988), is capable of accepting tyrosine (Hall 1979). It is a structure which might be expected to play a role in replicase recognition, although the significance of aminoacylation remains to be determined. Regions close to the 3'-terminus, for example, between nucleotides 130 and 277 of CMV Q RNA 1, are also able to form stem-loop structures (Rezaian *et al.* 1985).

The 5'-terminus of all four genomic RNAs has an m^7G cap structure (Symons 1975). There is a significant degree of homology between the 5'-terminal sequences of CMV Q RNAs 1 and 2 (Fig. 2.5), extending beyond the initiating codon at position 98. It is a degree of homology which does not, however, extend to RNA 3. RNAs 1 and 2 of CMV Fny also have a significant 5'-terminal homology; 84 per cent between the first 86 nucleo-

Table 2.1 Structure of CMV RNAs

RNA species	Nucleotide number						Reference
	Total	5'-untranslated	Coding*	Intercistronic	Coding*	3'-untranslated	
Q RNA 1	3389	97	2973	—	—	319	Rezaian *et al.* 1985
Fny RNA 1	3357	94	2979	—	—	284	Rizzo and Palukaitis 1989
Q RNA 2	3035	92	2517	—	—	426	Rezaian *et al.* 1984
Fny RNA 2	3050	86	2571	—	—	393	Rizzo and Palukaitis 1988
Q RNA 3	2197	95	837	287	654	324	Davies and Symons 1988
Q RNA 3	2217	122	837	301	654	303	Hayakawa *et al.* 1989
Q RNA 4	1031	53	—	—	654	324	Davies and Symons 1988

*Coding regions are exclusive of stop codons

```
       1                                                        50
CON.  GUCCGAAGAC GUUAAACUAC gCUCUCuuuA UuGCGAGUGC UGAGUUGGUA
RNA 1 GUCCGAAGAC GUUAAACUAC GCUCUCUUUA UUGCGAGUGC UGAGUUGGUA
RNA 2 GUCCGAAGAC GUUAAACUAC GCUCUCUUUA UUGCGAGUGC UGAGUUGGUA
RNA 3 GUCCGAAGAC GUUAAACUAC ACUCUC..AA UCGCGAGUGC UGAGUUGGUA

       51                                                       100
      guuuUGCUcu AAACUauCUG AAGUCgCUAA A.uccauuau ugGuuGcGAA
      GUUUUGCUUU AAACUAUCUG AAGUCGCUAA A.UCCAGUAU UGGUUGCGAA
      AGUUUGCUCU AAACUAUCUG AAGUCGCUAA A.UCCAUUAC UGGUUGCGAA
      GUGCUGCUCC AAACUGCCUG AAGUCCCUAA ACGUGUUGUU GCGCGGGGAA

       101                                                      150
      CGGGUUGUCC AUCCAGCUUA CGGCUAAAAU GGUCAGUauG cccCaaaggc
      CGGGUUGUCC AUCCAGCUUA CGGCUAAAAU GGUCAGUAUG CCCCAAAGGC
      CGGGUUGUCC AUCCAGCUUA CGGCUAAAAU GGUCAGUAUG CCCCAAAGGC
      CGGGUUGUCC AUCCAGCUUA CGGCUAAAAU GGUCAGU.CG UGUC.UUUCA

       151                                                      200
      a.uGCCGAca uCcUACA.gG uUGUCGAGcU ACCCUUGAAA UCAuCUCCUA
      A.UGCCGACA UCCUACAAGG UUGUCGAGCU ACCCUUGAAA UCAUCUCCUA
      AGUGCCGACA .CCUACAGGG UUGUCGAGCU ACCCUUGAAA UCAUCUCCUA
      CACGCCGAUG UCUUACA.AG AUGUCGAGGU ACCCUUGAAA UCACCUCCUA

       201                                                      250
      GAUUUCUUCG GAAGgGCUUC GUGAGAAGCU CGUGCACGGU AAUACACUGA
      GAUUUCUUCG GAAGGGCUUC GUGAGAAGCU CGUGCACGGU AAUACACUGA
      GAUUUCUUCG GAAGAGCUUC GUGAGAAGCU CGUGCACGGU AAUACACUGA
      GAUUUCUUCG GAAGGGCUUC GUGAGAAGCU CGUGCACGGU AAUACACUGA

       251                                                      300
      UAUUACCAAG AGUGCGGGUA UCGCCUGUGG UUuUCCACAG GUUCUCCAUA
      UAUUACCAAG AGUGCGGGUA UCGCCUGUGG UUUUCCACAG GUUCUCCAUA
      UAUUACCAAG AGUGCGGGUA UCGCCUGUGG UUCUCCACAG GUUCUCCAUA
      UAUUACCAAG AGUGCGGGUA UCGCCUGUGG UUUUCCACAG GUUCUCCAUA

       301
      AGGAGACCA
      AGGAGACCA
      AGGAGACCA
      AGGAGACCA
```

Fig. 2.2. Comparison of the 3'-terminal nucleotide sequences of CMV Q RNAs 1, 2 and 3; a consensus sequence is indicated in the top row. (Data from Rezaian *et al.* 1984, 1985; Davies and Symons 1988.)

tides (Rizzo and Palukaitis 1989). As indicated for the Q strain, there is an even greater homology between the first 50 nucleotides of these two species. The 5'-terminus is also capable of forming a stem-loop structure (Fig. 2.6).

Nucleotide sequence homology between strains

Nucleotide sequence comparisons between the 3'-termini of the genomic RNAs of different strains, indicates that the RNAs of some strains (e.g. Q

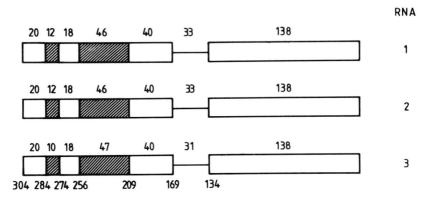

CMV Q : 3′ – TERMINAL HOMOLOGY

NUCLEOTIDE NUMBER FROM 3′– END OF RNA 3

Fig. 2.3. Homology between the 3′-terminal nucleotide sequences of CMV Q RNAs 1, 2, and 3. Open boxes correspond to regions of complete homology; hatched areas, partial homology; and single lines, little or no homology. (Modified from Symons 1979.)

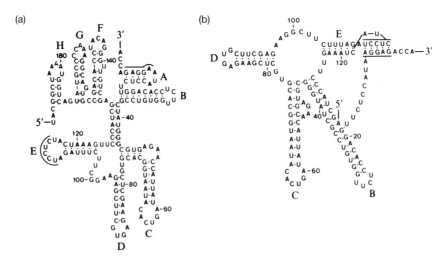

Fig. 2.4. Secondary structures predominating at the 3′-terminus of CMV Q RNA 2 in the absence (a) and in the presence (b) of magnesium. (a is from Rizzo and Palukaitis 1988, modified from Ahlquist *et al.* 1981a. b is from Joshi *et al.* 1983.)

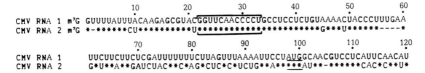

```
                10            20            30            40            50            60
                •             •             •             •             •             •
CMV RNA 1  m'G GUUUUAUUUACAAGAGCGUACGGUUCAACCCCUGCCUCCUCUGUAAAACUACCCUUUGAA
CMV RNA 2  m'G *-******CU************U*************,***********G***U******----*
                70            80            90           100           110           120
                •             •             •             •             •             •
CMV RNA 1      UUCUUCUUCUCGAUUUUUUUCUUAGUUUAAAAUUCCUAUGGCAACGUCCUCAUUCAACAU
CMV RNA 2      G*U**A**GAUCUAC**C*AG*CUC*C*UCUG**A*****AU**-******CAC*C**U*
```

Fig. 2.5. Nucleotide sequence homology at the 5'-termini of RNAs 1 and 2 of CMV Q. Homology is indicated (*), and initiating AUGs are underlined. Sequences in parentheses are common to BMV RNAs 1 and 2. (From Rezaian *et al.* 1985.)

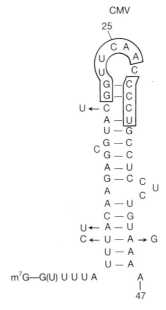

Fig. 2.6. Possible secondary structure at the 5'-terminus of CMV Q RNA 1. (From Rezaian *et al.* 1985.)

and P) share regions of strong homology, and so would appear to be closely related. The RNAs of other strains (e.g. T and M) do not share comparable regions of homology at the 3'-terminus (Symons 1985).

RNAs 1 and 2 of the two strains for which data are available, Q and Fny, also have regions of significant homology. There is, for example, 62 per cent homology between the 3'-terminal 180 nucleotides, and 80 per cent homology between the 5'-terminal 86 nucleotides of the RNA 1 species. The degrees of homology between the corresponding sections of the RNA

Table 2.2 Translation products of Q and Fny CMV RNAs

RNA species	Protein	Size of translation product			Reference
		Amino acid number	M_r from		
			Nucleotide sequence	Electrophoresis*	
Q RNA 1	P1a	991	110 791	95 000	Rezaian *et al.* 1985
Fny RNA 1	P1a	993	111 404	—	Rizzo and Palukaitis 1989
Q RNA 2	P2a	839	94 333	110 000	Rezaian *et al.* 1984
Fny RNA 2	P2a	857	96 720	—	Rizzo and Palukaitis 1988
Q RNA 3	P3a	279	30 353	35 000	Davies and Symons 1988
Q RNA 4	CP	218	24 247	24 500	Davies and Symons 1988

*Data from Gordon *et al.* 1982

2 species of the two strains are 64 and 81 per cent respectively (Rizzo and Palukaitis 1989). Similarities in coding regions are referred to in following sections.

Genome expression

CMV Q RNA 1 has a long open reading frame (ORF) extending from nucleotide 98 to 3070, encoding a protein 991 amino acids in length (P1a), that of the Fny strain extends from 95 to 3074 (Tables 2.1 and 2.2). Both begin at the first AUG from the 5'-terminus. Comparison of the nucleotide sequences of these ORFs indicates a 76 per cent homology between them, whilst comparison of predicted amino acid sequences indicates 85 per cent homology. RNA 2 also has a long ORF, that of the Q strain extending from nucleotide 93 to 2609 and encoding an 839 amino acid protein (P2a), that of the Fny strain extends from 87 to 2657. The degree of homology in this coding region is 71 per cent at the nucleotide level and 73 per cent at amino acid sequence level. The nucleotide sequence homology between the central regions (Table 2.3) is 77 per cent, but rather less at the N- and C-terminal regions. Amino acid sequence homology is 89 per cent in the central region. The ORF of CMV Q RNA 2 begins at the second AUG from the 5'-terminus; the first AUG, at positions 19–21, is followed by an in-phase stop at 43–45. The ORF of CMV Fny RNA 2 begins at the first AUG from the 5'-terminus.

CMV Q RNA 3 contains two major ORFs, the 5'-proximal sequence extending from nucleotide 96 to 932, and encoding a protein (P3a) 279 amino acids in length. The ORF encoding the capsid protein extends from nucleotide 1220 to 1873, a sequence also contained, of course, within RNA 4, from which CP is translated. The corresponding ORFs of CMV strain O begin at nucleotides 123 and 1261 and also encode proteins of 279 and 218 amino acids in length (Hayakawa *et al.* 1989). There is 83 per cent homology between the amino acid sequences of P3a of CMVs O and Q, and 80 per cent homology between their CP. There is also 81 per cent homology between the CP amino acid sequences of CMVs Q and strain Y, but 97 per cent between strains D and O (Hayakawa *et al.* 1988; 1989).

In vitro, each RNA species of the Q strain translated into a protein with the apparent size, assessed by electrophoresis in 12 per cent polyacrylamide gels, indicated in Table 2.2 (Schwinghamer and Symons 1977; Gordon *et al.* 1982). Although of the correct order, the apparent sizes do not quite correspond with those expected from nucleotide sequence data. Translation of RNAs from other strains, and analysis of products in other gel systems (Gordon *et al.* 1982; Kaminski and Wood unpublished) indicates a degree of variability; apparent sizes for the translation products of the W strain separated in 6–18 per cent polyacrylamide gels are perhaps the closest to those expected (Fig. 2.7).

Demonstration of the biosynthesis of these CMV genome encoded proteins (or indeed any others) *in vivo* has proved more difficult to achieve; several earlier attempts to identify them, either in plants, or in protoplasts have not been particularly convincing (Ziemiecki and Wood 1975; Gonda and Symons 1979; Roberts and Wood 1981; Gould *et al.* unpublished).

Table 2.3 Sequence homologies between nucleotide coding regions and encoded proteins of Q-CMV RNA 2 and Fny-CMV RNA 2

	Region		
Sequence	Amino terminus	Central	Carboxy terminus
Nucleotide position			
Q–CMV RNA 2	93–950	951–2030	2031–2609
Fny–CMV RNA 2	87–953	954–2033	2034–2657
Amino acid position			
Q–CMV RNA 2	1–286	287–646	647–849
Fny–CMV RNA 2	1–289	290–649	650–857
Homology (%)			
Nucleotide coding region	68	77	66
Protein sequence	64	89	56

(From Rizzo and Palukaitis 1988)

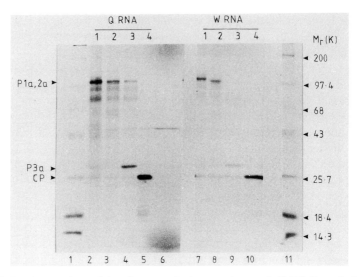

Fig. 2.7. Separation of *in vitro* translation products of CMV Q and CMV W RNAs in 6–18 per cent polyacrylamide gradient gels. Lanes 2–5 contain the translation products of CMV Q RNAs 1–4 respectively; lanes 7–10 contain products from CMV W RNAs 1–4; lane 6 contains the products of a translation mixture to which no RNA was added. Protein standards with M_r indicated down the right hand side are in lanes 1 and 11. Positions of major viral RNA encoded proteins are indicated down the left hand side. (Kaminski and Wood unpublished.)

However, several techniques are now readily available for production of antisera that will permit detection of putative virus-encoded proteins both *in situ* and in cell fractions. Infection of cucumber protoplasts with CMV Q or W, and radiolabelling of those proteins synthesized between 4 and 9 h after inoculation has revealed biosynthesis of a protein which is apparently not produced in uninfected cells (Kaminski and Wood unpublished; Fig 2.8). Antiserum raised to *in vitro* translation products, and which reacts with the *in vitro* translation product of RNA 2, also reacts with this 105k protein produced *in vivo*. This would suggest that the protein observed *in vivo* is P2a, the translation product of RNA 2, although it is not impossible that P1a is also produced and is co-migrating with P2a. Detection of P3a *in situ* has also been reported; these experiments are discussed later in the context of protein function. It is clear, however, that much more needs to be done to characterize the virus-encoded proteins that are produced *in vivo*, and to establish their relation to those protein structures predicted from nucleotide sequence data and *in vitro* translation experiments.

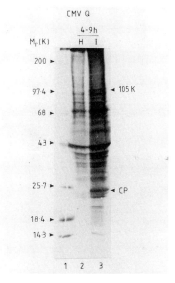

Fig. 2.8.   Polyacrylamide gel (6–18 per cent) separation of proteins synthesized in CMV Q infected (I) and in uninfected (H) cucumber protoplasts pulse-labelled between 4 and 9 h after infection. Protein standards are in lane 1 and positions of putative viral RNA encoded proteins indicated down the right hand side. (Kaminski and Wood unpublished.)

Functions of genome-encoded proteins

Proteins Pla and P2a

The complete nucleotide sequences of the genomes of several viruses are now available and often there is rather more information available on the nature and function of the proteins which they encode than is the case with CMV. Sequence comparisons can, therefore, provide valuable information not only with respect to protein function, but also with respect to evolutionary relationships. The genome of CMV has many similarities with that of other tricornaviruses, such as alfalfa mosaic ilarvirus (A1MV) and particularly brome mosaic bromovirus (BMV). Tobacco mosaic tobamovirus (TMV), with a ss(+) RNA genome consisting of a single species of about 6400 nucleotides, is superficially quite different. The TMV genome does, however, encode two proteins, $M_r 183k$ and 126k, both of which have been detected *in vivo*. They share a common AUG initiation codon, with the 183k protein produced by readthrough of the UAG codon which terminates translation of the 126k protein. Importantly, however, there are significant regions of homology between these coding regions of this virus, CMV, and other tricornaviridae (Cornelissen and Bol 1984; Haseloff *et al.* 1984;

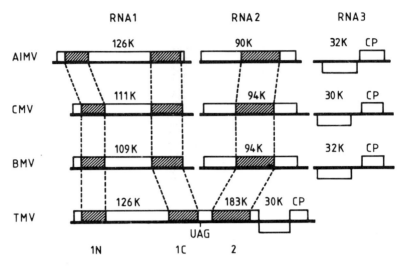

Fig. 2.9. The ss(+) RNA genomes of A1MV, BMV, CMV, and TMV. Genomic RNAs are indicated by solid lines and coding regions by boxes, with protein size above. Regions of partial homology are indicated by hatching. (Modified from Cornelissen and Bol 1984.)

Kamer and Argos 1984; Goldbach 1986). In particular, there is homology between the central sections of the tricornaviridae RNA 2 coding regions and the readthrough section of the TMV genome; and also between the N- and C-terminal parts of the RNA 1 coding regions and parts of the RNA encoding the TMV 183k protein (Fig. 2.9).

This degree of homology is sufficient not only to suggest an evolutionary relationship between the viruses, but also a functional relationship between the encoded proteins. There is evidence suggestive that the TMV 183k and 126k proteins are involved in virus replication (e.g. Ishikawa *et al.* 1986). There is also good evidence that the protein encoded by BMV RNA 1 forms an essential part of the replication complex involved in viral genome transcription. Antiserum to this protein inhibits replicase activity (Horikoshi *et al.* 1988; Quadt *et al.* 1988). It is likely that the BMV RNA 2- encoded protein is also required for RNA replication. RNAs 1 and 2 of A1MV replicate in protoplasts in the absence of RNA 3 (Nassuth and Bol 1983). No other virus-encoded proteins are required, although the necessity for a host contribution is not precluded. There are also sequence homologies with animal viruses with ss(+) RNA genomes. P2a of both Q and Fny CMV (and of BMV) contain the Gly-Asp-Asp sequence expected in a replicase (Kamer and Argos 1984; Rizzo and Palukaitis 1988), and P1a has sequences in common with nucleotide bind-

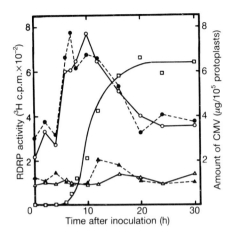

Fig. 2.10. Time course of appearance of soluble (S) and membrane-bound (M) RNA-dependent RNA polymerse (RDRP) activity in tobacco mesophyll protoplasts inoculated with CMV. Graphs represent extracted S-RDRP (▲,△) and M-RDRP (●.○) activity in the presence (▲,●) or absence (△,○) of exogenous CMV RNA template. Accumulation of CMV, as indicated by ELISA, is also shown (□). (From Nitta *et al.* 1988.)

ing proteins (Hodgman 1988; Rizzo and Palukaitis 1989). Both might therefore be expected to form part of the viral replication complex.

Earlier experiments on CMV had, however, led to the conclusion that virus-encoded proteins did not form part of the replication complex (Gordon *et al.* 1982; Jaspars *et al.* 1985). Membrane-bound replicase was isolated and purified from CMV-infected cucumber plants, and was found to consist principally of a protein of M_r about 100k, similar to those encoded by genomic RNAs 1 and 2. This 'replicase' protein was isolated from plants infected with any one of the three strains, Q, P, and T, and found to have approximately the same electrophoretic mobility in each case (Gordon *et al.* 1982). In contrast, the products of *in vitro* translation of RNAs 1 and 2 from these three strains differed between strains in their mobility. This led the authors to the view that the principal component of the viral replicase complex and virus-encoded proteins P1a and P2a were different species, a conclusion supported by proteolytic digestion experiments. More recent evidence (Nitta *et al.* 1988) is somewhat at variance with these conclusions. When tobacco protoplasts are inoculated with CMV, there is a significant stimulation of membrane-bound RNA-dependent RNA polymerase (M-RDRP) activity (Fig. 2.10) which occurs in the early stages of virus replication. Activity *in vitro* is virtually independent of added CMV RNA template, indicating that viral RNA is associated with the replication complex. Activity of the soluble polymerase (S-RDRP), which is stimulated in infected tobacco and cucumber leaves (Kumarasamy and Symons 1979; Takanami and Fraenkel-Conrat 1982) remains virtually constant in this system. This enzyme is presumed by Nitta and colleagues not to participate in viral genome replication.

When M-RDRP is isolated from protoplasts inoculated with the three genomic RNAs, it has the ability to produce, *in vitro*, double-stranded (ds) RNAs 1–4. Significantly, however, M-RDRP produced in protoplasts inoculated with RNAs 1 and 2 only is able to produce ds RNAs 1 and 2. Polymerase isolated from protoplasts inoculated with RNA 1 alone is unable to produce ds RNA 1. These observations are taken to indicate that proteins encoded by RNAs 1 and 2 form a necessary part of the replication complex, a conclusion similar to that derived from studies on the replication of BMV RNAs in protoplasts. However, more direct evidence for this association awaits confirmation; the presence of viral genome-encoded proteins in the replication complex remains to be demonstrated.

Protein 3a

The CMV P3a protein, predicted from nucleotide sequence data and observed as a product of *in vitro* translation of RNA 3, is similar in size to proteins encoded by the genomes of other viruses including those encoded by RNA 3 of A1MV and BMV, and to the 30k protein encoded by the genome of TMV. Again, it might be expected that these proteins would have functional similarities. There is now good evidence that the 30k TMV protein is involved in intercellular virus transport, a subject dealt with comprehensively by A.J. Maule in Chapter 1. When transgenic plants which express normal 30k protein are infected with TMV which has a mutation in the 30k coding region and which therefore, does not have the ability to spread from cell to cell, the mutation is complemented and progeny virus are able to spread and produce a systemic infection (Deom *et al.* 1987). In addition, *in situ* immunogold labelling using antisera to some of these 30k proteins has indicated their association with plasmodesmata and/or cell walls, a location which might at least suggest a potential for involvement in intercellular transport. However, immunogold labelling using a monoclonal antibody to P3a (MacKenzie and Tremaine 1988) of sections of tobacco tissue infected with CMV indicates a localization in the nucleolus (Fig. 2.11); an association of the TMV 30k protein with the nucleus has also been reported (Watanabe *et al.* 1986). It may be that this situation truly reflects the site of action of P3a. If so, the protein would appear to be influencing events at some distance from the site at which transport is taking place. As far as CMV is concerned, the conclusions to be derived from these experiments remain unclear.

An alternative approach, already referred to in relation to the function

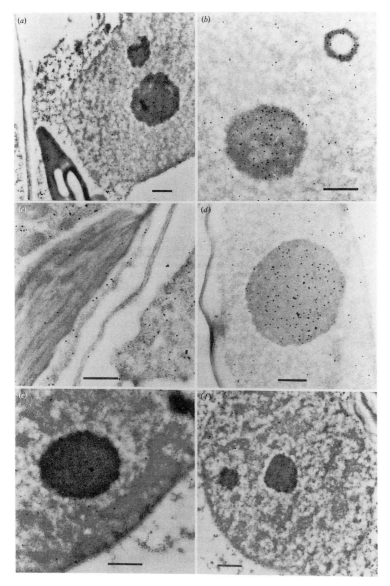

Fig. 2.11. Immunogold labelling of the 3a protein with monoclonal antibody in ultrathin sections of tobacco tissue systemically-infected with CMV (a, b, c) or in sections of CMV-inoculated leaves (d). Labelling of tissue systemically-infected with TAV is indicated in (e), and of healthy tissue in (f). Bar markers represent 0.5 mμ. (From MacKenzie and Tremaine 1988.)

```
CMV  1 .....MAFQGPSRT...LTQQSSAASSDDLQKILFSPDAIKKMATECDLG 42
         : :  ::        :       : :   ::      ::    :: ::
BMV  1 MSNIVSPFSGSSRTTSDVGKQAGGTSDEKLIESLFSEKAVKEIAAECKLG 50

    43 RHHWMRADNAISVRPLVPQVTSNNLLPFFKSGYDAGELRSKGYMSVPQVL 92
         ::           :::       :       :: :::  ::::: :: :
   51 CYNYLKSNEPRNYIDLVPKSHVSAWLSWATSKYDKGELPSRGFMNVPRIV 100

    93 CAVTRTVSTDAEGSLKIYLADLGDKE....LSPIDGQCVTLHNHELPALI 138
        :   ::    :  : : :  :        : ::  :       :: :: :
   101 CFLVRTTDSAESGSITVSLCDSGKAARAGVLEAIDNQEATIQLSALPALI 150

   139 SFQPTYDCPMELVG...NRHRCFAVVVERHGYIGYGGTTASVCSNWQAQF 185
         : :::::: ::     : :::  :     : :: :          :: :
   151 ALTPSYDCPMEVIGGDSGRNRCFGIATQLSGVVGTTGSVAVTHAYWQANF 200

   186 SSKNNNYTHAAAAKTLVLPYNRLAEHSKPSAVARLLKSQLNNVSSSRYLL 235
         :   :       ::::: ::   :         :             :::
   201 KAKPNNYKLHGPATIMVMPFDRLRQLDKKS.LKNYIRGISNQSVDHGYLL 249

   236 PNVALNQNASGHESEILKESPPIAIGSPSASRNNSFRSQVVNGL...... 279
          ::         :                          : ::
   250 GRPLQSVDQVAQEDLLVEESESPSALGRGVKDSKSVSASSVAGLPVSSPT 299

   280 .... 280

   300 LRIK 303
```

Fig. 2.12. Comparison between the predicted amino acid sequences of the 3a proteins encoded by CMV RNA 3 (top) and BMV RNA 3 (bottom). Identity and similarity between the two is indicated (:). (Data from Ahlquist *et al.* 1981b; Davies and Symons 1988.)

of P1a and P2a, is to compare sequences. The virus most closely related to CMV and for which appropriate data are available, is BMV. The amino acid sequences for the P3a proteins of BMV and CMV are aligned in Fig 2.12. Clearly a limited degree of homology exists between the two proteins (see also Murthy 1983), a similarity enhanced by the Diagon plot of Fig. 2.13. There is also some similarity with the comparable protein encoded by A1MV RNA 3, a protein also observed in association with the cell wall in tobacco (Stussi-Garaud *et al.* 1987).

Homology with the TMV 30k protein, the one whose involvement in transport is most clearly demonstrated, is slight (Davies and Symons 1988). However, although the lack of sequence homology is perhaps disappointing (and largely inconclusive) a functional relationship between the proteins is certainly not precluded. In view of other similarities between the viruses in question, it would be surprising if they did not also extend to similarities in P3a function.

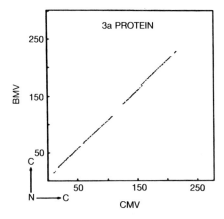

Fig. 2.13.   Diagon comparison of CMV and BMV 3a protein sequences. (Davies and Symons 1988.)

Other open reading frames

All three genomic RNAs have ORFs in addition to those already discussed. There are, for example, ORFs of 285, 126 and 99 nucleotides in length in CMV Q RNA 3 (Gould and Symons 1982). A second ORF in CMV Q RNA 2 begins at nucleotide 2410 and extends for 300 nucleotides (Rezaian *et al.* 1984), while that in Fny RNA 2 begins at nucleotide 2419 and extends for 330 nucleotides. There is a short ORF of 174 nucleotides on the negative strand of Fny RNA 2, but not on the corresponding Q RNA (Rizzo and Palukaitis 1988). In CMV Q RNA 1, there is a second ORF beginning at nucleotide 2328 and extending for 207 nucleotides; the corresponding ORF in Fny RNA 1 begins at the same site, but is only 165 nucleotides in length, and the homology between the polypeptides which would be encoded by these ORFs is only 30 per cent. There is also an ORF on the negative strand of CMV Q RNA 1, beginning at nucleotide 1093 of the positive strand and extending for 198 nucleotides; the comparable Fny ORF begins at nucleotide 1090 (+ve strand) and extends for 411 nucleotides. Here, homology is only 23 per cent (Rezaian *et al.* 1985; Rizzo and Palukaitis 1989). The significance of these potential coding regions remains unclear.

Non-translated regions

There is a marked homology between those nucleotide sequences near the 5'-terminus of CMV RNAs 1–3 and the corresponding regions of RNAs 1–3 of BMV (e.g. Davies and Symons 1988; Table 2.4). This sequence appears also in the intercistronic regions of the RNA 3 of both viruses. It

Table 2.4 Sequence homologies in the 5'-untranslated and intercistronic regions of Q–CMV and BMV

		Sequence	Reference
5'-untranslated region	CMV RNA 1 (20)[*]	A C G G U U C A A C C C C U G C C	Rezaian et al. 1985
	BMV RNA 1 (15)	G A G G U U C A A A C C C U U G U	Ahlquist et al 1984
	CMV RNA 2 (19)	A U G G U U C A A C C C C U G C C	Rezaian et al. 1984
	BMV RNA 2 (15)	G A G G U U C A A U C C C U U G U	Ahlquist et al. 1984
Intercistronic region	CMV RNA 3 (1099)	A A G G U U C A A U U C C C U U U U	Davies and Symons 1988
	BMV RNA 3 (1098)	U G G G U U C A A U U C C C U U U A	Ahlquist et al. 1981b
Consensus sequence		— — G_6 G_6 U_6 U_6 C_6 A_6 A_6 u_4 c_4 C_6 C_6 u_5 u_4 —	
		c_2 u_2 $\qquad\qquad$ c_1 g_2	

*Numbers in parentheses indicate the distance of the first nucleotide from the 5'-terminus of the RNA
(From Davies and Symons 1988)

may be that, as in the case of BMV, RNA 4 from CMV does not replicate autonomously (e.g. Hull and Maule 1985; Nitta *et al.* 1988), but is produced from (−) strand RNA 3; these sequences could be involved in the initiation of (+) strand RNAs (but see also Marsh *et al.* 1988).

Virus–host interactions

Different isolates of CMV can cause markedly different symptoms in a wide range of host plants (Francki *et al.* 1979; Kaper and Waterworth 1981). For example, systemic symptoms in tobacco can vary from mild mottling, to severe yellowing, and although most strains elicit a hypersensitive response in cowpea (*Vigna unguiculata* cv. Blackeye), some produce a systemic infection (Whipple and Walker 1941; Marchoux *et al.* 1975; Edwards *et al.* 1983). The biochemical, physiological, and cytological changes occurring during the course of various types of host response are well documented (e.g. Goodman *et al.* 1986; Fraser 1987). There is, however, very little information on the mechanisms by which these changes are induced. In attempts to determine the genome species involved in the induction of various types of response, pseudorecombinants between several strains have been produced, have been inoculated on to various hosts, and responses to them monitored.

In some cases, these experiments have suggested that response in a given host is determined by one of the genomic RNAs. For example, when pseudorecombinants were constructed between strains B and Ls (Edwards *et al.* 1983) and between D and Ds (Marchoux *et al.* 1975), ability to infect cowpea systemically was shown to be detemined by B or D RNA 2. Pseudorecombinants containing CMV B RNA 2 could also infect bean and pea systemically (Edwards *et al.* 1983). Similarly, ability of CMV K to infect maize was determined by RNA 2, and RNA 2 of CMV U determined symptom severity in *Nicotiana Edwardsonii* (Rao and Francki 1982).

In other cases (e.g. Rao and Francki 1982; Rizzo and Palukaitis 1989), host response seemed to involve two, or even all three, genomic RNAs, and symptoms elicited by pseudorecombinants could differ from those elicited by either parent. These types of experiments have not yet permitted satisfactory conclusions to be drawn on the functions of the viral RNAs or their translation products; they need to be combined with more sophisticated techniques involving site-directed mutagenesis and construction of chimaeric viral genomes. Expression of cDNA sequences in transgenic plants could also provide useful information.

Satellites

When RNA preparations from purified virions are electrophoretically separated, RNA species additional to RNAs 1–4 are often revealed. Some represent fragments of the viral genome or host RNA sequences (Symons 1985), but there is often present in preparations of some (but not all) isolates, an RNA about 340 nucleotides in length with a nucleotide sequence quite different from that of viral genomic RNA (Diaz-Ruiz and Kaper 1977; Palukaitis and Zaitlin 1984). This species is satellite RNA (designated CARNA 5 by Kaper), which depends on the presence of a helper cucumovirus for its replication, and which is encapsidated in capsid protein encoded by the helper (Kaper and Waterworth 1981; Murant and Mayo 1982; Linthorst and Kaper 1985).

An important feature of satellite RNAs is that they are able to modify the symptoms produced by helper viruses. In some cases, symptom severity is reduced, a reduction which is usually, but not always, correlated with decreased accumulation of helper (e.g. Kaper and Tousignant 1977). This observation has been put to practical use by protective inoculation of plants with satellite-containing CMV. In addition, when transgenic tobacco plants which are expressing satellite sequences are infected with CMV, then both symptom severity and virus accumulation are reduced (Harrison et al. 1987). In other cases, symptom type may be quite different from that produced by interaction between the helper virus alone and the host. Some satellites have the additional ability to induce a more severe reaction, which can, in extreme cases, take the form of a lethal necrosis. This is perhaps best illustrated by infection of tomato with appropriate CMV strains containing necrosis-inducing satellites (Kaper and Waterworth 1977).

The effect on the host depends on the particular combination of satellite, helper virus and host (Garcia-Arenal et al. 1987; Kaper et al. 1988). The mechanisms by which these effects are mediated are the subject of active investigation, but remain currently unresolved. The nucleotide sequences of several satellite RNAs have been determined, revealing a significant degree of homology between them, particularly between variants which induce necrosis (Garcia-Arenal et al. 1987; Kaper et al. 1988). All have one or more ORFs, although there is no indication that any potential translation product is involved in necrogenesis (Collmer and Kaper 1988).

Although for several variants there did not appear to be a correlation between nucleotide sequence and pathogenicity (Garcia-Arenal et al. 1987), David Baulcombe and his colleagues (unpublished) have recently identified two domains in the CMV Y satellite sequence (Takamami 1981; Kaper et al. 1986); a central region which is involved in induction of the severe yellowing mosaic in tobacco, and a second domain close to the 3'-terminus which is involved in induction of necrosis. It is of interest that satellite RNAs are capable of hybridization with genomic RNAs (Rezaian et al. 1985; Rezaian and Symons 1986; Rizzo and Palukaitis 1988); it is not impossible that such interactions are involved in the induction of some of these effects. It may also be relevant that significant amounts of ds satellite

RNAs may accumulate during the course of infection (Piazzolla *et al.* 1979; Diaz-Ruiz *et al.* 1987).

It is evident, therefore, that although we now know quite a lot about both the structure of the virus genome and its expression, we know very little about the interactions between this virus (or indeed any other) and its hosts that lead to the expression of the disease. However, techniques are now available which should permit rapid progress to be made towards finding some answers.

Acknowledgements

I am indebted to Dave Ruffles for photography, to Pauline Hill for artwork, to Frances Harold for typing, and to Trevor Griffiths and Emma Wallington for critical reading of the manuscript.

References

Ahlquist, P., Dasgupta, R., and Kaesberg, P. (1981a). Near identity of 3' RNA secondary structure in Bromoviruses and cucumber mosaic virus. *Cell* **23**, 183–9.

Ahlquist, P., Luckow, V., and Kaesberg, P. (1981b). Complete nucleotide sequence of brome mosaic virus RNA 3. *Journal of Molecular Biology* **153**, 23–38.

Ahlquist, P., Dasgupta, R., and Kaesberg, P. (1984). Nucleotide sequence of the Brome mosaic virus genome and its implications for viral replication. *Journal of Molecular Biology* **172**, 369–83.

Collmer, C.W. and Kaper, J.M. (1988). Site-directed mutagenesis of potential protein-coding regions in expressible cloned cDNAs of cucumber mosaic viral satellites. *Virology* **163**, 292–8.

Cornelissen, B.J.C. and Bol, J.F. (1984). Homology between the proteins encoded by tobacco mosaic virus and two tricornaviruses. *Plant Molecular Biology* **3**, 379–84.

Davies, C. and Symons, R.H. (1988). Further implications for the evolutionary relationships between tripartite plant viruses based on cucumber mosaic virus RNA 3. *Virology* **165**, 216–24.

Deom, C.M., Oliver, M.J., and Beachy, R.N. (1987). The 30-kilodalton gene product of tobacco mosaic virus potentiates virus movement. *Science* **237**, 389–94.

Diaz-Ruiz, J.R. and Kaper, J.M. (1977). Cucumber mosaic virus-associated RNA 5. III. Little sequence homology between CARNA 5 and helper virus. *Virology* **80**, 204–13.

Diaz-Ruiz, J.R., Avila-Rincon, M.J., and Garcia-Luque, I. (1987). Subcellular localization of cucumovirus-associated satellite double-stranded RNAs. *Plant Science* **50**, 239–48.

Edwards, M.C., Gonsalves, D., and Provvidenti, R. (1983). Genetic analysis of cucumber mosaic virus in relation to host resistance: Location of determinants for pathogenicity to certain legumes. *Phytopathology* **73**, 269–73.

Francki, R.I.B., Mossop, D.W., and Hatta, T. (1979). In *Descriptions of plant viruses*, No 213. Association of Applied Biologists, Wellesbourne.

Fraser, R.S.S. (1987). *Biochemistry of virus-infected plants*. Research Studies Press, Letchworth.

Garcia-Arenal, F., Zaitlin, M., and Palukaitis, P. (1987). Nucleotide sequence analysis of six satellite RNAs of cucumber mosaic virus: primary sequence and secondary structure alterations do not correlate with differences in pathogenicity. *Virology* **158**, 339–47.

Goldbach, R.W. (1986). Molecular evolution of plant RNA viruses. *Annual Review of Phytopathology* **24**, 289–310.

Gonda, T.J. and Symons, R.H. (1978). The use of hybridization analysis with complementary DNA to determine the RNA sequence homology between strains of plant viruses: its application to several strains of cucumovirus. *Virology* **88**, 361–70.

Gonda, T.J. and Symons, R.H. (1979). Cucumber mosaic virus replication in cowpea protoplasts: Time course of virus, coat protein, and RNA synthesis. *Journal of General Virology* **45**, 723–36.

Goodman, R.N., Király, Z., and Wood, K.R. (1989). *Biochemistry and physiology of plant disease*. University of Missouri Press, Columbia.

Gordon, K.H.J., Gill, D.S., and Symons, R.H. (1982). Highly purified cucumber mosaic virus-induced RNA-dependent RNA polymerase does not contain any of the full length translation products of the genomic RNAs. *Virology* **123**, 284–95.

Gould, A.R. and Symons, R.H. (1982). Cucumber mosaic virus RNA 3. Determination of the nucleotide sequence provides the amino acid sequences of protein 3A and viral coat protein. *European Journal of Biochemistry* **126**, 217–26.

Hall, T.C. (1979). Transfer RNA-like structures in viral genomes. *International Review of Cytology* **60**, 1–26.

Harrison, B.D., Mayo, M.A., and Baulcombe, D.C. (1987). Virus resistance in transgenic plants that express cucumber mosaic virus satellite RNA. *Nature* **328**, 799–802.

Haseloff, J., Goelet, P., Zimmern, D., Ahlquist, P., Dasgupta, R., and Kaesberg, P. (1984). Striking similarities in amino acid sequence among nonstructural proteins encoded by RNA viruses that have dissimilar genome organisation. *Proceedings of the National Academy of Sciences USA* **81**, 4358–62.

Hayakawa, T., Hazama, M., Onda, H., Komiya, T., Mise, K., Nakayama, M., and Furusawa, I. (1988). Nucleotide sequence analysis of cDNA encoding the coat protein of cucumber mosaic virus: genome organization and molecular features of the protein. *Gene* **71**, 107–14.

Hayakawa, T., Mizukami, M., Nakajima, M., and Suzuki, M. (1989). Complete nucleotide sequence of RNA 3 from cucumber mosaic virus (CMV) strain 0: comparative study of nucleotide sequences and amino acid sequences among CMV strains O,Q,D, and Y. *Journal of General Virology* **70**, 499–504.

Hodgman, T.C. (1988). A new superfamily of replicative proteins. *Nature* **333**, 22–3.

Horikoshi, M., Mise, K., Furusawa, I., and Shishiyama, J. (1988). Immunological analysis of brome mosaic virus replicase. *Journal of General Virology* **69**, 3081–7.

Hull, R. and Maule, A.J. (1985). In *The plant viruses, Vol.1*, (ed. R. I. B. Francki), p. 83. Plenum Press, New York.

Ishikawa, M., Meshi, T., Motoyoshi, F., Takamatsu, N., and Okada, Y. (1986). *In vitro* mutagenesis of the putative replicase genes of tobacco mosaic virus. *Nucleic Acids Research* **14**, 8291–305.

Jaspars, E.M.J., Gill, D.S., and Symons, R.H. (1985). Viral RNA synthesis by a particulate fraction from cucumber seedlings infected with cucumber mosaic virus. *Virology* **144**, 410–25.

Johnson, J.E. and Argos, P. (1985). In *The plant viruses, Vol. I*, (ed. R.I.B. Francki), p. 19. Plenum Press, New York.

Joshi, R.L., Joshi, S., Chapeville, F., and Haenni, A.L. (1983). tRNA-like structures of plant viral RNAs: conformational requirements for adenylation and aminoacylation. *EMBO Journal* **2**, 1123–7.

Kamer, G. and Argos, P. (1984). Primary structural comparison of RNA-dependent polymerases from plant, animal and bacterial viruses. *Nucleic Acid Research* **12**, 7269–82.

Kaper, J.M. and Tousignant, M.E. (1977). Cucumber mosaic virus-associated RNA 5. 1. Role of host plant and helper strain in determining amount of associated RNA 5 with virions. *Virology* **80**, 186–95.

Kaper, J.M. and Waterworth, H.E. (1977). Cucumber mosaic virus associated RNA 5: causal agent for tomato necrosis. *Science* **196**, 429–31.

Kaper, J.M. and Waterworth, H.E. (1981). In *Handbook of plant virus infections*, (ed. E. Kurstak), p. 257. Elsevier, Amsterdam.

Kaper, J.M., Duriat, A.S., and Tousignant, M.E. (1986). The 368-nucleotide satellite of cucumber mosaic virus strain Y from Japan does not cause lethal necrosis in tomato. *Journal of General Virology* **67**, 2241–6.

Kaper, J.M., Tousignant, M.E., and Steen, M.T. (1988). Cucumber mosaic virus-associated RNA 5. XI. Comparison of 14 CARNA 5 variants relates ability to induce tomato necrosis to a conserved nucleotide sequence. *Virology* **163**, 284–92.

Kumarasamy, R. and Symons, R.H. (1979). Extensive purification of the cucumber mosaic-virus-induced RNA-replicase. *Virology* **96**, 622–32.

Linthorst, H.J.M. and Kaper, J.M. (1985). Cucumovirus satellite RNAs cannot replicate autonomously in cowpea protoplasts. *Journal of General Virology* **66**, 1839–42.

Lot, H., Marchoux, G., Marrou, J., Kaper, J.M., West, C.K., van Vloten-Doting, L., and Hull, R. (1974). Evidence for three functional RNA species in several strains of cucumber mosaic virus. *Journal of General Virology* **22**, 81–93.

MacKenzie, D.J. and Tremaine, J.H. (1988). Ultrastructural location of non-structural protein 3A of cucumber mosaic virus in infected tissue using monoclonal antibodies to a cloned chimeric fusion protein. *Journal of General Virology* **69**, 2387–95.

Marchoux, G., Marrou, J., Devergne, J.C., Quiot, J.B., Douine, L., and Lot, H. (1975). Cucumber mosaic virus hybrids constructed by exchanging RNA compo nents. *Mededelingen van de Fakulteit Landbouwwetenschappen Rijksuniversiteit Gent* **40**, 59–72.

Marsh, L.E., Dreher, T.W., and Hall, T.C. (1988). Mutational analysis of the core and modulator sequences of the BMV RNA 3 subgenomic promoter. *Nucleic Acids Research* **16**, 981–95.

Murant, A.F. and Mayo, M.A. (1982). Satellites of plant viruses. *Annual Review of Phytopathology* **20**, 49–70.

Murthy, M.R.N. (1983). Comparison of the nucleotide sequences of cucumber mosaic virus and brome mosaic virus. *Journal of Molecular Biology* **168**, 469–75.

Nassuth, A. and Bol, J.F. (1983). Altered balance of the synthesis of plus-and minus-strand RNAs induced by RNAs 1 and 2 of alfalfa mosaic virus in the absence of RNA 3. *Virology* **124**, 75–85.

Nitta, N., Takanami, Y., Kuwata, S., and Kubo, S. (1988). Inoculation with RNAs 1 and 2 of cucumber mosaic virus induces viral RNA replicase activity in tobacco mesophyll protoplasts. *Journal of General Virology* **69**, 2695–700.

Owen, J. and Palukaitis, P. (1988). Characterisation of cucumber mosaic virus. I. Molecular heterogeneity mapping of RNA 3 in eight CMV strains. *Virology* **166**, 495–502.

Palukaitis, P. and Zaitlin, M. (1984). Satellite RNAs of cucumber mosaic virus: Characterization of two new satellites. *Virology* **132**, 426–35.

Peden, K.W.C. and Symons, R.H. (1973). Cucumber mosaic virus contains a functionally divided genome. *Virology* **53**, 487–92.

Piazzolla, P., Diaz-Ruiz, J.R., and Kaper, J.M. (1979). Nucleic acid homologies of eighteen cucumber mosaic virus isolates determined by competition hybridization. *Journal of General Virology* **45**, 361–9.

Piazzolla, P., Tousignant, M.E., and Kaper, J.M. (1982). Cucumber mosaic virus-associated RNA 5. IX. The overtaking of viral RNA synthesis by CARNA 5 and dsCARNA 5 in tobacco. *Virology* **122**, 147–57.

Quadt, R., Verbeek, H.J.M., and Jaspars, E.M.J. (1988). Involvement of a non-structural protein in the RNA synthesis of brome mosaic virus. *Virology* **165**, 256–61.

Rao, A.L.N. and Francki, R.I.B. (1982). Distribution of determinants for symptom production and host range on the three RNA components of cucumber mosaic virus. *Journal of General Virology* **61**, 197–205.

Rezaian, M.A. and Symons, R.H. (1986). Anti-sense regions in satellite RNA of cucumber mosaic virus form stable complexes with the viral coat protein gene. *Nucleic Acids Research* **14**, 3229–39.

Rezaian, M.A., Williams, R.H.V., Gordon, K.H.J., Gould, A.R., and Symons, R.H. (1984). Nucleotide sequence of cucumber-mosaic-virus RNA 2 reveals a translation product significantly homologous to corresponding proteins of other viruses. *European Journal of Biochemistry* **143**, 277–84.

Rezaian, M.A., Williams, R.H.V., and Symons, R.H. (1985). Nucleotide sequence of cucumber mosaic virus RNA 1. *European Journal of Biochemistry* **150**, 331–39.

Rizzo, T.M. and Palukaitis, P. (1988). Nucleotide sequence and evolutionary relationships of cucumber mosaic virus (CMV) strains: CMV RNA 2. *Journal of General Virology* **69**, 1777–87.

Rizzo, T.M. and Palukaitis, P. (1989). Nucleotide sequence and evolutionary relationships of cucumber mosaic virus (CMV) strains: CMV RNA 1. *Journal of General Virology* **70**, 1–11.

Roberts, P.L. and Wood, K.R. (1981). Protein synthesis in cucumber mosaic virus-infected tobacco leaves, pre-infected cells and protoplasts. *Archives of Virology* **70**, 115–22.

Schwinghamer, M.W. and Symons, R.H. (1977). Translation of the four major RNA species of cucumber mosaic virus in plant and animal cell-free systems and in toad oocytes. *Virology* **79**, 88–108.

Stussi-Garaud, C., Garaud, J-C., Berna, A., and Godefroy-Colburn, T. (1987). *In situ* location of an alfalfa mosaic virus non-structural protein in plant cell walls: correlation with virus transport. *Journal of General Virology* **68**, 1779–84.

Symons, R.H. (1975). Cucumber mosaic virus RNA contains 7-methylguanosine at the 5'-terminus of all four RNA species. *Molecular Biology Reports* **2**, 277–85.

Symons, R.H. (1979). Extensive sequence homology at the 3'-termini of the four RNAs of cucumber mosaic virus. *Nucleic Acids Research* **7**, 825–37.

Symons, R.H. (1985). In *The plant viruses, Vol. I*, (ed. R.I.B. Francki), p. 57. Plenum Press, New York.

Takanami, Y. (1981). A striking change on cucumber mosaic virus-infected tobacco plants induced by a satellite RNA. *Virology* **109**, 120–6.

Takanami, Y. and Fraenkel-Conrat, H. (1982). Comparative studies on ribonucleic acid dependent RNA polymerases in cucumber mosaic virus infected cucumber and tobacco and uninfected tobacco plants. *Biochemistry* **21**, 3161–7.

Watanabe, Y., Ooshika, I., Meshi, T., and Okada, Y. (1986). Subcellular localization of the 30k protein in TMV-inoculated protoplasts. *Virology* **152**, 414–20.

Whipple, O.C. and Walker, J.C. (1941). Strains of cucumber mosaic virus pathogenic on bean and pea. *Journal of Argicultural Research* **62**, 27–60.

Ziemiecki, A. and Wood, K.R. (1975). Protein synthesis in CMV-infected cucumber plants. *Mededelingen van de Fakulteit Landbouwwetenschappen Rijksuniversiteit Gent* **40**, 89–99.

3 Organization of the genomes of geminiviruses which infect the Gramineae

P.M. MULLINEAUX, G. CREISSEN and
H.V. REYNOLDS

*John Innes Institute, Institute of Plant Science Research, Colney Lane,
Norwich NR4 7UH, UK*

G.P. ACCOTTO

*Istituto di Fitovirologia Applicata del CNR, Strada delle Cacce 73,
10135 Torino, Italy*

C.J. WOOLSTON

*Department of Applied Biology, University of Hull, Hull HU6 7RX,
UK*

Introduction

Historically, the study of the organization and expression of the genomes of DNA viruses which infect bacteria and animals has had ramifications beyond the confines of molecular virology, both in terms of development of techniques, and as a stimulus to the study of equivalent phenomena, that is organization and expression of the genomes in their hosts. In the next few years, as more is learned about the structure and function of the genomes of plant DNA viruses, it is possible that such work will have a similar impact on the field of plant molecular biology.

Only two plant DNA virus groups have been studied in any detail, the caulimoviruses and the geminiviruses. Since the first report that geminiviruses possess genomes of circular single-stranded DNA (Harrison *et al.* 1977), and the appearance of the first complete nucleotide sequence of a geminivirus genome (Stanley and Gay 1983), most of the effort in several laboratories over the last 6 years has been concentrated on characterization of the structure of geminivirus genomes. For example, determination and analysis of the nucleotide sequences of the genomes from several different geminiviruses covering the three different sub-groups (reviewed in Howarth and Vandemark 1989); establishment of infectivity techniques for cloned virus DNA (allowing their genomes to be defined), identification of major transcription units and virus encoded proteins; mutagenesis studies, and establishment of protoplast and allied plant transformation techniques which will allow work to proceed in future. Inevitably, progress has been faster with those geminiviruses, such as tomato golden mosaic virus (TGMV) or African cassava mosaic virus (ACMV), both of which are mechanically transmissible and infect the Solanaceae, the latter group including species amenable to tissue culture and transformation techniques. Progress with those geminiviruses which infect the Gramineae has been slower because transmission of cloned DNA to host plants has not been possible until recently, and this has hampered both a complete definition of their genomes and mutagenesis studies. Tissue culture and protoplast systems for the Gramineae are technically difficult to maintain, with very few protoplast systems capable of undergoing sustained cell division. Nevertheless, these problems have been overcome and the pace of research on the Gramineae-infecting geminiviruses is quickening.

The aim of this chapter is to present a view of the organization of this subgroup of geminiviruses and to highlight the work of our laboratory on the characterization of *Digitaria* streak virus (DSV) and wheat dwarf virus (WDV).

The geminivirus group

The geminiviruses have twinned particles, approximately 20×35 nm, comprising a dimer of quasi-icosahedral units (Fig. 3.1). The geminate or twinned structure is the most common and is typical of the group, although trimers and tetramers can also be formed by some members (Francki *et al.* 1985; Davies *et al.* 1987). The members of the group can be subdivided into three categories:

1. Those that infect dicotyledenous hosts, are transmitted by whiteflies, and have bipartite (two component) genomes. Members of this subgroup include ACMV (Stanley and Gay 1983) and TGMV (Hamilton *et al.* 1984) which are the best characterized of the geminiviruses. •
2. Those that infect dicotyledenous hosts, are transmitted by leafhopper insect vectors, and have a single component genome. The best characterized of this group is beet curly top virus (BCTV; Stanley *et al.* 1986).
3. Those that infect the Gramineae, are transmitted by leafhopper insect vectors, and have a single component genome. Both DSV and WDV belong to this category. Others include *Chloris* striate mosaic virus (CSMV; Andersen *et al.* 1988) and the well characterized maize streak virus (MSV; Mullineaux *et al.* 1984).

All four Gramineae-infecting geminiviruses are transmitted by different leafhopper insect species. For example, WDV is transmitted by *Psammotettix alienus* (Vacke 1972) and DSV by *Nesoclutha declivata* (quoted in Donson *et al.* 1987).

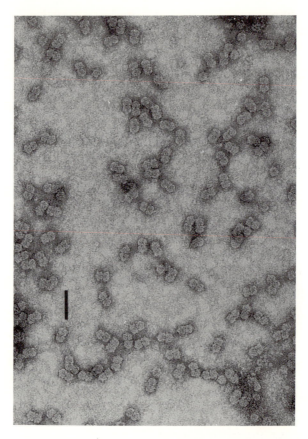

Fig. 3.1.   Electron micrograph of negatively stained geminate particles of WDV which have been purified from infected wheat tissue. The bar represents 50 nm.

WDV infects wheat (symptoms are shown in Fig. 3.2), barley, oats and several species of temperate grass, such as *Lolium* and *Poa* (Vacke 1972). Interestingly, the Czech isolate of WDV (Woolston *et al.* 1988) does not form symptoms on barley (Creissen *et al.* 1990). DSV infects *Digitaria setigera* in its native habitat (Vanuatu; Dollet *et al.* 1984) and, *via* 'agroinfection' (see below), can infect maize, oats, *Digitaria sanguinalis*, and barley (Donson *et al.* 1988, Creissen *et al.* 1990). The typical symptoms of a DSV infected *Digitaria* leaf are shown in Fig. 3.3.

None of this sub-group of geminiviruses are serologically related to those which infect dicotyledonous species (whether transmitted by whitefly or leafhopper) (Roberts *et al.* 1984). Within this subgroup there is also considerable diversity in serological relationships. For example, WDV and

Fig. 3.2.   Photograph of wheat plants (cv. Sappo). The plant on the left is uninoculated, whilst the plant on the right is infected with WDV-CJI.

MSV show no immunological cross-reactivity (Roberts *et al.* 1984), while DSV and MSV show a relationship although they are immunologically distinct (Dollet *et al.* 1984). These serological relationships have tended to be confirmed by the now more abundant nucleotide sequence data (Howarth and Vandemark 1989).

Geminivirus DNA from infected tissue

DNA forms

As with all geminiviruses, the nucleic acid from the virus particles of both DSV and WDV has been shown to be circular single-stranded (ss) DNA, of molecular weight about 900 000. (Dollet *et al.* 1984; Donson *et al.*

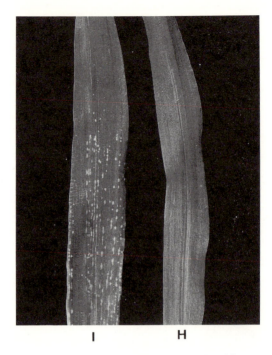

I   H

Fig. 3.3.   Leaves from uninoculated (H) and infected (I) *Digitaria sanguinalis* plants. The photograph was taken 12 days post-inoculation.

1987; 1988; Woolston *et al.* 1988). The major non-encapsidated virus DNA form from infected leaf material is a double-stranded (ds), supercoiled species which is presumed to replicate in the nucleus of the infected cell (Donson *et al.* 1988; Woolston *et al.* 1988; Creissen *et al.* 1990). Lesser amounts of multimeric ds DNA forms are frequently observed in, for example, DSV infected material (Fig. 3.4. and Donson *et al.* 1988), as well as variable amounts of open-circular and linear ds DNA species that are likely to be artefacts of the isolation procedures used.

Infectivity systems

A considerable effort has been, and continues to be, expended in development of methods for studying the infectivity of Gramineae-infecting geminiviruses in whole plants, in protoplasts, and more recently, in selected organs and tissues.

Agroinfection

The major breakthrough which allowed defined (i.e. DNA-sequenced) geminivirus genomes to be transmitted to whole plants was the technique

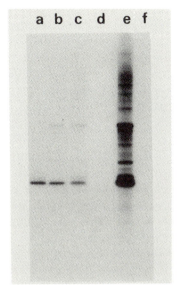

Fig. 3.4. Southern transfer hybridization of a 1.2 per cent (w/v) neutral agarose gel probed with *in vitro* labelled cloned DSV DNA. DSV virion DNA (lane a) and cellular DNA (lane b) from naturally DSV infected *Digitaria sanguinalis*. Cellular DNA from agroinfected (lane c) and healthy (lane d) *D. sanguinalis* plants 21 days post-inoculation. Lanes (e) and (f) are an increased exposure of lanes (c) and (d).

called agroinfection, first developed for transmission of cloned cauliflower mosaic virus DNA (Grimsley *et al.* 1986). Agroinfection is defined as the transmission of viral DNA to plant cells that is mediated by *Agrobacterium*. The observation that dimers of MSV DNA, cloned between the right and left borders of Ti plasmid T-DNA, could be transmitted to maize when a suspension of the bacteria were injected into the meristematic region of the plant, was an important one (Grimsley *et al.* 1987). Since then, agroinfection of DSV (Donson *et al.* 1988) and WDV (Woolston *et al.* 1988; Hayes *et al.* 1988a) has been achieved. Agroinfection allows studies to be performed on plants infected with a defined geminivirus genome (e.g. Accotto *et al.* 1989). In addition, this procedure can make use of a geminivirus as a sensitive marker to study the interaction between *Agrobacterium* and cereal plants (Boulton *et al.* 1989a; Dale *et al.* 1989), or with anther-culture derived barley embryos (Creissen *et al.* 1990).

Protoplast studies

We have been able to show that WDV DNA can replicate in protoplasts prepared from suspension culture cells of *Triticum monococcum*

(Woolston *et al.* 1989). The non-regenerable cell-suspension culture of *T. monococcum* was first described for use in gene transfer experiments by Lorz *et al.* (1985). The rapid rates of growth of the suspension cultures, and the relatively high percentage of protoplasts which will undergo sustained cell division (compared with many suspension cultures of the Gramineae) have allowed us to develop procedures for observing replication and expression of the WDV genome. It should be noted that any WDV based construct which can replicate in *T. monococcum* protoplasts can be used subsequently in other whole-cell inoculation procedures.

One of the early experiments conducted in our laboratory, using the plasmid pWDVK10D, is typical of the response that can be obtained. pWDVK10D is a tandem dimer, of the wild-type WDV-CJI genome, in pIC20R (Woolston *et al.* 1989). Inoculation of pWDVK10D DNA into protoplasts of *T. monococcum* leads to production of novel WDV specific DNA forms. These show the same electrophoretic mobilities as the double-stranded forms of the viral DNA that are found in WDV infected plants, and can be first detected 24 hours after inoculation (Fig. 3.5). An increase in abundance is observed during the course of the experiment (Fig. 3.5 and 3.6). Digestions of these WDV-specific forms with restriction enzyme can be used to confirm that they are dsDNA forms (open-circular, linear, and supercoiled respectively) of the circular genome. Inoculated protoplasts start to divide approximately three days after inoculation. By five to six days post-inoculation, sustained cell division reaches a maximum of about 10 per cent of the original protoplast population. After eight days in culture, protoplast viability shows a significant decline.

Degradation of the inoculum DNA appars to occur immediately follow-ing inoculation, and gives rise to the heterogeneous smear of WDV-specific DNA in the zero-day timepoint on gel electrophoresis. However, about five per cent of the input DNA is maintained apparently intact in subsequent timepoints (Fig. 3.5 and 3.6). This probably represents the amount of DNA which is protected from degradation by delivery into the protoplasts, and from which WDV-specific forms are derived.

Additional data is provided by Woolston *et al.* (1989) which demon-strates that the novel ds WDV specific DNA forms are products of replica-tion in the protoplasts and not accumulated products of homologous recombination of the input DNA. Restriction enzyme isoschizomers, which recognize different DNA methylation patterns on the same sequence, have been used to show that the novel ds forms of WDV DNA are generated in the protoplasts. Further indirect evidence that the novel ds forms undergo replication rather than recombinational escape from the plasmid input DNA, is provided by the observation that DSV DNA does not replicate in *T. monococcum* protoplasts. DSV DNA ought to be as

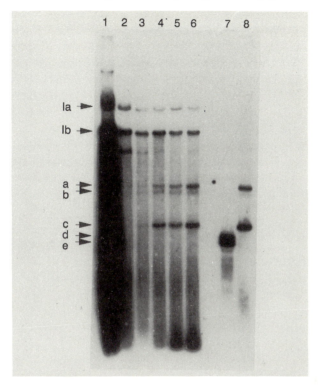

Fig. 3.5. Autoradiograph of a Southern blot, using a full length WDV-specific probe, of DNA samples extracted from inoculated protoplasts at 0 days (1), 1 day (2), 2 days (3), 3 days (4), 4 days (5), and 5 days (6) post-inoculation (dpi) with pWDVK10D. Lane 7 is viral ssDNA and lane 8 is dsDNA extracted from infected plants. Bands indicated are inoculum DNA (1a and 1b), progeny dsDNA; open circular (a), linear (b), supercoiled (c), viral ssDNA (d, e; lane 7 only).

capable of recombinational escape from plasmid based tandem sequences as WDV (Woolston *et al.* 1989).

There is no evidence for replication of the input DNA in the protoplasts. Recent experiments in our laboratory suggest that the failure of such a construct to replicate is due to sequences on the plasmid pIC20R, rather than the size of the insert DNA *per se* (H. Reynolds and E. Dekker unpublished observations). Currently, this protoplast system can only be used for WDV.

Alternative methods of delivering geminivirus DNA to plant cells

An alternative approach is the 'biolistic process', in which high velocity microprojectiles are used to deliver DNA to cells, tissues and organelles (Sanford 1988). High success rates were obtained using the particle

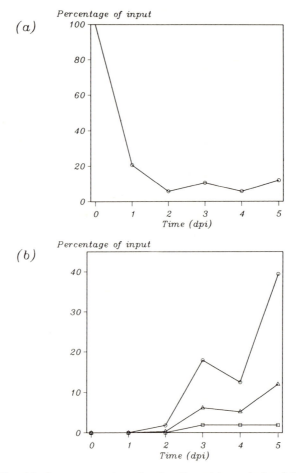

Fig. 3.6. Graphical representation of a densitometric analysis of the autoradiograph shown in Fig. 3.2. Panel (a) shows the changes in the amount of inoculum DNA over the course of the experiment. Panel (b) shows the changes in amounts of the progeny DNA forms (open-circular△, linear □, and supercoiled ○). Results are expressed as the percentage of the input DNA amount at t=0.

bombardment approach with barley anther-culture derived embryos. For example, following bombardment with the plasmid pWDVK10D Southern analysis of the DNA prepared from the cultures confirmed the presence of replicative forms of the virus in all of the replicate treatments (Criessen *et al.* 1990). Representative samples are shown in Fig. 3.7. Again methylation tests show that the WDV DNA has been synthesized *de novo* in the barley embryos.

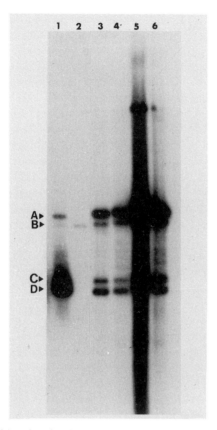

Fig. 3.7. Southern blot showing WDV infection following particle bombardment with pWDVK10D. Lanes 1 and 2, WDV markers; 3–6, DNA extracted from four separate replicates of barley microspore derived material 1 week after bombardment. The different WDV forms indicated by arrows are A, open circular ds DNA; B, linear ds DNA; C, covalently closed circular ds DNA; D, viral ssDNA.

Even simpler approaches are possible. Recently, Topfer *et al.* (1989) has shown that DNA can be delivered to dried cereal embryos by imbibition, and productive expression of a reporter gene was subsequently obtained. This expression was shown to be partly dependent on replication of the virus DNA.

Sequencing of geminivirus genomes

Over the last 5 years the nucleotide sequences of the four geminiviruses that infect the Gramineae have been determined. This includes sequence data from wild type virus populations, different strains of the same virus and more recently, from infectious cloned DNA (reviewed by Howarth and Vandemark 1989). Before the advent of an inoculation system for the Gramineae-infecting geminiviruses, it was the accommodation into one circle of DNA of the sequence data from 'shotgun' cloning experiments which provided the best evidence for their single component genome (Mullineaux *et al.* 1984; Donson *et al.* 1987; Andersen *et al.* 1988). This approach has now been superceded by the agroinfection system (see below) for this subgroup of geminiviruses (Grimsley *et al.* 1987), a technique which has been used to confirm that their genomes consist of a single circular molecule of DNA (Donson *et al.* 1988; Lazarowitz 1988; Woolston *et al.* 1988).

The nucleotide sequences from infectious clones of a Swedish isolate (WDV-S; MacDowell *et al.* 1985; Hayes *et al.* 1988a) and a Czechoslovak isolate (WDV-CJI; Woolston *et al.* 1988) have been determined for WDV. Although, in contrast to the normal infection caused by the WDV-CJI cloned DNA (Woolston *et al.* 1988), the WDV-S clone does not appear to elicit symptoms on wheat (Hayes *et al.* 1988a). For DSV, the sequence of the wild type virus DNA population, and of an infectious clone of a single isolate, has been determined (Donson *et al.* 1987; 1988). No differences between the two sequences were detected (Donson *et al.* 1988).

At the nucleotide sequence level, there is no homology between those geminiviruses which infect the monocots and those which infect dicots, except for a nononucleotide sequence (5' TAATATTAC 3') which is present in the head of the most energetically favourable stem-loop structures from all the geminiviruses sequenced so far (Davies *et al.* 1987). This hairpin and its conserved sequence is always located in the large intergenic region (Fig. 3.8), or common region of the bipartite geminiviruses (Davies *et al.* 1987). There is sequence homology among the different Gramineae-infecting geminiviruses. For example, 46 per cent between WDV and MSV (MacDowell *et al.* 1985), although most of this homology is between the sequences coding for the complementary sense C2 ORF.

The circular genome maps of WDV-CJI (Woolston *et al.* 1988) and DSV (Donson *et al.* 1987) which have been derived from their nucleotide sequences are shown in Fig. 3.8. Open reading frames (ORFs) are numbered clockwise from the zero coordinate if encoded by the virion (V) strand (i.e. that strand which is encapsidated as ssDNA), and numbered anticlockwise if encoded by the complementary (C) strand (only present in unencapsidated ds DNA). Following the convention established by Stanley and Gay (1983), only those ORFs which code for a polypeptide greater than 10 kDa when measured from the first in-frame AUG translation initiation codon, are included. The exception is the C2 ORF of WDV and DSV, whose derived amino acid sequence shares extensive homology with those of other geminiviruses, including that of MSV which does have an

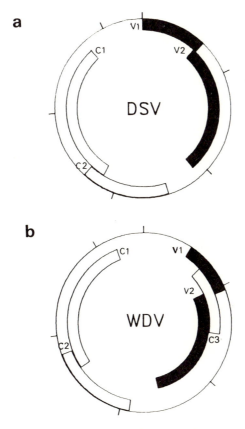

Fig. 3.8. Potential coding regions within DSV DNA (a) and WDV DNA (b). The position and orientation of regions with the capacity to code for polypeptides >10 kDa from the first in-frame ATG triplet are shown. The exception is C2 which does not have a 5' proximal ATG. The product of this open region was calculated from the first codon. Virion (+) sense DNA runs in a clockwise direction.

AUG codon near the beginning of its C2 ORF (Mullineaux *et al.* 1984). Following these criteria, only four ORFs seem to be common to the Gramineae-infecting geminiviruses, as typified by DSV (Fig. 3.8a). Extra ORFs located on other geminivirus sequences, e.g. C3 on WDV (Fig. 3.8b), often show no conservation of position and considerable variation of sequence, even within different isolates of the same virus such as the case with the Nigerian, Kenyan, and South African isolates of MSV (Mullineaux *et al.* 1984; Howell 1984; Lazarowitz 1988).

Detection and possible functions of virus encoded proteins

Virion sense encoded products

The capsid of geminiviruses appears to be made of a single protein encoded by the virus. For the Gramineae-infecting geminiviruses, the coat protein was shown to be encoded by the V2 ORF of MSV by performing hybrid arrest translation experiments on poly (A)+ RNA prepared from infected maize tissue, followed by immuneprecipitation of the coat protein polypeptide using an antiserum raised against purified virus particles (Morris-Krsinich *et al.* 1985). The size of the coat protein monomer, as measured on SDS polyacrylamide gels (about 28 kDA), and the experimentally determined amino acid content are in agreement with the predicted values derived from sequence data. The predicted amino acid sequences of the V2, ORF products of DSV, WDV, and CSMV show good homology to that of MSV, and therefore must also code for their respective capsid proteins (MacDowell *et al.* 1985; Donson *et al.* 1987; Woolston *et al.* 1988, Andersen *et al.* 1988). In addition to their structural role, the coat protein may also determine insect vector specificity (Stanley *et al.* 1986; Briddon *et al.* 1989), although the precise sequences involved have yet to be elucidated.

Immediately upstream of the coat protein ORF, there is the V1 ORF (Fig. 3.8) which could encode for a polypeptide of about 11 kDa (Mullineaux *et al.* 1988). The product of the V1 ORF of MSV has been detected both in protein extracts and from among the *in vitro* translation products from RNA which had been prepared from infected tissue, using an antiserum raised against a β-galactosidase-V1 fusion protein which was expressed in, and purified from, *E. coli* (Mullineaux *et al.* 1988). Mutagenesis studies have shown that the sequences of the V1 ORF are essential for infection, and may also be involved in symptom formation and cell-to-cell spread (Mullineaux *et al.* 1988; Boulton *et al.* 1989b; Lazarowitz *et al.* 1989). As with the V2 ORF, there is no direct information on the nature of the V1 product from WDV and DSV, although there is sufficient homology (Fig. 3.9) to suggest an identical role. One further feature worthy of comment is the unusual number of proline residues present for polypeptides of this size (Fig. 3.9). The precise function of the V1 product remains to be elucidated, but any proposal must take into account the fact that the V1 product and its sequences are found in only those geminiviruses which infect the Gramineae.

Complementary sense encoded products

The potential polypeptides of M_r approximately 31 kDa and 17 kDa, that are coded by the C1 and C2 ORFs respectively (Fig. 3.8), contain amino

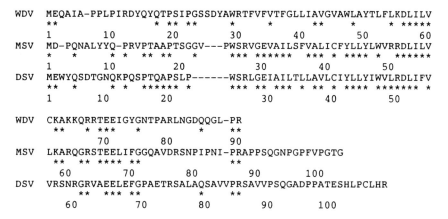

Fig. 3.9. Comparisons of the amino acid sequences of the V1 encoded polypeptides of WDV, MSV, and DSV respectively. Asterisks indicate the position of
amino acids which are identical in at least two of the three polypeptides. The amino
acid residues are contiguous, however gaps (indicated by dashed lines) have been
introduced in the sequences for maximum alignment.

acid sequences which are the best conserved across the geminivirus sub-
groups (Mullineaux *et al.* 1985; Donson *et al.* 1987). Apart from the
conserved sequences in the C1 ORF, which are indicative of a function fundamental to the geminivirus life-cycle (Mullineaux *et al.* 1985), there are no
further clues to a possible function for the product of this ORF. It is
unlikely that the product of the C2 ORF is expressed as a 17 kDa protein,
since in both DSV and WDV there is no translation initiation codon present in the 5' region of the ORF. However, C2 is expressed as a fusion to
C1, brought about by the splicing-out from the nascent virus encoded
RNA, of an intron located at the junction between the two ORFs (see
below and Accotto *et al.* 1989; Schalk *et al.* 1989). Since both spliced and
unspliced RNAs cover both complementary sense ORFs (see below and
Accotto *et al.* 1989), two protein products which share common N-termini
but differ at their C-termini, can be made from one transcription unit. The
product of a spliced mediated fusion of C1 and C2 would be approximately
41 kDA, retaining all the previously identified conserved sequences, and
would be equivalent to the AC1 (or AL 1) product of dicot-infecting geminiviruses (Accotto *et al.* 1989; Schalk *et al.* 1989). Mutants of WDV which
do not express functional C2 sequences cannot replicate in protoplasts of
Triticum monococcum suspension-culture cells. However, the C1:C2 product seems to be the only virus encoded product required for replication of
its ds DNA (Schalk *et al.* 1989). The involvement of the C1:C2 protein in
DNA replication is underscored by the recent observation that the most

conserved sequences of the C2 ORF contain a motif which is indicative of their ability to bind dNTP, possibly DNA (Gorbalenya and Koonin. 1989). It cannot be ruled out that the C1:C2 protein, like its counterpart the large T antigen of SV40, has more than one function (Depamphilis and Wasserman 1982). For example, it could be involved in the modulation of virion sense promoter activity or activation of host genes involved in DNA replication.

Transcription

Bidirectional transcription of MSV and DSV has been reported. Using S1 nuclease mapping techniques, Morris-Krsinich *et al.* (1985) were able to map the 5' and 3' ends of the virion sense transcripts, and despite being barely detectable, gave approximate coordinates to the complementary sense transcript of MSV. A more recent study of the transcripts of DSV which employed the same techniques (Accotto *et al.* 1989), revealed a similar pattern of transcripts encoded by the virion strand, but the synthesis of the complementary strand encoded RNAs was more complex than had been suspected from the data obtained from MSV (Fig. 3.10). The virion strand encoded transcripts (termed 1+ and 2+; Fig. 3.10), have 3Δ ends which are coterminal and distal to conventional polyadenylation sequences (AATAAA; Proudfoot and Brownlee 1976). The 5' ends of the WDV virion strand encoded transcript, and the 1+ versions of the DSV and MSV encoded RNAs are located downstream of a consensus eukaryotic promoter 'TATA' box (Breathnach and Chambon 1981). However, in contrast to DSV and MSV, preliminary mapping data (C.J. Woolston unpublished data) for WDV virion sense RNAs indicate that only one transcript (equivalent to 1+) is present in infected wheat leaves.

No consensus promoter sequences can be located upstream of the 5' ends of the 2+ RNAs of MSV and DSV (see Fig. 3.10). It is possible that the synthesis of these RNAs is directed by a promoter which does not have a recognisable eukaryotic consensus sequence. Alternatively, it may be that the 2+ transcripts are derived from a specific cleavage of the 1+ transcripts. The recent work of Fennol *et al.* (1988) may shed some light on this problem. They have observed that deletion of sequences around co-ordinate 0 in the MSV large intergenic region, including the 5' end of transcript 1+, abolished expression of a V2 (coat protein)-*cat* gene fusion, assayed in maize protoplasts derived from a suspension-culture line, using a transient expression system. These data could be interpreted as removal of sequences necessary for conversion of a 1+ transcript to a 2+ transcript. These workers also showed that the virion sense promoter may be composed of a core 'TATA' element and an enhancer-like sequence (Upstream Activating Sequence or UAS), in an arrangement similar to

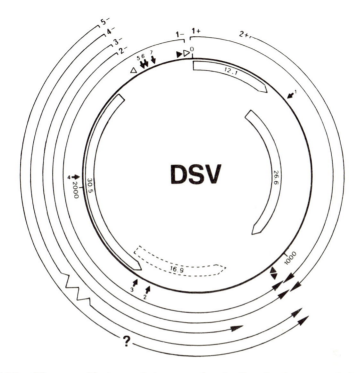

Fig. 3.10. Virus specific transcripts mapped onto the circular map of the DSV genome. The positions of the ORFs, consensus promoters (▷), polyadenylation sequences (►), and potential hairpin structures ($\Delta G > -14$ kcal/mol; arrows numbered 1–7) are shown. The consensus sequences can be found in the text. The zigzag (W) in transcript 4– indicates the approximate position of the intron in the DSV sequence. The question mark (?) in transcript 5– indicates the approximate position of an internal S1 nuclease sensitive site (Accotto *et al*. 1989) which may have been caused by hairpin 2.

that of the cauliflower mosaic virus (CaMV) 35S promoter (Ow *et al*. 1987). The core 'TATA' sequence of the CAMV 35S promoter was activated by the fusion of one or more copies of UAS. Interestingly, using this fusion, the levels of transiently expressed MSV RNA were elevated, but appeared to be only of the 2+ form, no 1+ transcript being detected. However, the differences between the data derived from infected plant material and that from suspension-culture derived protoplasts which are expressing transiently, may be due to the abnormal cell types encountered in the latter system, and the use of only a portion of the MSV genome in their study.

Attempts to align sequences upstream from the conserved 'TATA' sequence revealed a surprising lack of homology in this region (C.J. Woolston unpublished data). The next conserved sequence upstream from the 'TATA' box was the nonanucleotide sequence of the head of the most stable stem loop structure. This failure to align stretches of sequence in the large intergenic regions of these viruses places more significance on those sequences which are conserved, and suggests that functional analysis of this region will be required for all of the viruses.

A detailed analysis of the complementary sense encoded transcripts, using S1 nuclease mapping techniques, is available only for DSV (Fig. 3.10, Accotto *et al.* 1989) and it raises as many questions as it provides answers.

A consensus eukaryotic promoter sequence (TATAa/tAa/t; Breathnach and Chambon 1981) is located about 20 nucleotides upstream of the mapped 5' terminus in each of RNAs 2−, 3− and 4− from DSV (Fig. 3.11). This sequence is conserved in the other viruses. The region upstream of the 5' terminus of RNA 1− contains no conserved sequences in DSV (Fig. 3.10 and 3.11), nor a recognisable promoter sequence in any of the other Gramineae infecting geminiviruses.

Splicing of complementary sense RNAs

Analysis of S1 nuclease resistant complementary sense RNA:DNA hybrids in two dimensional gel systems, revealed the presence of a single internal S1 nuclease sensitive site in both RNA 4− and 5− (Fig. 3.10, Accotto *et al.* 1989). The internal S1 nuclease sensitive site of RNA 4− was located approximately between coordinates 1750 and 1650 on the DSV genome. A search of the sequence in this region revealed good matches to splice donor and acceptor sites at coordinates 1759 and 1666 respectively (Accotto *et al.* 1989). The position of the 3' end of the 5' exon, corresponding to the splice donor site in the nascent RNA, was confirmed by high resolution S1 nuclease mapping (Accotto *et al.* 1989). Simulated splicing, to give RNA 4−, and subsequent translation (Fig. 3.12) shows that the removal of a 92 base intron results in a fusion of the C1 and C2 ORFs to give a 41 kDA polypeptide (see above).

The internal S1 nuclease sensitive site in RNA 5− (Fig. 3.10) is located approximately between coordinates 1550 and 1500 (Accotto *et al.* 1989). In contrast to the situation with RNA 4−, no acceptable matches to splice donor and acceptor sequences were located in the region of this internal S1 nuclease sensitive site. However, a stem loop structure (hairpin No.2; Fig. 3.10) is located at that site, which could account for the observed S1 nuclease sensitivity.

In addition to a spliced DSV specific transcript, there is also the more abundant unspliced RNA 2− which contains the spliced sequences which

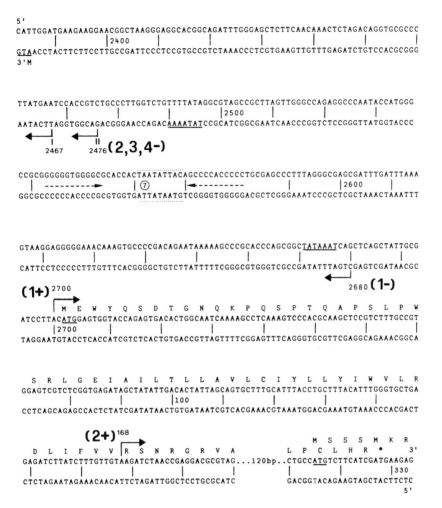

Fig. 3.11. Location of the 5' ends of virus specific transcripts on the DSV sequence (2383–334). The 5' termini and directions of transcription are marked (⌐→). Consensus promoter sequences and the initiator codons of the ORFs shown in Fig. 3.10 are underlined. The conserved nonanucleotide sequence in the head of stem-loop 7 (see text) is marked by a dotted line.

are absent in RNA 4− (Fig. 3.10). This suggests that the maturation of virus specific nascent RNAs, during the infection cycle can be differentially regulated (Accotto *et al.* 1989). Selective processing of RNAs occurs in mammals (Leff *et al.* 1986), and among their viruses, particularly in the adenovirideae and the papovavirideae (Depamphilis and Wasserman 1982; Broker 1984). It occurs also in *Drosophila* (Leff *et al.* 1986). However, we

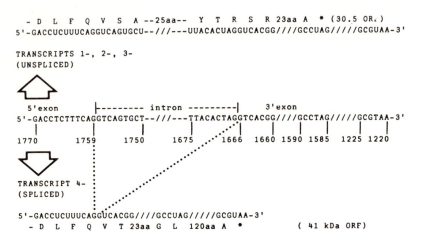

```
   - D   L   F   Q   V   S   A  --25aa--   Y   T   R   S   R 23aa A   *  (30.5 OR.)
5'-GACCUCUUUCAGGUCAGUGCU--///---UUACACUAGGUCACGG////GCCUAG/////GCGUAA-3'

TRANSCRIPTS 1-, 2-, 3-
(UNSPLICED)
```

```
   5'exon        |--------- intron ---------|     3'exon
5'-GACCTCTTTCAGGTCAGTGCT--///---TTACACTAGGTCACGG////GCCTAG/////GCGTAA-3'
   |         |        |         |         |     |    |    |    |    |
   1770      1759     1750      1675      1666  1660 1590 1585 1225 1220

TRANSCRIPT 4-
(SPLICED)

5'-GACCUCUUUCAGGUCACGG////GCCUAG/////GCGUAA-3'
   - D   L   F   Q   V   T 23aa G   L   120aa A   *          ( 41 kDa ORF)
```

Fig. 3.12. Computer generated transcription and translation of the complementary sense strand of the DSV genome between coordinates 1770 and 1220.

are not aware of any reported cases of selective processing of the transcripts of a single gene in plants.

A similar observation of the presence of a spliced and unspliced RNA has been made by analysis of polymerase-chain-reaction (PCR)-amplified cDNAs, synthesized using primers, flanking a putative intron in the WDV genome (Schalk *et al.* 1989). The presence of an equivalent unspliced transcript was also likely, although the techniques employed could not distinguish between a PCR amplified product derived from a cDNA or from contaminating virus DNA.

The problem of multiple minor complementary sense transcripts

Although two proteins could be synthesized by the RNAs from the complementary sense transcription unit, it does not explain the presence of a minimum of four RNAs which may be coded by the complementary sense strand of DSV (Fig. 3.10). Could these RNAs have an alternative or additional role?

RNA 1− has a 5' terminus at coordinate 2680 (Fig. 3.11) and does not code for any additional polypeptides (P.M. Mullineaux unpublished data). However, the most stable stem loop structure and its conserved nonanucleotide sequence (see above) are transcribed as RNA 1−. The free energy of formation of this stem loop structure in DSV is −48.9 kcal/mol (Donson *et al.* 1987) and if it is formed in RNA 1− it could be inhibitory to the translation of the C1 ORF. The introduction of a hairpin structure of $\triangle$G for synthesis = 50 kcal/mol, into the untranslated 5' leader sequence of a pre-proinsulin RNA, completely abolished translation of the downstream cistron (Kozak 1986). Below this value translation of the preproinsulin coding sequence was not significantly inhibited, presumably because scanning ribosomes were able to 'melt out' any secondary structure encountered (Kozak 1986). Therefore, the free energy value for formation of the stable hairpin of DSV could be of some significance. The values for free energy of formation of the equivalent hairpin structures of MSV, WDV and CSMV are similar. To follow this argument, if RNA 1− cannot be translated to produce a C1 product, then what is its role? Possible functions could include involvement in the replication of virus DNA, or a regulatory role.

The presence of RNA 3− is difficult to explain. Its 3' end maps at about coordinate 1200 (Fig. 3.10), different from the rest of the complementary sense transcripts. The nearest possible polyadenylation sequence (AATAA) is at coordinate 1350, well away from its 3' end. In addition, it potentially codes for the C1 and C2 ORFs, the same as RNA 2− does. However, it is interesting to note that its 3' end maps in the vicinity of the 5' end of the RNA moiety of the virion-associated primer that is a feature of all the geminiviruses which infect the Gramineae (see below).

Evidence for regulation of transcription in DSV

There is extensive overlapping of transcripts in DSV (Fig. 3.10) and in MSV. Transcript 2+ is internal to the sequences which code for transcript 1+, the complementary sense RNAs 2−, 3−, 4−, and 5− are transcribed from the same part of the DSV genome and are internal to the longer RNA 1−, and the 3' termini of the virion sense and complementary sense RNAs overlap. This extensive overlapping of transcripts may be interpreted as indirect evidence for temporal regulation of gene expression, if the assumption is made that when analysis of RNAs is performed in infected plants all stages of the virus life cycle are represented. In addition, it is difficult to postulate that the same stretch of DNA could simultaneously serve as a promoter and be transcribed from a promoter further upstream, either on the same or opposite strand. For example, the potential promoter(s) of RNAs 2−, 3−, 4−, and 5− are transcribed as RNA 1−. As well as the complexities of transcript synthesis by the viruses there is also the regulation of RNA maturation. If temporal regulation of transcription does occur it is likely to be highly complex and is not likely to be clearly defined until synchronized single-cell or protoplast systems become available.

DNA replication

Many aspects of the interconversion between single-stranded and double-stranded forms of geminivirus DNAs, and the presence of replicative intermediates remain to be elucidated. However, some data are available.

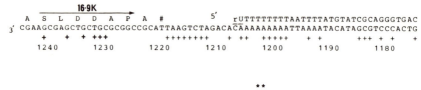

Fig. 3.13. Nucleotide sequence of the small DNA molecules associated with DSV. The sequence is shown aligned with the viral strand sequence. All the bases are deoxyribonucleotides except for the U residue and an indeterminate number of other undefined ribonucleotides (r) at the 5' terminus of the small DNA molecules. The four most predominant 3' termini ($_*$* followed by *) are indicated. Sequences conserved with MSV are indicated (+). Alignment of DSV and MSV sequences for the homology shown required the insertion of five padding characters (−) in the DSV sequence at position 1161 on the virion sense strand. The position of the stop codon (#) and the single letter codes of the eight carboxy terminal amino acids of the product of the C2 ORF (16.9k) are given.

Conversion of circular single-stranded (ss) DNA to double-stranded (ds) DNA

Virion DNA which has been prepared from purified virions of the Gramineae-infecting geminiviruses, and which appears to be primarily circular and single-stranded, is capable of directing synthesis of complementary strand DNA without the addition of an exogenous primer providing that nucleotides, Mg^{2+} and DNA polymerase I are supplied (Donson *et al.* 1984, 1987; Hayes *et al.* 1988b; Andersen *et al.* 1988). This self-priming ability results from the presence of a population of 'nested' DNA molecules found in association with the circular ssDNA. These molecules have ribonucleotides covalently linked to the discrete 5' deoxyribonucleotide terminus of the DNA species. The major species has a DNA sequence of about 80 nucleotides and is located almost exclusively in the small intergenic region of these viruses (Fig. 3.8 and 3.13). No primer-like molecule has been found associated with the geminiviruses that infect dicotyledonous hosts (Stanley and Townsend 1985).

The ribonucleotides which have been detected at the 5' end of these small molecules could represent the vestiges of a larger RNA primer laid down prior to encapsidation. In the four geminiviruses investigated in this respect, there appear to be no unifying features immediately 5' to the ribonucleotide moieties suggestive of a possible recognition site for a primase complex. It is conceivable that a priming RNA species could be long enough to have a 5' end several hundred bases away from the recognition site and this may be an explanation of some of the minor RNAs encoded by the complementary strand of the DSV genome (Accotto *et al*. 1989; Fig. 3.10).

Possession of a primer would allow synthesis of a ds DNA species to proceed immediately upon entry of the virus into the nucleus of a new host cell. However, Donson *et al*. (1984) have suggested that the encapsidated DNA represents essentially a replicative intermediate, which if it had not been packaged would have been converted to a ds DNA form.

Possible routes of synthesis of ssDNA and ds DNA

A further implication of the model above is that the production of mainly ssDNA for the virions is brought about by an interruption of DNA polymerase activity by encapsidation. The observed heterogeneity of the 3' ends of these small primer molecules would support this view (Donson *et al*. 1984, 1987). Further circumstantial evidence for this model comes from studies on replication of WDV DNA in protoplasts of *Triticum monococcum* suspension-culture cells (Woolston *et al*. 1989). In this system, WDV ds DNA replicates, but no virus particles or free coat protein is detected. The levels of ssDNA are very low, compared with infected leaf tissue. It could be argued that ssDNA does not accumlate because it cannot be encapsidated and therefore is converted to ds DNA. However, much more experimentation is needed to establish whether these views are correct.

While nothing is known about the mechanism of ds DNA replication, some authors have suggested that the presence in infected tissues of concatameric DNA forms may be indicative of a rolling circle mode of replication. From the various studies on replication of WDV ds DNA in protoplasts of *Triticum monococcum*, it is clear that only one virus encoded protein is required, namely the C1:C2 ORFs fusion product (see above; Schalk *et al*. 1989). It is also presumed that a cis acting origin of replication is required and may be located in the large intergenic region (Figure 8) (Mullineaux *et al*. 1984; MacDowell *et al*. 1985; Donson *et al*. 1987; Andersen *et al*. 1988).

It is worth noting that transcription and DNA replication may be tightly coupled processes in the geminiviruses, subject to a complex mode of control. For example, the C1:C2 product of DSV and WDV is clearly required for DNA replication. The C1:C2 protein can only be synthesized when the RNA transcribing these ORFs is spliced and in turn, this latter process may itself be controlled (Accotto *et al*. 1989).

Future prospects

The work presented above, represents our preliminary efforts towards understanding the genome organisation of this important group of DNA viruses. We are in a position to identify some areas which are worthy of detailed study because they ought to advance our understanding of similar processes which are certain to go on in the host. For example, differential splicing, the arrangement of transcription regulatory sequences and DNA replication. The possibility that the supercoiled forms of the geminiviruses may exist as a minichromosome (Abouzid *et al.* 1988), opens up the exciting prospect of examining the interaction between host and viral proteins, and defining host genes which may be involved in processes such as DNA replication.

References

Abouzid, A.M., Frischmuth, T., and Jeske, H. (1988). A putative replicative form of the abutilon mosaic virus (gemini group) in a chromatin-like structure. *Molecular and General Genetics* **212**, 252–8.

Accotto, G.P., Donson, J., and Mullineaux, P.M. (1989). Mapping of *Digitaria* streak virus transcripts reveals different RNA species from the same transcription unit. *EMBO Journal* **8**, 1033–9.

Andersen, M.T., Richardson, K.A., Harbison, S-A, and Morris, B.A.M. (1988). Nucleotide sequence of the geminivirus chloris striate mosaic virus. *Virology* **164**, 443–9.

Breathnach, R. and Chambon, P. (1981). Organization and expression of eucaryotic split genes coding for proteins. *Annual Review of Biochemistry* **50**, 349–83.

Boulton, M.I., Buchholz, W.G., Marks, M.S., Markham, P.G., and Davies, J.W. (1989a). Specificity of *Agrobacterium* mediated delivery of maize strak virus DNA to members of the Gramineae. *Plant Molecular Biology* **12**, 8711–21.

Boulton, M.I., Steinkellner, H., Donson, J., Markham, P.G., King, D.I., and Davies, J.W. (1989b). Mutational analysis of the virion sense genes of maize streak virus. *Journal of General Virology* **70**, 2309–23.

Briddon, R.W., Watts, J., Markham, P.G., and Stanley, J. (1989). The coat protein of beet curly top virus is essential for infectivity. *Virology* **172**, 628–33.

Broker, T.R. (1984). Animal virus RNA processing. In *Processing of RNA*, (ed. D. Apirion), pp. 182–203. CRC Press, Boca Raton FL, USA.

Creissen, G., Smith, C., Francis, R., Reynolds, H., and Mullineaux, P. (1990). *Agrobacterium*- and microprojectile- mediated viral DNA delivery into barley microspore-derived cultures. *Plant Cell Reports* **8**, 680–3.

Dale, P.J., Marks, M.S., Brown, M.M., Woolston, C.J., Gunn, H.V., Mullineaux, P.M., Lewis, D.M., Kemp, J.M., Chen, D.F., Gilmour, D.M., and Flavell, R.B. (1989). Agroinfection of wheat: Inoculation of in vitro grown seedlings and embryos. *Plant Science* **63**, 237–45.

Davies, J.W., Stanley, J., Donson, J., Mullineaux, P.M., and Boulton, M.I. (1987). Structure and replication of geminivirus genomes. *Journal of Cell Science Supplement* **7**, 95–107.

Depamphilis, M.L. and Wasserman, P.M. (1982). Organisation and replication of papovavirus DNA. In Kaplan, A.S. (ed.), *Organisation and replication of viral DNA*, (ed. A.S. Kaplan), pp. 38–114. CRC Press, Boca Raton FL, USA.

Dollet, M., Accotto, G.P., Lisa, V., Menissier, J., and Boccardo, G. (1984). A geminivirus, serologically related to maize streak virus, from *Digitaria sanguinalis* from Vanuatu. *Journal of General Virology* **67**, 933–7.

Donson, J., Morris-Krsinich, B.A.M., Mullineaux, P.M., Boulton, M.I., and Davies, J.W. (1984). A putative primer for second-strand DNA synthesis of maize streak virus is virion-associated. *EMBO Journal* **3**, 3069–73.

Donson, J., Accotto, G.P., Boulton, M.I., Mullineaux, P.M., and Davies, J.W. (1987). The nucleotide sequence of a geminivirus from *Digitaria sanguinalis*. *Virology* **161**, 160–9.

Donson, J., Gunn, H.V., Woolston, C.J., Pinner, M.S., Boulton, M.I., Mullineaux, P.M., and Davies, J.W. (1988). *Agrobacterium* mediated infectivity of cloned *Digitaria* streak virus DNA. *Virology* **162**, 248–50.

Fenoll, C., Black, D.M., and Howell, S. (1988). The intergenic region of maize streak virus contains promoter elements involved in rightward transcription of the viral genome. *EMBO Journal* **7**, 1589–96.

Francki, R.I.B., Milne, R.G., and Hatta, T. (eds) (1985). The geminivirus group. In *Atlas of plant viruses* Vol.1, pp. 33–46. CRC Press, Boca Raton FL, USA.

Gorbalenya, A.E. and Koonin, E.V. (1989). Viral proteins containing the putative NTP-binding sequence pattern. *Nucleic Acids Research* **17**, 8413–30.

Grimsley, N., Hohn, B., Hohn, T., and Walden, R. (1986). 'Agroinfection', an alternative route for viral infection of plants by using the Ti plasmid. *Proceedings of the National Academy of Sciences USA* **83**, 3282–6.

Grimsley, N., Hohn, T., Davies, J.W., and Hohn, B. (1987). *Agrobacterium*-mediated delivery of infectious maize streak virus into maize plants. *Nature* **325**, 177–9.

Hamilton, W.D.O., Stein, V.E., Coutts, R.H.A., and Buck, K.W. (1984). Complete nucleotide sequence of the infectious cloned DNA components of tomato golden mosaic virus: Potential coding regions and regulatory sequences. *EMBO Journal* **3**, 2197–205.

Howarth, A.J. and Vandemark, G.J. (1989). Phylogeny of geminiviruses. *Journal of General Virology* **70**, 2717–27.

Harrison, B.D., Barker, H., Bock, K.R., Guthrie, E.J., Meredith, G., and Atkinson, M. (1977). Plant viruses with circular single-stranded DNA. *Nature* **270**, 760–2.

Hayes, R.J., MacDonald, H., Coutts, R.H.A., and Buck, K.W. (1988a). Agroinfection of *Triticum aestivum* with cloned DNA of wheat dwarf virus. *Journal of General Virology* **69**, 891–6.

Hayes, R.J., MacDonald, H., Coutts, R.H.A., and Buck, K.W. (1988b). Priming of complementary DNA synthesis *in vitro* by small DNA molecules tightly bound to virion DNA of wheat dwarf virus. *Journal of General Virology* **69**, 1345–50.

Howell, S.H. (1984). Physical structure and genetic organisation of the genome of maize streak virus (Kenyan isolate). *Nucleic Acids Research* **12**, 7359–75.

Kozak, M. (1986). Influences of mRNA secondary structure on initiation by eukaryotic ribosomes. *Proceedings of the National Academy of Sciences USA* **83**, 2850–4.

Lazarowitz, S.G. (1988). Infectivity and complete nucleotide sequence of the genome of a South African isolate of maize streak virus. *Nucleic Acids Research* **16**, 229–50.

Lazarowitz, S.G., Pinder, A.J., Damsteegt, V.D., and Rogers, S.G. (1989). Maize streak virus genes essential for systemic spread and symptom development. *EMBO Journal* **8**, 1023–32.

Leff, S.E., Rosenfeld, M.G., and Evans, R.M. (1986). Complex transcriptional units: Diversity in gene expression by alternative RNA processing. *Annual Review of Biochemistry* **55**, 1091–117.

Lorz, H., Baker, B., and Schell, J. (1985). Gene transfer to cereal cells mediated by protoplast transformation. *Molecular and General Genetics* **199**, 178–82.

MacDowell, S.W., MacDonald, H., Hamilton, W.D.O., Coutts, R.H.A., and Buck, K.W. (1985). The nucleotide sequence of cloned wheat dwarf virus DNA. *EMBO Journal* **4**, 2173–80.

Morris-Krsinich, B.A.M., Mullineaux, P.M., Donson, J., Boulton, M.I., Markham, P.G., Short, M.N., and Davies, J.W. (1985). Bidirectional transcription of maize streak virus DNA and identification of the coat protein gene. *Nucleic Acids Research* **13**, 7237–56.

Mullineaux, P.M., Donson, J., Morris-Krisinich, B.A.M., Boulton, M.I., and Davies, J.W. (1984). The nucleotide sequence of maize virus DNA. *EMBO Journal* **3**, 3063–8.

Mullineaux, P.M., Donson, J., Stanley, J., Boulton, M.I., Morris-Krsinich, B.A.M., Markham, P.G., and Davies, J.W. (1985). Computer analysis identifies sequence homologies between potential gene products of maize streak virus and those of cassava latent virus and tomato golden mosaic virus. *Plant Molecular Biology* **5**, 125–31.

Mullineaux, P.M., Boulton, M.I., Bowyer, P., van der Vlugt, R., Marks, M., Donson, J., and Davies, J.W. (1988). Detection of a non-structural protein of M_r 11 000 encoded by the virion DNA of maize streak virus. *Plant Molecular Biology* **11**, 57–66.

Ow, D., Jacobs, J.D., and Howells, S.H. (1987). Functional regions of the cauliflower mosaic virus 35S RNA promoter determined by use of the firefly luciferase gene as a reporter of promoter activity. *Proceedings of the National Academy of Sciences USA* **84**, 4870–4.

Proudfoot, N.J. and Browlee, G.G. (1976). *Nature* **263**, 211

Roberts, I.M., Robinson, D.J., and Harrison, B.D. (1984). Serological relationships and genome homologies among geminiviruses. *Journal of General Virology* **65**, 1723–30.

Sanford, J.C. (1988). The biolistic process. *Trends in Biotechnology* **6**, 299–302.

Schalk, H.-J., Matzeit, V., Schiller, B., Schell, J., and Gronenborn, B. (1989). Wheat dwarf virus, a geminivirus of graminaceous plants needs splicing for replication. *EMBO Journal* **5**, 1761–7.

Stanley, J. and Gay, M.R. (1983). Nucleotide sequence of cassava latent virus DNA. *Nature* **301**, 260–2.

Stanley, J. and Townsend, R. (1985). Characterization of DNA forms associated with cassava latent virus infection. *Nucleic Acids Research* **13**, 2189–206.

Stanley, J., Markham, P.G., Callis, R.J., and Pinner, M.S. (1986). The nucleotide sequence of an infectious clone of the geminivirus beet curly top virus. *EMBO Journal* **5**, 1761–7.

Topfer, R., Gronenborn, B., Schell, J., and Steinbiss, H-H. (1989). Uptake and transient expression of chimaeric genes in seed-derived embryos. *The Plant Cell* **1**, 133–9.

Vacke, J. (1972). Host plants, range and symptoms of wheat dwarf virus. *Vedecke Prace Vyzk Ustavu Rostl Vroby, Praha-Ruzyne* **17**, 151–62.

Woolston, C.J., Barker, R., Gunn, H., Boulton, M.I., and Mullineaux, P.M. (1988). Agroinfection and nucleotide sequence of cloned wheat dwarf virus DNA. *Plant Molecular Biology* **11**, 35–43.

Woolston, C.J., Reynolds, H.V., Stacey, N.J., and Mullineaux, P.M. (1989) Replication of wheat dwarf virus DNA in protoplasts and analysis of coat protein mutants in protoplasts and plants. *Nucleic Acids Research* **17**, 6029–41.

4 Comovirus capsid proteins: synthesis, structure, and evolutionary implications

G.P. LOMONOSSOFF, M. SHANKS,
C.L. HOLNESS, and A.J. MAULE

Department of Virus research, John Innes Institute and AFRC Institute of Plant Science Research, Colney Lane, Norwich NR4 7UH, UK

D. EVANS

Department of Biological Sciences, University of Warwick, Coventry CV4 7AL, UK

Z. CHEN, C.V. STAUFFACHER, and
J.E. JOHNSON

Department of Biological Sciences, Purdue University, West Lafayette, Indiana 47907, USA

Introduction

Coat proteins play an important role in several aspects of the replication cycle of plant viruses. Their most obvious function is to provide a protective shell surrounding the relatively labile genetic material, which in the case of most plant viruses is single-stranded RNA. This protective function is most evident during transmission of the virus from host to host, when the virions are often exposed to harsh conditions. This is particularly true in the case of viruses such as comoviruses, which are beetle-transmitted. Beetle regurgitant, the natural medium for the virus inoculum, contains such high levels of ribonuclease that the RNA of many non-beetle transmitted viruses is degraded despite being encapsidated (Gergerich *et al.* 1986; Gergerich and Scott 1988). The protective function of coat proteins is not, however, limited to that stage of the replication cycle involving transmission from host to host; the presence of functional coat protein is often required for efficient spread of a virus within a host plant (see Maule Chapter 1). Thus mutants of tobacco mosaic virus (TMV) with defects in their coat protein spread only slowly in susceptible plants (Siegel *et al.* 1962; Dawson *et al.* 1988).

Recent work has shown that as well as providing protection for the viral nucleic acid, the coat proteins of a number of viruses can have a dramatic effect on the response of the host to infection. For example, a point mutation in the coat protein of TMV produces a virus that induces a hypersensitive response in a host that does not respond in this fashion when infected with the wild-type TMV (Knorr and Dawson 1988). In addition, mutants of TMV with deletions within the coat protein can induce symptoms that are dramatically different from those observed during a wild-type infection (Dawson *et al.* 1988). The mechanism whereby mutations in the coat protein of a virus affect the host response to infection is unclear at present.

Properties of comoviruses

Comoviruses form a group of fourteen plant viruses which predominantly infect legumes. Their genomes consist of two molecules of single-stranded, positive-sense RNA of different sizes that are separately encapsidated within isometric particles of approximately 28 nm diameter. As a consequence of their behaviour in caesium chloride density gradients the two types of nucleoprotein particles are termed middle (M) and bottom (B) component, the RNAs within the particles being known as M and B RNA, respectively. Both types of particle have an identical protein composition, consisting of 60 copies each of a large, (VP37) and a small, (VP23) coat protein. In addition to the nucleoprotein particles, comovirus preparations contain a variable amount of empty (protein-only) capsids which are known as top (T) component.

To date, much of the work on the molecular biology of comoviruses has been concentrated on the type member, cowpea mosaic virus (CPMV). In the case of CPMV, it is known that both M and B RNA are polyadenylated and have a small protein (VPg) covalently linked to their 5' termini. More limited studies on other comoviruses suggest that these features are shared by the RNAs of all members of the group. Both RNAs from CPMV have been sequenced and shown to consist of 3481 nucleotides in the case of M RNA and 5889 in the case of B RNA, excluding the poly(A) tails (van Wezenbeek *et al.* 1983; Lomonossoff and Shanks 1983). Both RNAs contain a single, long open reading frame, expression of the viral gene products occurring through synthesis and subsequent cleavage of large precursor polypeptides. Though both RNAs are required for infection of whole plants, the larger B RNA is capable of independent replication in protoplasts, though no virus particles are produced in this case (Goldbach *et al.* 1980). This observation, coupled with earlier genetic studies (Gopo and Frist 1977), established that the coat proteins are encoded by M RNA.

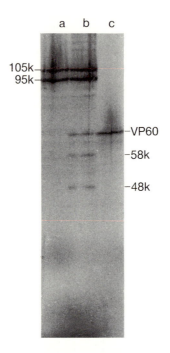

Fig. 4.1.   Proteolytic processing of the *in vitro* translation products of CPMV M RNA. Full-length CPMV M RNA was transcribed from a full-length cDNA clone using SP6 RNA polymerase and was then translated for 60 min at 30° in rabbit reticulocyte lysates in the presence of [^{35}S]-methionine. Lanes a and b show, respectively, the M RNA products before and after incubation with the translation products of CPMV B RNA. Lane c shows the products synthesized in response to SP6 transcripts from a cDNA clone that contains just the sequence of the two CPMV coat proteins. Fractionation of the products was on a 10 per cent polyacrylamide/SDS gel.

Expression of comovirus coat proteins

Genetic map of CPMV M RNA

Translation of CPMV M RNA in either wheat germ (Davies *et al.* 1977) or rabbit reticulocyte lysates (Pelham 1979), leads to synthesis of two proteins of apparent molecular weights 105 and 95 kDa (the 105k and 95k proteins, see Fig. 4.1). A similar pattern of synthesis is observed when the M RNAs of other comoviruses are translated *in vitro*, the exact sizes of the products being virus-specific (Goldbach and Krijt 1982). Since the combined size of the 105 and 95k proteins clearly exceeded the coding capacity of CPMV M

RNA, Pelham (1979) suggested that the two proteins must be related, and demonstrated that they were, in fact, carboxy co-terminal.

Once synthesized, the primary products of CPMV M RNA are processed by a protease that is encoded by B RNA to give the mature viral proteins. *In vitro* the processing reaction can be mimicked, at least in part, by treating the M RNA-specific 105 and 95k proteins with the translation products of B RNA (Fig. 4.1). Three products arise during such a treatment, the 58k protein, the 48k protein, and VP60. Peptide mapping and immunological experiments have shown that the 58 and 48k proteins correspond to the amino terminal portions of the 105 and 95k proteins respectively, and that VP60 is the precursor to the two viral coat proteins (Pelham 1979; Franssen *et al.* 1982). At some stage during the viral replication cycle, VP60 must be cleaved to give the mature virus coat proteins, VP37 and VP23. The little that is known about the timing and mechanism of this process will be discussed in a later section.

Comparison both of the sizes of the *in vitro* translation products, and of the terminal sequences of the cleavage products with the amino acid sequence of the long open reading frame, predicted from the M RNA sequence, has led to the derivation of a genetic map for CPMV M RNA (Fig. 4.2). Although all of the cleavage products shown in Fig. 4.2 have now been detected in CPMV-infected cowpea plants and/or protoplasts (Wellink *et al.* 1987; Holness *et al.* 1989), the 105 and 95k proteins have yet to be observed *in vivo*, presumably because they are processed rapidly.

The mechanism of translation of CPMV M RNA

The fact that the 105 and 95k proteins were carboxy co-terminal and that the kinetics of their syntheses were similar, led Pelham (1979) to conclude that they arose as a result of alternative initiation events. Examination of the nucleotide sequence of CPMV M RNA shows that the 5' proximal AUG codon occurs at position 115 in the sequence (Fig. 4.3). However, this AUG precedes an open reading frame of only 60 residues and is in a poor context i.e. it is flanked by nucleotides which do not correspond to the consensus sequence believed to be important for efficient initiation of translation (Lutcke *et al.* 1987; Kozak 1986, 1989). The out-of-phase AUG at 115 probably does not play a significant role in the translation properties of M RNA since it is not conserved in the M RNA sequence of the related comovirus, red clover mottle virus (RCMV) (Shanks *et al.* 1986). The next AUG encountered in the sequence occurs at position 161 and is in-phase with the only long open reading frame found on CPMV M RNA. Initiation at position 161 would lead to the synthesis of a protein of 116kDa—a plausible candidate for the 105k *in vitro* translation product (van Wezenbeek *et al.* 1983). From inspection of the M RNA sequence, van

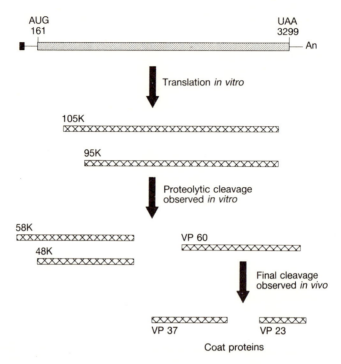

Fig. 4.2. Model for the expression of CPMV M RNA. The top line represents the RNA with the long open reading frame shown as a shaded bar. The positions of the translational initiation and termination codons at the extremities of this open reading frame are indicated. The box at the 5' end of the RNA represents the VPg, and An represents the poly(A) tail. The proteins produced as a result of translation and processing are shown as cross-hatched boxes.

Wezenbeek *et al.* (1983) identified AUGs at 512 and 524 as the most likely initiators for the synthesis of the 95k protein.

To investigate further the translation of CPMV M RNA, we have made mutants of a full-length cDNA copy of M RNA that had been inserted in the transcription vector pPMl (Ahlquist and Janda 1984). RNA transcribed from the full-length clone, pPMM2902, has properties identical to virion M RNA, including the ability to multiply in protoplasts when co-inoculated with virion B RNA (Holness *et al.* 1989). Four site-directed mutants of pPMM2902 were constructed: pGM161 with the AUG at 161 changed to AGC (coding for serine), pGM512 with the AUG at 512 changed to CUC (coding for leucine), pGM524 with the AUG at 524 changed to ACU (coding for threonine), and pGM512+524 with the two AUGs at positions 512

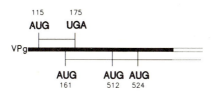

5′ REGION OF CPMV M–RNA

Sequences around the AUGs

115	U U G C **A U G** A G	
161	U A C A **A U G** U U	
512	U G A A **A U G** G A	
524	C A U U **A U G** A G	
Plant consensus	A A C A **A U G** G C	
Kozak consensus	C R C C **A U G** G –	

Fig. 4.3. Model of the 5′ terminal region of CPMV M RNA. The heavy line represents the RNA with the positions of the first 4 AUGs indicated. The short open reading frame which follows the AUG at 115 is shown above the RNA, together with the position of the in-phase terminator. The long open reading frame in-phase with AUGs at 161, 512, and 524 is shown below the RNA. The 'Plant consensus' is derived from Lutcke *et al.* (1987) and the 'Kozak consensus' is from Kozak (1989).

and 524 changed to CUC and ACU, respectively. Transcripts were made from these mutants and the biological effect of the mutation were assessed.

The effects of the mutations outlined above on the *in vitro* translation properties of the pPMM2902-based transcripts, are shown in Table 4.1. While transcripts from the wild-type pPMM2902 directed synthesis of both the 105 and 95k proteins in reticulocyte lysates, transcripts from pGM161 gave only the 95k protein. This result identifies unambiguously the AUG at position 161 as the initiator for the synthesis of the 105k protein, at least *in vitro*. The situation regarding the initiation site for synthesis of the 95k protein is somewhat less clear. Transcripts from pGM512 and pGM524 can still direct synthesis of both the 105 and 95k proteins, synthesis of the 95k product only being abolished in the case of the double AUG mutant, pGM512+524. These observations suggest that either of the codons at 512 and 524 can act as the initiator for 95k protein synthesis in reticulocyte lysates.

In order to confirm that the results obtained *in vitro* were relevant *in vivo*, transcripts from the four mutants were electroporated into cowpea protoplasts. Transcripts from pGM512, pGM524, and pGM512+524 all multiplied in the presence of B RNA, and all directed synthesis of viral coat protein (Table 4.1). The properties of the double mutant pGM512+524 imply that synthesis of the 95k protein is neither essential for the virus multiplication in protoplasts nor a pre-requisite for coat protein production. The only transcripts which failed to replicate were from pGM161. The most probable reason for this failure is that the mutations that were introduced affected not only the translational properties of the RNA but also its ability to be recognized by the B RNA-encoded replication machinery. This idea is supported by the fact that some transcripts with different mutations in the 5' region also failed to multiply (Holness, unpublished).

While the failure of pGM161 to replicate precluded a direct assessment of the role of AUG at 161 in translation *in vivo*, analysis of the functions of the AUGs at 512 and 524 was possible. Western blot analysis of the proteins produced in protoplasts that had been electroporated with transcripts from pGM512 and pGM524, showed that both these mutants could still direct synthesis of the 48k as well as the 58k protein (Table 4.1). Since the 48k protein is derived from the 95k precursor (see Fig. 4.2), these results confirm the conclusion of the *in vitro* translation experiments that 95k synthesis can initiate from the AUG at either position 512 or 524. The fact that transcripts from the double mutant pGM512+524 did not produce any 48k protein is also consistent with the AUG at either 512 or 524 being the initiator for 95k synthesis.

The results obtained with the mutant transcripts confirm that the two

Table 4.1 Properties of site-directed mutants of CPMV M RNA

Name of construct	*In vitro* translation products[a]	Multiplication in protoplasts[b]	Production of 48k and 58k proteins[c]	Production of coat proteins[d]
pPMM2902	105k, 95k	YES	48k, 48k	YES
pGM161	95k	NO	*	*
pGM512	105k, 95k	YES	48k, 58k	YES
pGM524	105k, 95k	YES	48k, 58k	YES
pGM512+524	105k	YES	58k	YES

[a] Determined in rabbit reticulocyte lysates.
[b] Determined by Northern blot analysis of nucleic acids extracted from electroporated protoplasts.
[c] Determined by Western blots of the proteins extracted from electroporated protoplasts. The blots were probed with anti-58/48k serum.
[d] Determined by immunofluorescent staining of electroporated protoplasts using anti-capsid serum.
* Not detected as transcripts failed to replicate.

primary translation products of CPMV M RNA arise as a result of the use of alternative initiation codons. This occurs most probably as a result of 'leaky' scanning, many ribosomes failing to initiate at position 161 because of the relatively poor context of this AUG (Fig. 4.3). The ribosomes which fail to initiate at 161 would carry on 'scanning' the M RNA sequence until they reach a suitable initiation codon. Inspection of the sequences around the next two AUGs (at positions 512 and 524) suggests that of the two the AUG at 512 is in the better context. Though both AUGs have a purine at the −3 position, only the AUG at 512 has a G at the +4 position, a feature regarded as important for initiation in both plants and animals (Lutcke *et al.* 1987; Kozak 1986, 1989). The results with mutant pGM524 clearly show that the AUG at 512 can act as an initiator and since ribosomes will encounter this AUG before the one at 524, at least some 95k protein synthesis will initiate here. Though the properties of the mutant pGM512 demonstrate that the AUG at 524 can also act as an initiator, whether it is actually used when the AUG at 512 is present is uncertain. For the AUG at 524 to be used in wild-type M RNA, the AUG at 512 would have to be 'leaky', a possibility rendered less likely, but not ruled out, by its favourable context.

Though the function of the double initiation event on CPMV M RNA is not clear, the fact that a similar phenomenon occurs with the M RNAs of all the comoviruses that have been examined to date suggests that it is important. However, it is unlikely that double initiation has a significant role in production of the coat proteins since transcripts of the mutant pGM512+524, which direct the synthesis of only the 105k protein, still produce functional coat protein in protoplasts. The significance of the phenomenon is more likely to be that it allows the production of two proteins, the 58 and 48k proteins, which differ at their N-termini. One or both of these proteins are believed to play a role in the cell-to-cell spread of comoviruses (see Maule, Chapter 1).

Virus assembly

The actual process of assembly of the comoviruses is poorly understood. The first step of the process is presumably release of VP60 from the 105k and 95k precursors. This cleavage takes place at a glutamine–methionine dipeptide and appears to involve two B RNA-encoded proteins (Vos *et al.* 1988). VP60 appears to have only a transient existence in plant cells, accumulation to detectable amounts in cowpea protoplasts occurring only when they have been treated with Zn^{2+} ions to inhibit further proteolysis (Wellink *et al.* 1987). The mechanism whereby VP60 is cleaved to give the mature coat proteins, VP37 and VP23, is uncertain as this cleavage has yet to be achieved *in vitro*. In view of the precise 1:1 stoichiometry of VP37 and VP23 in CPMV virions, it seems plausible that the cleavage takes place after the initial stages of capsid formation.

Little is known about the initial stages of the interaction between the viral RNAs and the coat proteins. A notable feature of preparations of comoviruses is that approximately 20 per cent of the particles consist of empty capsids i.e. particles devoid of RNA. These empty particles are similar in overall structure to the RNA-containing ones and are only slightly less stable. The existence of capsids that are empty shows that strong interactions which are independent of the presence of RNA, occur between the coat proteins, and it also suggests that the RNA may be packaged once much of the capsid has been formed.

The influence of particle assembly on accumulation of the viral RNA in infected cells is marked. When cowpea protoplasts are inoculated with isolated B RNA, RNA replication occurs but the unencapsidated RNA fails to accumulate to a significant extent (de Varennes and Maule 1985). A similar effect is observed when M RNA-specific transcripts with deletions or frame-shifts in the viral coat proteins are co-inoculated with B RNA (Holness, unpublished).

The structure of comovirus capsids

Studies on cowpea mosaic virus

Investigations into the structure of comoviruses initially focused on CPMV. The first studies involved image reconstruction from electron micrographs and established the icosohedral symmetry of the virus (Crowther *et al.* 1974). However, it was only after the crystallization of CPMV (White and Johnson 1980) that a detailed picture of the arrangement of the protein subunits began to emerge, culminating in the production of a 3.5Å electron density map of the virus (Stauffacher *et al.* 1987).

The 3.5Å map showed that there was a clear relationship between CPMV and the T=3 plant viruses such as tomato bushy stunt virus (TBSV) and southern bean mosaic virus (SBMV). The capsids of these latter viruses are composed of 180 identical coat protein subunits, each consisting of a single β-barrel domain. These can occupy three different positions, A, B, and C, within the virions (Harrison *et al.* 1978; Abad-Zapatero *et al.* 1980; Fig. 4.4). The two coat proteins of CPMV were shown to consist of three distinct β-barrel domains, two being derived from VP37 and one from VP23. Thus, in common with the T=3 viruses, each CPMV particle is made up of 180 β-barrel structures. The single domain from VP23 occupies a position analogous to that of the A type subunits of TBSV and SBMV, whereas the N- and C-terminal domains of VP37 occupy the positions of the C and B type subunits respectively (Fig. 4.4).

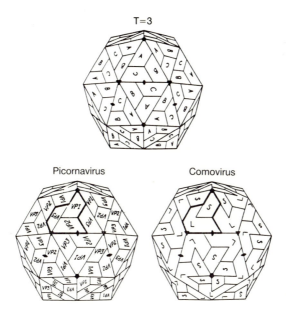

Fig. 4.4. Comparison of the structures of simple T=3 virus, picornavirus and comovirus capsids. The icosohedral asymmetric unit of the T=3 shell contains three identical subunits, labelled A, B, and C. The asymmetric unit of picornaviruses (shown in heavy outline) contains three different subunits known as VP1, VP2, and VP3. The comovirus capsid is similar to the picornavirus capsid except that VP2 and VP3 are covalently linked to form the large (L) coat protein (referred to in the text as VP37), the small coat protein (S or VP23) corresponding to VP1.

As well as demonstrating the structural relationship between CPMV and the T=3 plant viruses, the 3.5Å map also established a link between the (plant) comoviruses and the (animal) picornaviruses. The capsids of picornaviruses consist of 60 copies each of three different coat proteins VP1, VP2, and VP3, each one consisting of a single β-barrel domain. As in the case of comoviruses, these coat proteins are released by cleavage of a precursor polyprotein and are synthesized in the order VP2–VP3–VP1. Comparison of the 3-dimensional structure of CPMV with that of picornaviruses established that the N- and C-terminal domains of VP37 are equivalent to VP2 and VP3 respectively, and that VP23 is equivalent to VP1 (Fig. 4.4). The equivalence between structural position and gene order suggests that VP37 corresponds to an uncleaved form of the two picornavirus capsid proteins, VP2 and VP3.

Studies on bean pod mottle virus

Though the 3.5Å electron density map of CPMV enabled the polypeptide backbone of both coat protein subunits to be traced, a detailed interpretation of the side-chain densities was not possible (Stauffacher *et al*. 1987). However, X-ray diffraction analysis of crystals of another member of the group, bean pod mottle virus (BPMV), has enabled the structure of a comovirus to be determined at atomic resolution (Chen *et al*. 1989). The structure of BPMV has been solved using crystals containing only purified middle components and used the co-ordinates obtained from CPMV as an aid to the structure determination. The 3Å electron density map so produced for BPMV was of very high quality and allowed all of the 374 amino acids of VP37, and 185 out of the 198 residues of VP23 to be located. The missing 13 residues of VP23 are from the C-terminus and are probably absent from the particles as a result of proteolytic cleavage.

The 3Å structure of BPMV has both confirmed the general features of the quarternary structure of comoviruses first observed for CPMV, and has provided additional detail concerning the structure of the individual subunits and the interactions between them. The N-terminal domains of VP37, which occupy the C positions in the structure (Fig. 4.4), consist of the first 182 amino acids of VP37. The bulk of this domain consists of an eight-stranded anti-parallel β-barrel. However, the 20 or so amino terminal residues are in an extended conformation and make intersubunit contacts across the two-fold axes on the inside of the virus particles. The extreme end of this 'arm' also makes contact with the viral RNA (see below). The B domains consist of the C-terminal 192 residues of VP37, once again arranged in the form of an eight-stranded β-barrel. The five C-terminal residues of this domain can follow two different routes. The route of lower occupancy is the one which brings the C-terminus of VP37 closer to the N-terminus of VP23, though the two termini are still 12Å apart. This finding implies that there is a substantial rearrangement in the structure of the coat proteins when VP60 is cleaved. The VP23 molecules, which occupy the A positions around the 5-fold axes are ten-stranded β-barrels of somewhat streamlined appearance.

As well as revealing details of the structure of the protein subunits, the 3Å map of BPMV contains an unexpected feature—some prominent density corresponding to ordered RNA. A chain of seven ribonucleotides can be fitted to this density at all of the 60 icosohedral asymmetric units. The bases of the RNA in these segments are stacked, and the sugar-phosphate backbone stereochemistry is similar to that found in one strand of an A-type RNA helix. The seven well-ordered ribonucleotides are located near the three-fold axes and lie in a pocket formed by the N- and C-terminal domains of VP37, the surface of VP37 being rich in hydrophilic residues.

The interactions between the protein and the RNA are mainly through van der Waals with only a few specific contacts.

Further examination of the 3Å map has revealed that the 3' terminal ribonucleotide of each of the seven ordered nucleotide segments is connected to the 5' terminal ribonucleotide of a three-fold related segment by electron density that is one third the height of the ordered RNA density. Four ribonucleotides in an extended conformation can be built into this density though the fit is ambiguous. This second segment of RNA makes no close contact with the coat protein and is somewhat mobile. The total RNA modelled from both the well-ordered and the disordered segments corresponds to 660 nucleotides per particle—about 20 per cent of the RNA that is packaged in middle components.

Comparison of comovirus coat protein sequences

Alignment of the primary amino acid sequences of the coat protein precursors (VP60), of the three comoviruses for which this information is currently available (CPMV, BPMV, and RCMV), demonstrates that a high degree of overall homology exists between the coat proteins of these closely related viruses. However, the degree of homology is not uniform but depends markedly on that portion of the sequence being compared. This is illustrated in Fig. 4.5, where the percentage homology between the CPMV and BPMV coat proteins is plotted as a function of amino acid number. The highest degree of conservation is found in the C-terminal domain of VP37 (the B domain) and the least in the VP23 (the A domain). It is particularly noticeable that the homology drops to an extremely low value in two places, around the cleavage site between the VP37 and VP23 (the border between the B and A domains) and at the C-terminus of VP23. The low degree of sequence conservation in these two regions suggests that they do not play an important role in the structure of the virions. Indeed the C-terminal portions of VP23 are known to lie on the surface of the virions and can be removed by proteolysis without affecting capsid stability. It is also probable that the sequence around the VP37–VP23 cleavage site is relatively unstructured and is near the surface of the uncleaved VP60 molecule. These features may be important for proteolytic cleavage of VP60 to take place to form VP37 and VP23.

An intriguing feature of the relative conservation of primary sequences in the three comovirus domains is that the situation is exactly mirrored in picornaviruses. Thus VP3 is the most conserved protein among picornaviruses while the VP1 is the most variable.

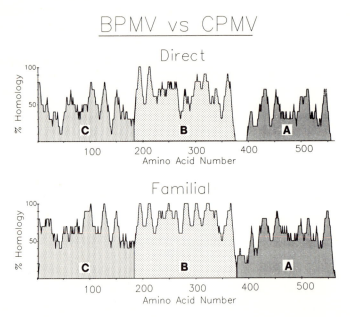

Fig. 4.5. Homology plot comparing the amino acid sequences of the coat proteins of CPMV and BPMV. The percentage homology is shown as a function of position in the sequence and is calculated either by scoring only identical amino acids (Direct), or by allowing substitution of related amino acids (Familial) as defined by Schwartz and Dayhoff (1978). The A, B, and C domains, as defined by the 3Å structure of BPMV, are shown in different shading.

Evolutionary implications of structure of comovirus capsids

With the single exception of satellite tobacco necrosis virus (Liljas *et al.* 1982), the protein shells of all simple spherical viruses examined to date consist of 180, eight-stranded anti-parallel β-barrels (ESABs). In the case of simple plant viruses, these 180 ESABs are provided by 180 identical subunits. In contrast, the 180 ESABs in picornaviruses are provided by 60 copies each of three different coat proteins. Since the first option is more economical in terms of the coding capacity required by the virus, Hosur *et al.* (1987) suggested that the added complexity displayed by picornaviruses gives them a selective advantage in terms of flexibility in the capsid structure. This flexibility is probably important in the ability of picornaviruses to cope with immune surveillance. The fact that comoviruses use a strategy in their capsid structure very reminiscent of picornaviruses supports the idea that they (the comoviruses) may be derived from a picorna-like ancestor that passed from animals to plants (Goldbach 1986).

Although the quarternary structures of comoviruses clearly resemble those of picornaviruses, it appears that comoviruses are structurally less sophisticated. Thus the protein subunits of comoviruses lack the extended N- and C-termini and the large insertions between the strands of the β-barrels that are found in picornaviruses. The surface loops are the principal antigenic determinants in picornaviruses and one of them probably plays a role in receptor binding (Rossmann *et al*. 1985). The simplification of comovirus capsids probably represents an adaptation of these viruses to their plant hosts, since plants do not possess an immune system, and virus infection is mediated by mechanical damage to the host cells rather than through binding to specific receptors.

Conclusion

In the last few years, the coat proteins of comoviruses have received considerable attention from several points of view. These viewpoints include their mode of synthesis, their role in the transmission of the viruses and their 3-dimensional structures. For technical reasons, these aspects have been studied more or less independently of each other. However, development of cDNA clones from which infectious RNA can be obtained suggests that a more integrated approach to an understanding of the functions of the coat proteins may be possible in the future. Using the cDNA clones, it should be possible to introduce defined mutations into the coat protein region of M RNA with a view to an examination of the effect of the mutations on the phenotype of the resulting virus. In this way it should be possible to gain an insight into such diverse phenomena as nucleic acid-protein interactions, the mechanism of proteolytic cleavage and the mechanism of virus transmission. All in all, it seems comoviruses have a lot to contribute to the future.

References

Abad-Zapatero, C., Abdel-Meguid, S.S., Johnson, J.E., Leslie, A.G.W., Rayment, I., Rossmann, M.G., Suck, D., and Tsukihara, T. (1980). Structure of southern bean mosaic virus at 2.8Å resolution. *Nature* **286**, 33–9.

Ahlquist, P. and Janda, M. (1984). cDNA cloning and *in vitro* transcription of the complete brome mosaic virus genome. *Molecular and Cellular Biology* **4**, 2876–82.

Chen, Z., Stauffacher, C., Li, Y., Schmidt, T., Bomu, W., Kamer, G., Shanks, M., Lomonossoff, G.P., and Johnson, J.E. (1989). Protein-nucleic acid interactions in a spherical virus: the structure of bean pod mottle virus at 3.0Å resolution. *Science* **245**, 154–9.

Crowther, R.A., Geelen, J.L.M.C., and Mellema, J.E. (1974). A three-dimensional reconstruction of cowpea mosaic virus. *Virology* **57**, 20–7.

Davies, J.W., Aalbers, A.M.J., Stuick, E.J., and Van Kammen, A. (1977). Translation of cowpea mosaic virus RNA in a cell-free extract from wheat germ. *FEBS Letters* **77**, 265–9.

Dawson, W.O., Bubrick, P., and Grantham, G.L. (1988). Modifications of the tobacco mosaic virus coat protein gene affecting replication, movement and symptomatology. *Phytopathology* **78**, 783–9.

De Varennes, A. and Maule, A.J. (1985). Independent replication of cowpea mosaic virus bottom component RNA: *in vivo* instability of the viral RNAs. *Virology* **144**, 495–501.

Franssen, H., Goldbach, R., Broekhuijsen, M., Moerman, M., and Van Kammen, A. (1982). Expression of middle-component RNA of cowpea mosaic virus: *in vitro* generation of a precursor to both capsid proteins by a bottom-component RNA-encoded protease from infected cells. *Journal of Virology* **41**, 8–17.

Gergerich, R.C. and Scott, H.A. (1988). The enzymatic function of ribonuclease determines plant virus transmission by leaf-feeding beetles. *Phytopathology* **78**, 270–2.

Gergerich, R.C., Scott, H.A., and Fulton, J.P. (1986). Evidence that ribonuclease in beetle regurgitant determines the transmission of plant viruses. *Journal of General Virology* **67**, 367–70.

Goldbach, R. (1986). Molecular evolution of plant RNA viruses. *Annual Review of Phytopathology* **24**, 289–310.

Goldbach, R. and Krijt, J. (1982). Cowpea mosaic virus-encoded protease does not recognise primary translation products of M-RNA from other comoviruses. *Journal of Virology* **43**, 1151–4.

Goldbach, R., Rezelman, G., and van Kammen, A. (1980). Independent replication and expression of B-component RNA of cowpea mosaic virus. *Nature* **286**, 297–300.

Gopo, J.M. and Frist, R.H. (1977). Location of the gene specifying the smaller protein of the cowpea mosaic virus capsid. *Virology* **79**, 259–66.

Harrison, S.C., Olson, A.J., Schutt, C.E., Winkler, F.K., and Bricogne, G. (1978). Tomato bushy stunt virus at 2.9Å resolution. *Nature* **276**, 268–373.

Holness, C.L., Lomonossoff, G.P., Evans, D., and Maule, A.J. (1989). Indentification of the initiation codons for translation of cowpea mosaic virus middle component RNA using site-directed mutagenesis of an infectious cDNA clone. *Virology* **172**, 311–20.

Hosur, M.V., Schmidt, T., Tucker, R.C., Johnson, J.E., Gallagher, T.M., Selling, B.H., and Rueckert, R.R. (1987). Structure of an insect virus at 3.0A resolution. *Proteins* **2**, 167–76.

Kozak, M. (1986). Point mutations define a sequence flanking the AUG initiator codon that modulates translation by eukaryotic ribosomes. *Cell* **44**, 283–92.

Kozak, M. (1989). The scanning model for translation: an update. *Journal of Cell Biology* **108**, 229–41.

Knorr, D.A. and Dawson, W.O. (1988). A point mutation in the tobacco mosaic virus capsid protein gene induces hypersensitivity in *Nicotiana sylvestris. Proceedings of the National Academy of Sciences USA* **85**, 170–4.

Liljas, L., Unge, T., Jones, T.A., Friborg, K., Lovgren, S., Skoglund, U., and Strandberg, B. (1982). Structure of satellite tobacco necrosis virus at 3.0Å resolution. *Journal of Molecular Biology* **159**, 93–108.

Lomonossoff, G.P. and Shanks, M. (1983). The nucleotide sequence of cowpea mosaic virus B RNA. *EMBO Journal* **2**, 2153–8.

Lutcke, H.A., Chow, K.C., Mickel, F.S., Moss, K.A., Kern, H.F., and Scheele, G.A. (1987). Selection of AUG initiation codons differs in plants and animals. *EMBO Journal* **6**, 43–8.

Pelham, H.R.B. (1979). Synthesis and proteolytic processing of cowpea mosaic virus proteins in reticulocyte lysates. *Virology* **96**, 463–77.

Rossmann, M.G., Arnold, E., Erickson, J.W., Frankenberger, E.A., Griffith, J.P., Hecht, H.J., Johnson, J.E., Kamer, G., Luo, M., Mosser, A.G., Rueckert, R.R., Sherry, B., and Vriend, G. (1985). Structure of a human common cold virus and functional relationship to other picornaviruses. *Nature* **317**, 145–53.

Schwartz, R.M. and Dayhoff, M.O. (1978). In *Atlas of protein sequence and structure*, (ed. M.O. Dayhoff), Vol. 5, Supplement 3, pp. 353–8. National Biomedical Foundation, Washington DC.

Shanks, M., Stanley, J., and Lomonossoff, G.P. (1986). The primary structure of red clover mottle virus middle component RNA. *Virology* **155**, 697–706.

Siegel, A., Zaitlin, M., and Sehgal, O.P. (1962). The isolation of defective tobacco mosaic virus strains. *Proceedings of the National Academy of Sciences USA* **48**, 1845–51.

Stauffacher, C.V., Usha, R., Harrington, M., Schmidt, T., Hosur, M.V., and Johnson, J.E. (1987). The structure of cowpea mosaic virus at 3.5Å resolution. In *Crystallography in molecular biology*, (ed. D. Moras, J. Drenth, B. Strandberg, D. Suck, and K. Wilson), pp. 293–308. Plenum Publishing Corporation, New York.

van Wezenbeek, P., Verver, J., Harmsen, J., Vos, P., and van Kammen, A. (1983). Primary structure and gene organisation of the middle component RNA of cowpea mosaic virus. *EMBO Journal* **2**, 941–6.

Vos, P., Verver, J., Jaegle, M., Wellink, J., van Kammen, A., and Goldbach, R. (1988). Two viral proteins involved in the proteolytic processing of cowpea mosaic virus polyproteins. *Nucleic Acids Research* **16**, 1967–85.

Wellink, J., Jaegle, M., Prinz, H., van Kammen, A., and Goldbach, R. (1987). Expression of the middle component RNA of cowpea mosaic virus *in vivo. Journal of General Virology* **68**, 2577–85.

White, J.M. and Johnson, J.E. (1980). Crystalline cowpea mosaic virus. *Virology* **101**, 319–24.

5 Interplay between genes and gene products of alfalfa mosaic virus and its hosts

MARIANNE J. HUISMAN

MOGEN International NV, Einsteinweg 97, 2333 CB Leiden, The Netherlands

Introduction to alfalfa mosaic virus and other ilarviruses

Alfalfa mosaic virus (AlMV) causes a mild chlorosis on *Nicotiana tabacum* cv., Samsun NN, and necrotic local lesions on *Phaseolus vulgaris* cv. Berna. The virus causing these symptoms is a multipartite virus. The genome of AlMV is contained within three bacilliform particles which differ from all other types of virions known. They are unlike the tobamovirus rigid rods, or the flexuous rods of, for example, the potex- and potyviruses, or the icosahedrally shaped virions characteristic of tombus- and several of the tricornaviridae (Matthews 1982). The deviant morphology has been used as a reason to classify AlMV in a separate class of viruses within the alfalfa mosaic virus group. However, increasing knowledge of the molecular organization of viruses, has lead to a more sophisticated view of the classification of AlMV. Although, AlMV has a deviant morphology, it has a unique feature in common with the ilarviruses; notably dependence upon the coat protein to initiate infection (Van Vloten-Doting *et al.* 1981 and references therein). So, AlMV can be considered to be a member of the ilarvirus group.

AlMV infectious virus consists of three components, called bottom, middle, and top b. Each of these particles contains a unique RNA molecule. The RNA molecules are numbered one to three in order of decreasing lengths and are contained within the bottom, middle, and top b particles, respectively. The combination of the three genomic RNAs is not infectious. Both AlMV RNA and other ilarvirus RNAs, need the presence of coat protein, *or* the messenger RNA for the coat protein (RNA 4), to be able to infect host plants successfully (Fig. 5.1) (Bol *et al.* 1971; Gonsalves and Garnsey 1975). This holds for infection of whole plants as well as for protoplasts (Alblas and Bol 1978). It is intriguing that a virus needs its coat protein not only for its protective function, but also for an active function in RNA replication (see the section on early events in replication).

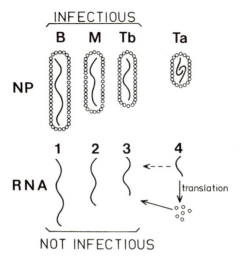

Fig. 5.1. The genomic RNAs (1, 2, and 3) of AlMV are not infectious as such. The viral B (bottom), M (middle), and Tb (top b) particles are. Addition of coat protein to the genomic RNAs makes them infectious.

Genome organization of AlMV

The three genomic RNAs 1, 2, 3, and the subgenomic RNA 4 (Fig. 5.2) all contain a 5' end cap structure and a 3'-terminal region of 145 nucleotides which is homologous (Koper-Zwarthoff *et al.* 1979; Cornelissen *et al.* 1983b). The homology is 87% and whilst this might seem low, the 20 nucleotides which are different within the 145 nucleotide stretch are located at loop structures or at two pairing bases in stem structures. This secondary structure is conserved (Fig. 5.3) not only between different genomic RNAs of AlMV, but also between ilarviruses. The untranslated 5'-terminal sections vary in size from 37 nucleotides for RNA 4 to 346 nucleotides for RNA 3 (Jaspars 1987). The untranslated leader sequences of RNA 1 and 2 are 100 and 55 nucleotides, respectively. RNA 4 is a very efficient messenger RNA. Its short untranslated leader sequence cannot form any secondary structure, and this may facilitate initiation of translation. Ribosomes probably protect the RNA from degradation because they are continuously translating RNA 4.

Coding capacity of the AlMV RNAs

The two long RNAs are 3644 nucleotides (RNA 1; Cornelissen *et al.* 1983a) and 2593 nucleotides long (RNA 2; Cornelissen *et al.* 1983b). Each

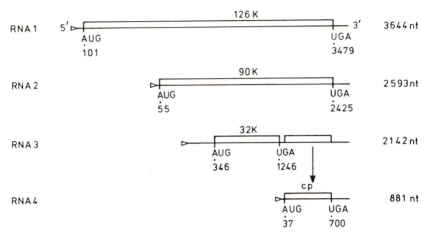

Fig. 5.2. Genomic organization of AlMV. RNA 4 is the subgenomic mRNA for the coat protein (CP) gene. ▷, cap structure; nt, nucleotides; K, kDa.

of these RNAs contains one large open reading frame, encoding proteins with calculated molecular weights of 126 kDa, and 90 kDa, respectively. The smallest genomic RNA, RNA 3 (2142 nucleotides; Langereis *et al.* 1986), contains two open reading frames, of which the 3'-terminal one represents the coding region for the viral coat protein (CP) of 24 kDa. The coat protein is translated from the subgenomic messenger RNA 4 (881 nucleotides; Brederode *et al.* 1980). The 5'-proximal open reading frame on RNA 3 encodes a protein with a calculated molecular weight of 32 kDa. The viral RNAs *in vitro* direct proteins with molecular weights close to the calculated values and, therefore, it is highly likely that these open reading frames encode functional proteins.

How to analyse virally encoded functions

Neither the primary structure of the RNAs itself, nor the primary structure of the proteins that the RNAs code for, can tell us anything about the way the virus multiplies, spreads, or produces symptoms within its host. To unravel the functions of the virally encoded products various strategies have been used. Among them isolation of the viral 'replicase' from infected tissues, analysis of replication of viral RNAs in isolated protoplasts, construction of a library of mutants of this virus, and, more recently, expression of parts of the viral genome within the host plant.

Replication cycle

It has been proposed that the replication cycle of AlMV may be divided into the following steps (Nassuth and Bol 1983). The first step is the uncoating of the virions. The second represents the start of minus-strand RNA synthesis. The third step is the switch to plus-strand RNA synthesis. After this step the plus-strand RNA molecules can start the replication cycle all over again, or they can be encapsidated into virions. In some way the virus has to be transported from cell-to-cell and throughout the whole plant. Dorokhov and coworkers (1983) have shown that the viral genome spreads through plants as part of a viral ribonuclear protein complex (vRNP). Unfortunately, this work was never followed up, so that to date, the general form in which viruses spread throughout whole plants is unknown. It might be as a complex of viral RNA with host proteins or as virions, or as a combination of both.

At the end of the replication cycle an RNA 3 encoded product is involved in the regulatory switch from a further round of RNA replication to production of mainly plus-stranded progeny and their encapsidation into virions. This product might be the coat protein. The function of the RNA 3 encoded product that is involved is either a negative controlling effect on minus-stranded RNA synthesis, or a positive controlling effect on plus-stranded RNA synthesis.

Early events in RNA replication

For the early events in AlMV replication two factors seem of paramount importance. One is the presence of the viral coat protein or its messenger RNA (Van Vloten-Doting 1978). The other is the presence of the so-called high-affinity sites located at the 3' terminus of the RNA (Koper-Zwarthoff and Bol 1980). The observation that the coat protein can be substituted by its messenger RNA, both in whole plants as well as in protoplasts, has led to the conclusion that for entrance into the cells the coat protein is not necessary. So the need for coat protein is at a stage later than that of entry into cells.

A unique phenomenon of ilarviruses as opposed to all other viruses is that naked ilarvirus RNAs are capable of uncoating ilarviral virions (Van Vloten-Doting and Jaspars 1972). This uncoating of virions by naked RNAs suggests a very important role for coat protein on the ilarvirus RNAs. As mentioned before, 3' termini of the RNAs of AlMV and several other ilarviruses, for example tobacco streak virus (TSV) contain a highly conserved secondary structure; specific stem-loop structures, spaced by the conserved AUGC element (Fig. 5.3; Koper-Zwarthoff and Bol 1980; Bol *et al.* 1985), which *in vitro* bind coat protein with high-affinity (Zuidema

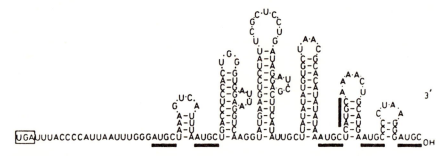

Fig. 5.3.   Secondary structure of the 3'-terminal region of the AlMV genomic RNAs. Stem-loop structures interspersed with AUGC sequences.

1983). Weaker binding sites have also been observed, positioned internally in the viral RNAs. *In vitro* the TSV high-affinity sites can bind either TSV or AlMV coat protein (Zuidema and Jaspars 1984). The reverse experiment with AlMV RNAs reveals the same phenomenon. Ilarvirus coat protein molecules are interchangeable on binding-sites *in vitro* and also in genome activation (Van Vloten-Doting 1975; Gonsalves and Fulton 1977). Whether the coat protein binds to these sites *in vivo* is unknown. However, their presence at the conserved 3' terminus of all genomic RNAs is most suggestive of a functional property.

It has been speculated that binding of a few coat protein molecules to the 3' end is necessary for recognition of the viral template by the virally encoded replicase complex (Houwing and Jaspars 1978). This would render the replicase complex highly specific for the AlMV templates. Another explanation for the need of ilarviruses for coat protein in genome activation might be simply a lack of compact folding of the 3'-terminal region of the RNA which would otherwise stabilize the genomic RNAs. Some viruses e.g. tobacco mosaic-, brome mosaic virus, and others, contain a 3'-terminal structure which can be folded into a tRNA-like structure. This compact folding of the RNA may protect against exonuclease degradation. Other viruses like potexviruses contain a poly(A) tail at their 3' end. Poly(A) tails have been shown to add stability to viral RNAs (Huez *et al.* 1983). The secondary structure of ilarviruses in itself may be too limited to stabilize the RNAs. Through the binding of coat protein molecules to these sites extra stability is provided. This extra stability might be necessary to extend the lifetime of the genomic RNAs, so that they can be translated and replicated.

Coat-protein expressing transgenic Samsun NN tobacco plants

Recently, transgenic plants have been produced which express either the coat protein gene of TSV, or that of AlMV (Van Dun *et al.* 1987; Loesch-Fries *et al.* 1987; Van Dun *et al.* 1988). These plants show coat protein-mediated cross-protection only against the corresponding virus. Thus, AlMV coat protein transgenic plants show coat protein-mediated cross-protection against AlMV but not against TSV. And *vice versa*: TSV coat protein transgenic plants show cross-protection against TSV but not against AlMV. Following inoculation with a mixture of AlMV RNAs 1, 2, and 3, virus replication is observed in AlMV and TSV coat protein expressing plants (Van Dun *et al.* 1987; 1988). So, the two viral coat proteins can be interchanged for activation of the respective genomes, but not for coat protein-mediated cross-protection. Apparently, different interactions and/ or domains of the coat protein are responsible for coat protein-mediated cross-protection phenomenon and the genome activation.

Protoplasts experiments reveal the replicative ability of the AlMV RNA 1 and 2 products

Inoculations of protoplasts, with one, or a combination of two components of the tripartite genome of AlMV have given indications as to which components can give rise to viral RNA replication and/or infectious virus production (Nassuth and Bol 1983). Only protoplasts inoculated with bottom and middle components, i.e. the monocistronic RNAs 1 and 2, plus coat protein, show replication of the two viral RNAs. However, no infectious virus production can be detected. None of the other combinations of purified components shows any replication of RNA and infectious virus production can only be detected after inoculation with bottom, middle and top b particles. Interestingly, the production ratio between the minus- and plus-stranded RNAs is different with inoculations using bottom and middle particles only, versus inoculations with bottom and middle combined with top b particles (Nassuth and Bol 1983).

Replication of temperature-sensitive mutants in protoplasts

The functions of the 126 and 90 kDa products have been studied extensively by the use of a set of temperature-sensitive mutants (Sarachu *et al.* 1985; Huisman *et al.* 1985). A mutant with a defect which was located in the bottom component (RNA 1) showed severe reduction in (I) minus-strand RNA synthesis, (II) production of coat protein, and (III) infectious virus production at the restrictive temperature compared to the wild type synthesis at that temperature. Similar results could be obtained with a temperature-sensitive mutant in which the mutation was located in the middle component (RNA 2). Moreover, temperature shift-up experiments have indicated that these proteins are necessary during the first six hours after inoculation. An alternative explanation could be that these proteins are very stable. Once formed at the permissive temperature they are able to retain their functional properties after transfer to higher temperatures, whereas native protein formation may not take place at the restrictive temperature. Curiously enough, all efforts to find a mutant in the library which was disturbed in the plus-strand RNA synthesis were unsuccessful. Although involvement of RNA 1 and 2 products in minus- as well as plus-strand RNA synthesis is predicted, to date there is no evidence for the involvement of virally encoded gene products in plus-strand RNA synthesis.

One specific temperature-sensitive mutant in greater detail

The origin of Tbts 7

One specific mutant called Tbts 7, from the library of temperature-sensitive mutants present at the plant virus group of State University of Leiden, will be described in greater detail. Tbts 7 was produced by ultra-violet irradiation of purified top b particles (Van Vloten-Doting *et al.* 1980). The irradiated top b particles were mixed with purified bottom and middle particles and infection sites spotted on inoculated Samsun NN tobacco plants, were passaged through several single lesion transfers to achieve apparent homogeneity. Afterwards the stock was maintained as dried frozen leaf material at -20°. One of the series of mutants obtained is Tbts 7. The symptoms induced on *Phaseolus vulgaris* cv. Berna by this mutant are indistinguishable from those induced by the parent strain. However, on *Nicotiana tabacum* cv. Samsun NN the mutant causes chlorotic spots which, in contrast to those induced by the parent strain are partly surrounded by a necrotic ring, usually described as half-moon-shaped necrotic lesions (Roosien *et al.* 1983; Huisman *et al.* 1987; 1989). Apparently, symptoms arise through interaction between viral and host encoded proteins. The interaction between *P. vulgaris* and AlMV or this specific mutant, is equal, but the interaction between Tbts 7 and *N. tabacum* is disturbed compared to that between AlMV and the same host. How the hypersensitive response in Samsun NN is evoked by Tbts 7 is not clear. Progeny from mixed infections of parent strain and Tbts 7 virions fail to induce mutant symptoms on Samsun NN tobacco plants. This suggests a dominance of the parent strain over the mutant (Roosien *et al.* 1983).

Tbts 7 virus preparation complemented with purified parent strain top b particles show the original parent strain symptoms on Samsun NN, confirming that the Tbts 7 mutation(s) are located on RNA 3 (Van Vloten-Doting *et al.*1980; Sarachu *et al.* 1985). As indicated below Tbts 7 deviates from the parent strain in a number of other phenotypic aspects. It is remarkable that this mutant is extremely stable. Tbts 7 has been harvested 6 times in a period of 9 years. For each harvest the virus had to be cultivated for 3–6 weeks. Each time the batch was started with inoculum from the stock supply of dried frozen infected leaf material, and each time the batch of Tbts 7 showed the same phenotype. This is definitely not true for all temperature-sensitive mutants. Many of them change in character in less than a month of culture.

Composition of the Tbts 7 virus preparations

Tbts 7 virus preparations show an exceptionally low amount of mutant RNA 3 containing particles. This reduced amount of top b components coincides with increased amounts of components which migrate in a position between top b and top a particles in a sucrose density gradient (Roosien 1983).

Electrophorectic mobility difference of Tbts 7 coat protein

The coat protein of Tbts 7 shows a retarded electrophoretic mobility on SDS-PAGE compared to the wild type (Fig. 5.4). It has been shown that the change in mobility is not due to a C-terminal nor to an N-terminal protein extension (Smit 1981). Apparently, the mutation(s) results in amino acid residue changes in the coding region. Comparative uncoating experiments with mutant and parent strain virions have shown that in the presence of parent AlMV RNAs the mutant virions uncoat less easily than the parent strain virions (Smit 1981). Apparently, the mutant coat protein has enhanced protein-protein interactions.

Temperature-sensitive phenotype of Tbts 7

Comparison of the amount of Tbts 7 produced in tobacco Samsun NN leaf discs at 23° and 30° showed that infectious virus production of Tbts 7 is temperature-sensitive. This temperature-sensitivity can be removed by addition of parent strain top b components. The coat protein of Tbts 7 besides having an altered electrophoretic mobility, is unable to activate the genome of AlMV at 30°C, the non-permissive temperature (Smit *et al.* 1981).

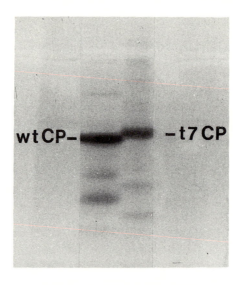

Fig. 5.4. Migration of translation products directed by wild type RNA 4 transcripts (wt CP) and Tbts 7 RNA 4 transcripts (t7 CP).

Differences in inoculation procedures can give rise to different observations

At the permissive temperature Tbts 7 multiplies in whole plants and leaf discs of Samsun NN tobacco. However, in cowpea protoplasts inoculated with this mutant in the presence of poly-L-ornithine (PLO) no infectious virus production can be observed at any temperature (Sarachu *et al.* 1985). Analysis on Northern blots of the RNA synthesized revealed that these protoplasts contained both plus- and minus-stranded RNAs 1 and 2 in the same ratio that was observed in protoplasts infected only with bottom and middle particles of the parental strain (Huisman *et al.* 1987). Protoplasts treated with PLO and inoculated with Tbts 7 and parental strain top b particles produced normal amounts of virus. Moreover, the ratio of plus- and minus-stranded RNAs produced was similar to that observed in an infection with parent strain bottom, middle, and top b particles. However, when protoplasts were inoculated with Tbts 7 in the presence of polyethylene glycol (PEG) virus was produced at the permissive temperature, and the mutant behaved as a temperature-sensitive mutant with a selective defect in replication of RNA 3 (Huisman *et al.* 1987). Thus a difference in inoculation procedure may result in differences in the data observed. Therefore, several inoculation methods have to be compared when using protoplasts in order to allow definite observations after virus inoculations.

The uncoating experiments (Smit 1981) may give a clue towards explaining the observations described above. Tbts 7 virions are known to uncoat less easily than parent strain virions. This difficulty in uncoating of Tbts 7 virions might be enhanced by specific interference of PLO on the virus particles which is different from that resulting from PEG. Knowing that top b particles of the parent strain loose their coat protein slightly quicker than the larger particles (Van Boxsel 1976), problems especially in uncoating of Tbts 7 top b particles, might be the reason for the infection type observed in Tbts 7 PLO inoculated protoplasts, where it seems as if only bottom and middle particles are present.

As indicated before an RNA 3 encoded product is involved in the regulation of minus- versus plus-stranded RNA production. The results with Tbts 7 provide circumstantial evidence for the involvement of coat protein, and not of the 32k protein, in the regulation of minus- and plus-stranded RNA production.

Nucleotide sequence elucidation of an RNA 3 mutant

Determination of the nucleotide sequence of mutants of an RNA virus such as AlMV has been considered with great caution, because of the plasticity of RNA genomes. Because of the high mutation rate many mutations may be present which do not affect the viral phenotype, but which will make a direct correlation between observed mutations and phenotypes very difficult (see below). One mutant analysed at the level of nucleotide sequence is Tbts 7. As mentioned before the mutation in this Tbts 7 is located in RNA 3, and the phenotype differs in several characteristics from that of the parent strain. cDNAs of the mutant have been cloned and the nucleotide sequence has been analysed.

The results with Tbts 7 are highly interesting (Huisman *et al.* 1989). In the coat protein cistron two point mutations are present of which only one changes an amino acid residue; asparagine 126 is changed into an aspartate. This point mutation was engineered into the parent strain RNA 4 transcript, and this transcript directed *in vitro* the synthesis of a product with the same aberrant electrophoretic mobility as the translation product directed by the native RNA 4 molecule of Tbts 7 (Huisman *et al.* 1989). Hence, the temperature-sensitivity of the Tbts 7 coat protein in regard to genome activation has to be due to the change of asparagine to aspartate in the coding region of the coat protein.

The 32 kDa protein cistron also contained two point mutations, one of which was a silent mutation whereas the other resulted in a change which is also present in two other AlMV strains neither of which are temperature-sensitive. Therefore, it is unlikely that this mutation has anything to do with the temperature-sensitive phenotype or any other aberrant features of this mutant.

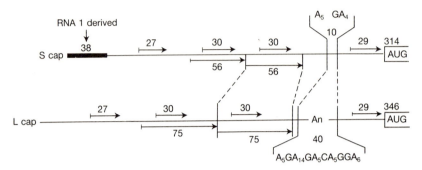

Fig. 5.5. Schematic diagram of a comparison between 5'-terminal leader sequence of AlMV RNA 3 of the Strasbourg (S) strain and the Leiden (L) strain. Identical stretches of nucleotides are depicted as arrows with the lengths denoted above. The 75 base repeated segment in the L strain is present as a shorter 56 nucleotide repeated segment in the S strain. The oligo(A) stretches present in both leaders are of different lengths. The bold bar at the extreme 5' terminus of RNA 3 S represents RNA 1 derived sequences.

RNA recombination at the 5' terminus of Tbts 7 RNA 3

To investigate whether the one point mutation in the coat protein is responsible for all the deviations observed the mutant RNA 3 had to be sequenced completely. The data revealed changes in the 5'-terminal part. The untranslated leader sequence of 346 nucleotides from the parent strain RNA 3 shows a few remarkable characteristics (Fig. 5.5): two direct repeats of 75 nucleotides, a poly(A) stretch of 40 bases interspersed with four Gs and one C residue, and two stretches of 27 and 29 nucleotides both of which are also present in the untranslated leader sequences of AlMV 425 Madison and AlMV S strain (Langereis *et al.* 1986). Moreover the S strain RNA 3 has an extra characteristic compared to the two 425 strains. At the extreme 5' end of the S strain RNA 3, 38 nucleotides are found which correspond to the very 5' end of the untranslated leader sequence of RNA 1 (Fig. 5.5). The nucleotide sequence of the 5' terminus of Tbts 7 RNA 3 deviates from the parent RNA 3 over approximately the 5'-proximal sequence of 100 nucleotides. The sequence of clones of Tbts 7 cDNA which have been analysed and which represent the 5'-terminal 270 nucleotides, fall into three different groups (Fig. 5.6). One group, the largest, contains molecules which are about 100 nucleotides shorter than the parent strain RNA 3. The 5' termini of those particular RNA 3 molecules would be the start of the 75 nucleotide repeated segment. Another group contains this 75 nucleotide repeated segment, not twice, but three times. These molecules are about parent RNA size, although with a different sequence.

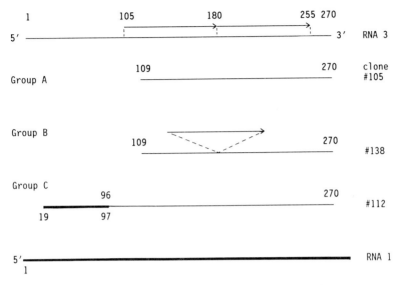

Fig. 5.6. Schematic representation of the three groups of 5' terminal leader sequences of RNA 3 of Tbts 7. The 75 base repeated segment is shown as an arrow. The extra arrow in group B represents the extra 75 base segment present in those cDNA clones. Group C clones contain RNA 1 derived sequences (bold bars). Numbers above the RNA represent lines corresponding to RNA 3 sequences. Numbers below those lines correspond to RNA 1 sequences.

The third group resembles the RNA 3 molecules of the Strasbourg strain in structure. These Tbts 7 RNA 3 molecules contain, at their extreme 5' termini, sequences which are colinear with those of the 5'-terminal nucleotides of the RNA 1. These molecules can only have arisen by RNA recombination. RNA recombination has always been suggested as a mechanism to account for observations like the occurrence of defective interfering particles. Bujarski and Kaesberg (1986) were the first to show that RNA recombination is not a rare event occurring only on evolutionary time scales, but can be observed in plants a few weeks after inoculation. The observation of two independent recombinants in Tbts 7 is in keeping with the suggestion that RNA recombination is a very common event, but that parent strain molecules will normally outcompete the mutant molecules formed by recombination. During mutant selection single lesion transfers were performed, so no parent strain virus is present. Since this mutant is less viable than the parent strain as demonstrated by competition experiments (Roosien *et al.* 1983), only when parent type RNA 3 is absent can RNA recombinant events be revealed. RNA recombination may be the mechanism by which over time RNA viruses are diverging. The viral

RNAs simply pick up genes from related systems and use these for their own life cycle. In this respect a comparison of tobamoviruses to the potexviruses is relevant. The latter group clearly contains two extra open reading frames which are absent in the tobamoviruses (Goelet *et el*. 1982; Huisman *et al*. 1989).

Nucleotide sequence analysis of an RNA 2 mutant

Another mutant analysed at the level of nucleotide sequence is $M_{syst(uv)b,c}$. This contains a mutation located in RNA 2 and was constructed by ultraviolet-irradiation of purified middle particles. The virus stock was prepared as described for Tbts 7. Its phenotypic characteristics are a systemic infection both in *Phaseolus vulgaris* cv. berna (bean) and in *Vigna unguiculata* (cowpea), whereas parent strain AlMV causes necrotic local lesions on both these hosts. Another deviation in $M_{syst(uv)b,c}$ is an extremely high production of middle particles in cowpea protoplast infections (Roosien 1983). Bean and cowpea plants inoculated with a mixture of $M_{syst(uv)b,c}$ and parent strain middle particles show mixed symptoms: both parent strain and mutant phenotypes can be observed in these hosts. In the case of $M_{syst(uv)b,c}$ more than 100 point mutations were present over the RNA 2 molecule resulting in 37 amino acid residue changes. Many, but not all, of these mutations have been observed to occur in another strain of AlMV called YSMV. Moreover, YSMV contains additional changes not present in $M_{syst(uv)b,c}$. This makes $M_{syst(uv)b,c}$ a true deviant of the parent strain, but it is not possible to make out which of all these mutations is responsible for the aberrant symptomology induced by this mutant. Consequently, the sequence of this mutant, although interesting in correlation to the RNA mutation rate, cannot be used to determine which part of the genome is involved in symptomology (L. Neeleman, unpublished result).

Spread of the virus through the plant

For tobacco mosaic virus Meshi and coworkers have elegantly demonstrated that the viral protein involved in spread of the virus through the plant is the 30 kDa protein (Meshi *et al*. 1987). In the case of ALMV a small number of mutants have been isolated which are temperature-sensitive with regard to infectious virus production in tobacco leaf discs, but not in tobacco protoplasts (Table 1 and Huisman *et al*. 1986). The conclusion seems justified that these mutants are disturbed in the spread of the virus and not in their replication cycle. Unfortunately, to date no further details are available of the mutations present in these mutants. The experiments of the plant virus group in Strasbourg have shown that the 32 kDa protein is mainly located in the outer cell wall (Stussi-Garaud *et al*. 1987).

Table 5.1

	Tobacco protoplasts	Tobacco leaf discs
Parent strain	558[a]	65
Mts 1[b]	746	2
Mts 10	158	2
Tbts 9	199	4
Tbts 18	1785	3

[a] numbers represent virus production at 30°C as a percentage of that at 25°C. The lesion numbers obtained at 25°C (i.e. 100%) ranged from 200–1200 per seven half-leaves.
[b] the mutation was determined to be located in the middle particle but additional mutations in e.g. top b particle cannot be ruled out.

This protein facilitates the movement of the virus to neighbouring cells. Whether this protein is also involved in the further spread of the virus throughout the vascular system of the plant is unknown.

Presently insufficient experimental data are available to allow an understanding of how symptoms are formed. Knorr and Dawson (1988) have shown that a point mutation in the coat protein gene of TMV results in necrotic lesions on the host, where as parent strain gives systemic mosaic type symptoms. This could be comparable to the situation with the mutant Tbts 7 which has a point mutation in the coat protein and deviant symptoms. However, in the case of Tbts 7 other mutations which are present make an unequivocal conclusion impossible. I hope that in the future more data on specific interactions between viral and host encoded proteins will become available, and that nucleotide sequence changes may be correlated with changes in phenotype.

Acknowledgements

Thanks are due to Dr Lous Van Vloten-Doting for her critical reading of this manuscript and for many useful discussions. Dinie Posthumus-Lutke Willink is gratefully acknowledged for her excellent help for the word-processing of this manuscript.

References

Alblas, F. and Bol, J.F. (1978). Coat protein is required for infection of cowpea protoplasts with alfalfa mosaic virus. *Journal of General Virology* **14**, 653–6.
Bol, J.F., Van Vloten-Doting, L., and Jaspars, E.M.J. (1971). A functional equivalence of top component a RNA and coat protein in the initiation of infection by alfalfa mosaic virus. *Virology* **46**, 73–85.
Bol, J.F., Cornelissen, B.J.C., Huisman, M.J., and Van Vloten-Doting, L. (1985). Structure and function of the tripartite RNA genome of ilarviruses. In *Molecu-

lar form and function of the plant genome, (ed. L. Van Vloten-Doting, G.S.P. Groot, and T.C. Hall).

Brederode, F.Th., Koper-Zwarthoff, E.C., and Bol, J.F. (1980). Complete nucleotide sequence of alfalfa mosaic virus RNA 4. *Nucleic Acids Research* **8**, 2213–23.

Bujarski, J.J. and Kaesberg, P. (1986). Genetic recombination between RNA components of a multipartite plant virus. *Nature* (London) **321**, 528–31.

Cornelissen, B.J.C., Brederode, F.Th., Moormann, R.J.M., and Bol, J.F. (1983a). Complete nucleotide sequence of alfalfa mosaic virus RNA 1. *Nucleic Acids Research* **11**, 1253–65.

Cornelissen, B.J.C., Brederode, F.Th., Veeneman, G.H., Van Boom, J.H., and Bol, J.F. (1983b). Complete nucleotide sequence of alfalfa mosaic virus RNA 2. *Nucleic Acids Research* **11**, 3019–25.

Dorokhov, Y.L., Alexandrova, N.M., Miroshnichenko, N.A., and Atabekov, J.G. (1983). Isolation and analysis of virus-specific ribonucleoprotein of tobacco mosaic virus-infected tobacco. *Virology* **127**, 237–52.

Goelet, P., Lomonossoff, G.P., Butler, P.J.G., Akam, M.E., Gait, M.J., and Karn, J. (1982). Nucleotide sequence of tobacco mosaic virus RNA. *Proceedings of the National Academy of Sciences USA* **79**, 5818–22.

Gonsalves, D., and Garnsey, S.M. (1975). Functional equivalence of an RNA component and coat protein for infectivity of citrus leaf rugose virus. *Virology* **64**, 23–31.

Gonsalves, D., and Fulton, R.W. (1977). Activation of Prunus necrotic ringspot virus and rose mosaic virus by RNA 4 component of some Ilarviruses. *Virology* **81**, 398–407.

Houwing, C.J. and Jaspars, E.M.J. (1978). Coat protein binds to the 3' terminal part of RNA 4 of alfalfa mosaic virus. *Biochemistry* **17**, 2927–33.

Huez, G., Cleuter, Y., Bruck, C., Van Vloten-Doting, L., Goldbach, R., and Verduin, B. (1983). Translational stability of plant viral RNAs microinjected into living cells. Influence of a 3'-poly(A) segment. *European Journal of Biochemistry* **130**, 205–9.

Huisman, M.J., Sarachu, A.N., Alblas, F., and Bol, J.F. (1985). Alfalfa mosaic virus temperature-sensitive mutants. II. Early functions encoded by RNAs 1 and 2. *Virology* **141**, 23–9.

Huisman, M.J., Sarachu, A.N., Alblas, F., Broxterman, H.J.G., Van Vloten-Doting, L., and Bol, J.F. (1986). Alfalfa mosaic virus temperature-sensitive mutants. III. Mutants with a putative defect in cell-to-cell transport. *Virology* **154**, 401–4.

Huisman, M.J., Lanfermeyer, F.C., Loesch-Fries, L.S., Van Vloten-Doting, L., and Bol, J.F. (1987). Alfalfa mosaic virus temperature-sensitive mutants. IV. Tbts 7, a coat protein mutant defective in an early function. *Virology* **160**, 143–50.

Huisman, M.J., Cornelissen, B.J.C., Groenendijk, C.F.M., Bol, J.F., and Van Vloten-Doting, L. (1989). Alfalfa mosaic virus temperature-sensitive mutants. V. The nucleotide sequence of Tbts 7 RNA 3 shows limited nucleotide changes and evidence for heterologous recombination. *Virology* **171**, 409–16.

Jaspars, E.M.J. (1987). Interaction of alfalfa mosaic virus nucleic acid and protein.

In *Molecular plant virology Vol. 1. Virus structure and assembly and nucleic acid–protein interactions*, (ed. J.W. Davies).

Knorr and Dawson (1988). A point mutation in the tobacco mosaic virus capsid protein gene induces hypersensitivity in *Nicotiana sylvestris*. *Proceedings of the National Academy of Sciences*, USA **85**, 170–4.

Koper-Zwarthoff, E.C. and Bol, J.F. (1980). Nucleotide sequence of the putative recognition site for coat protein in the RNAs of alfalfa mosaic virus and tobacco streak virus. *Nucleic Acids Research* **8**, 3307–18.

Koper-Zwarthoff, E.C., Brederode, F.Th., Walstra, P., and Bol, J.F. (1979). Nucleotide sequence of the 3'-noncoding region of alfalfa mosaic virus RNA 4 and its homology with the genomic RNAs. *Nucleic Acids Research* **7**, 1887–900.

Langereis, K., Mugnier, M., Cornelissen, B.J.C., Pinck, L., and Bol, J.F. (1986). Variable repeats and poly (A)-stretches in the leader sequence of alfalfa mosaic virus RNA 3. Virology **154**, 409–14.

Loesch-Fries, L.S., Merlo, D., Zinnen, T., Burhop, L., Hill, K., Krahn, K., Jarvis, N., Nelson, S., and Halk, E. (1987). Expression of alfalfa mosaic virus RNA 4 in transgenic plants confers virus resistance. *EMBO journal* **6**, 1845–51.

Matthews, R.E.F. (1982). Classification and nomenclature of viruses. Fourth report of the International Committee on Taxonomy of Viruses. *Intervirology* **17**, 1–199.

Meshi, T., Watanabe, Y., Saito, T., Sugimoto, A., Maeda, T., and Okada, Y. (1987). Function of the 30kd protein of tobacco mosaic virus: involvement in cell-to-cell movement and dispensability for replication. *EMBO Journal* **6**, 2557–63.

Nassuth, A., and Bol, J.F. (1983). Altered balance of the synthesis of plus- and minus-strand RNAs induced by RNAs 1 and 2 of alfalfa mosaic virus in the absence of RNA 3. *Virology* **124**, 75–85.

Roosien, J. (1983). Mutants of alfalfa mosaic virus. Ph.D. thesis, State University of Leiden.

Roosien, J., Van Klaveren, P., and Van Vloten-Doting, L. (1983). Competition between the RNA 3 molecules of wild type alfalfa mosaic virus and the temperature-sensitive mutant Tbts 7. *Plant Molecular Biology* **2**, 113–18.

Sarachu, A.N., Huisman, M.J., Van Vloten-Doting, L., and Bol, J.F. (1985). Alfalfa mosaic virus temperature-sensitive mutants. I. Mutants defective in viral RNA and protein synthesis. *Virology* **141**, 14–22.

Smit, C.H. (1981). Multiple activation of the genome of alfalfa mosaic virus. Ph.D. thesis, University of Leiden.

Smit, C.H., Roosien, J., Van Vloten-Doting, L., and Jaspars, E.M.J. (1981). Evidence that alfalfa mosaic virus infection starts with three RNA-protein complexes. *Virology* **112**, 169–73.

Stussi-Garaud, C., Garaud, J.-C., Berna, A., and Godefroy-Colburn, T. (1987). *In situ* location of an Alfalfa Mosaic Virus non-structural protein in plant cell walls: correlation with virus transport. *Journal of General Virology* **68**, 1779–84.

Van Boxsel, J.A.M. (1976). High-affinity sites for coat protein on alfalfa mosaic virus RNA. Ph.D. thesis, State University of Leiden.

Van Vloten-Doting, L. (1975). Coat protein is required for infectivity of tobacco streak virus: biological equivalence of the coat proteins of tobacco streak and alfalfa mosaic viruses. *Virology* **65**, 215–25.

Van Vloten-Doting, L. (1978). Early events in the infection of tobacco with alfalfa mosaic virus. *Journal of General Virology* **41**, 649–52.

Van Vloten-Doting, L. and Jaspars, E.M.J. (1972). The uncoating of alfalfa mosaic virus RNA. *Virology* **48**, 699–708.

Van Vloten-Doting, L., Hasrat, J.A., Oosterwijk, E., Van't Sant, P., Schoen, M.A., and Roosien, J. (1980). Description and complementation analysis of 13 temperature-sensitive mutants of alfalfa mosaic virus. *Journal of General Virology* **46**, 415–26.

Van Vloten-Doting, L., Francki, R.I.B., Fulton, R.W., Kaper, J.M., and Lane, L.C. (1981). Tricornaviridae—a proposed family of plant viruses with tripartite, single-stranded RNA genomes. *Intervirology* **15**, 198–203.

Van Dun, C.M.P., Bol, J.F., and Van Vloten-Doting, L. (1987). Expression of alfalfa mosaic virus and tobacco rattle virus coat protein genes in transgenic tobacco plants. *Virology* **159**, 299–305.

Van Dun, C.P.M., Overduin, B., Van Vloten-Doting, L. and Bol, J.F. (1988). Transgenic tobacco expressing tobacco streak virus of mutated alfalfa mosaic virus coat protein does not cross-protect against alfalfa mosaic virus infection. *Virology* **164**, 383–9.

Zuidema, D. (1983). Specific Binding Sites on RNAs and Coat Protein of Alfalfa Mosaic Virus Involved in Genome Activation, Ph.D. thesis, State University of Leiden.

Zuidema, D. and Jaspars, E.M.J. (1984). Comparative investigations on the coat protein binding sites of the genomic RNAs of alfalfa mosaic and tobacco streak viruses. *Virology* **135**, 43–52.

Zuidema, D., Bierhuizen, M.F.A., Cornelissen, B.J.C., Bol, J.F., and Jaspars, E.M.J. (1983). Coat protein binding sites on RNA 1 of alfalfa mosaic virus. *Virology* **125**, 361–9.

6 Molecular responses of tobacco to virus infection

JOHN F. BOL, HUUB J.M. LINTHORST, and RALPH L.J. MEUWISSEN

Department of Biochemistry, Leiden University, Gorlaeus Laboratories, Einsteinweg 5, 2333 CC Leiden, The Netherlands

Introduction

Resistance of plants to virus infection can manifest itself in a number of ways. The most common form is 'non-host resistance', i.e. all cultivars of a given plant species are resistant to infection by a particular virus. For instance, cowpea plants are not susceptible to infection by tobacco mosaic virus (TMV). The observation that cowpea protoplasts are able to support replication of TMV (Koike *et al.* 1976) indicates that this form of non-host resistance is due to a restriction of cell-to-cell spread of the virus. Possibly, the virus is able to replicate at an undetectable level in the primary-infected cells of cowpea plants.

A second defence mechanism is known as 'cultivar resistance'. It is the phenomenon in which some cultivars of a given plant species are susceptible to a particular virus whereas others are immune, and results from the presence of specific resistance genes. Again using cowpea plants as an example: most cowpea varieties are susceptible to infection with cowpea mosaic virus (CPMV) but the Arlington cultivar is not. This form of cultivar resistance has been correlated with the occurrence of an inhibitor of CPMV poly protein processing (Ponz *et al.* 1988). In many other cases (e.g. the Tm-1 and Tm-2 genes of tomato, the N-gene of tobacco) specific resistance genes are genetically well defined but their biochemical mode of action is unknown.

In contrast to non-host and cultivar resistance, both of which are manifested constitutively, 'induced resistance' is exerted in an individual plant only after pathogen attack. Two forms of resistance induced in response to virus infection are known, 'cross protection' and 'systemically acquired resistance'. Cross protection is the phenomenon in which infection of a plant with a mild strain of a given virus confers resistance to subsequent infection from a severe strain of the same virus. The mild strain must be able first to spread throughout the plant and then the infected parts become immune to a closely related challenging strain. Because of the strain specificity involved it is believed that gene products encoded by the first virus are responsible for the protection against the second. On the other hand, systemically acquired resistance has a broad specificity. It is induced by those pathogens, including viruses, fungi, and bacteria, that give rise to a necrotic infection of the plant. As a result of the hypersensitive response of the plant the pathogen remains localized at the site of infection, while the pathogen-free parts of the plant develop a broad resistance to subsequent infection by viruses, fungi, or bacteria. Induction of this resistance is paralleled by a large increase in *de novo* synthesis of many plant proteins, for example enzymes involved in biosynthesis of aromatic compounds, cell wall components, and the extracellular pathogenesis-related (PR) proteins (Bol and Van Kan 1988). Circumstantial evidence suggests that at least some of these proteins synthesized *de novo* play a role in the induced resistance. Here we report an analysis of the possible role in the mechanism of systemically acquired resistance, of PR proteins that are induced by TMV infection of Samsun NN tobacco.

Properties of PR proteins

General occurrence

Induction of PR proteins by pathogens or abiotic elicitors has been observed in over 20 plant species (for a review see Bol 1988, Bol *et al.* 1990). Available data on amino acid sequence homologies and serological relationships support the notion that PR's represent a class of proteins that are highly conserved throughout the plant kingdom. The group of approximately 10 acidic PR proteins induced by TMV infection of Samsun NN or Xanthi nc tobacco have been studied in most detail. On the basis of their molecular and serological properties a classification into five groups has been proposed. The situation is confused by the different nomenclatures used by the various laboratories studying tobacco PR proteins. These nomenclatures are listed in Table 6.1. In our studies we use the names proposed by Van Loon (1982).

The function of those proteins 1a, 1b and 1c that constitute group 1 of the tobacco PR's is not known. Groups 2 and 3 both contain hydrolytic enzymes that are able to degrade polysaccharide components from cell walls. The group 2 proteins, 2, N and O, have been shown to be 1,3-β-glucanases (Kauffmann *et al.* 1987) whereas the group 3 proteins, P and Q, have chitinase activity (Legrand *et al.* 1987). Little is known about the properties of group 4 proteins (PR-R). Two electrophoretically distinct components that are serologically related to each other and to group 2 proteins (Van Loon 1987) have been observed, with apparent Mr's of 13 and

Table 6.1 Nomenclature of tobacco PR proteins

Group	$M_r \times 10^{-3}$	Nomenclature used by				
		Van Loon 1982	Jamet and Fritig 1986	Pierpoint 1986	Asselin et al. 1985 Dumas et al. 1987	Van Loon et al. 1987
1	15	1a, 1b, 1c	1a, 1b, 1c	Ia, Ib, Ic	b1, b2, b3	1a, 1b, 1c
2	40	2, N, O	2, N, O	2, N, O	b4, b5, b6,	2a, 2b, 2c
3	29/30	P, Q	P, Q	P, Q	b7, b8	3a, 3b
4	13/15	R	—	R'	b9b	4a, 4b
5	23	S	R, S	R	b9a	5a, 5b

15k. Also two closely related components have been identified among the group 5 proteins (PR-S) by Pierpoint *et al.* (1987). A function for these proteins is suggested by their extensive amino acid sequence homology to a bifunctional inhibitor of amylase and protease activities of insects that has been isolated from maize (Richardson *et al.* 1987).

In addition to the acidic PR proteins listed in Table 2.1, infection of tobacco with TMV also induces synthesis of a number of basic proteins that show extensive amino acid sequence homology to proteins from groups 1, 2, 3, and 5. The acidic proteins are found almost quantitatively in the extra-cellular fluid from infected tobacco leaves (Parent and Asselin 1984), but little of the basic isoforms is excreted into the apoplast (L.C. van Loon, personal communication). In salt-stressed tobacco plants the basic equiva-lent of PR-S has been localized in the vacuoles (Singh *et al.* 1987) and this may also be the site of accumulation of other basic tobacco PR proteins.

Besides the hypersensitive response to pathogens and various stress con-ditions, many PR's are also induced by abiotic agents such as salicylic acid or polyacrylic acid (see Bol 1988). The observation that in most cases the chemical induction of various subsets of PR's is accompanied by induction of resistance to infection has lent support to the hypothesis that PR's are involved in defence mechanisms. The present state of knowledge about the function of the various groups of PR proteins is summarized below.

PR-1 proteins

The acidic PR-1 proteins of tobacco have been found to be serologically related to inducible proteins in cowpea, tomato, potato, *Solanum demis-sum*, *Gomphrena globosa*, *Chenopodium amaranticolor*, maize, and barley (Nassuth and Sänger 1986, White *et al.* 1987). The basic p14 protein, induced by TMV or viroid infection of tomato, was the first PR-1 type pro-tein to be sequenced (Lucas *et al.* 1985). Tobacco PR-1 sequences were initially derived from cDNA clones (Cornelissen *et al.* 1986a). Based on limited amino acid sequence data a full-length clone was assigned to PR-1b and incomplete clones were assigned to PR-1a and PR-1c. Recently, more extensive sequencing of proteins 1a and 1b showed these assignments to be correct (Payne *et al.* 1988). Sequencing of PR-1a and PR-1c at the cDNA level has been completed by Pfitzner and Goodman (1987), Matsuoka *et al.* (1987), and Cutt *et al.* (1988). The combined data showed that PR-proteins are synthesized as precursors with an N-terminal signal peptide of 30 amino acids attached to the mature protein (138 amino acids). Minor differences were observed between the sequences of the corresponding proteins from *Nicotiana tabacum* cvs. Samsun NN, Wisconsin-38, and Xanthi nc. Analy-sis of the 3'-ends of the cDNA clones indicated that various alternate polyadenylation sites are being used. The amino acid sequence homology between PR-1a, 1b, and 1c is about 90 per cent. A weak homology of these proteins to a venom allergen from the white face hornet has been reported (Fang *et al*. 1988).

By Northern blot analysis (Hooft van Huijsduijnen *et al*. 1985) and *in vitro* translation of poly (A) RNA from healthy and TMV-infected tobacco (Carr *et al*. 1985) it was shown that synthesis of PR-1 proteins is regulated at the level of mRNA accumulation. In primary infected leaves maximum levels of PR-1 mRNA are reached about 4 days after inoculation whereas PR-1 mRNA accumulation is detectable in systemically induced virus-free leaves from 8 days onwards. (Cornelissen *et al*. 1986a). Accumulation of PR-1 mRNA induced by spraying tobacco with a 5 mM salicylate solution reaches a maximum within two days (Hooft van Huijsduijnen *et al*. 1986a).

In addition to mRNAs for acidic PR-1 proteins, TMV infection of tobacco was found to induce mRNAs encoding basic PR-1-like proteins (Cornelissen *et al*. 1987). Compared to the acidic isoforms these proteins have a C-terminal extension of 36 amino acids. The amino acid sequence homology between the acidic and basic tobacco PR-1 proteins is 67 per cent; the homology between the tobacco PR-1 proteins and the tomato p14 is 61 per cent. In contrast to the acidic PR-1 protein, the basic PR-1 proteins are constitutively expressed in the roots of healthy plants (Memelink *et al*. 1987).

In TMV infected tobacco leaves the acidic PR-1 proteins accumulate initially around the local lesions (Antoniw and White 1986). Approximately 80 to 90 per cent of this protein is found in the extracellular spaces of the leaf (Ohashi and Matsuoka 1987). This extracellular location has been verified by immunofluorescence microscopy and immunogold labelling techniques (Carr *et al*. 1987, Dumas *et al*. 1988, Hosokawa and Ohashi 1988). Application of immunogold labelling to healthy tomato plants has shown p14 to be located in intercellular spaces as well as in the cytosol of those cells where this compartment is disorganized (Vera *et al*. 1988). Because of this latter observation, it has been suggested that p14 is involved in cell degeneration.

Glucanases and chitinases

A first clue to the function of virus-induced PR proteins came from the finding of the Strasbourg group that four tobacco PR's are 1,3-β-glucanases (Kaufmann *et al*. 1987) while four others have chitinase activity (Legrand *et al*. 1987). These two types of hydrolytic enzymes were known to be coordinately induced by ethylene, fungal infection, or fungal elicitors in several plant species (see, e.g., Vögeli *et al*. 1988). Plant chitinases have no known substrate in healthy plants, but are able to degrade chitin in fungal cell walls and to degrade the peptidoglycan in bacterial cell walls (Boller 1988). The coordinate induction of chitinases and glucanases may reflect a collective action of the two enzymes in the inhibition of fungal growth. Out of 18 fungi tested, 15 were inhibited by combinations of chitinase and 1,3-β-glucanase purified from pea, while the two enzymes used separately each inhibited only one of the fungi tested (Mauch *et al*. 1988). Induction of these hydrolases by virus infection may be responsible for the resistance of TMV-infected tobacco to subsequent infection by fungi and bacteria as reported by Gianninazzi (1983).

At present, about 25 1,3-β-glucanases from 18 different plant species, and 14 chitinases from 9 plant species have been described (see Boller 1988). Those characterized in most detail are the glucanases and chitinases from bean (Vögeli *et al*. 1988), pea pod (Mauch *et al*. 1988), tobacco (Shinshi *et al*. 1987, Kauffmann *et al*. 1987, Legrand *et al*. 1987, Shinshi *et al*. 1988), potato (Gaynor 1988, Kombrink *et al*. 1988 and chapter 15 this volume), and the chitinase from cucumber (Metraux *et al*. 1989). The major part of the basic glucanases and chitinases from bean are localized in the vacuoles whereas the acidic chitinase from cucumber accumulates in the apoplast (Mauch *et al*. 1988, Metraux *et al*. 1989). In tobacco three acidic 1,3-β-glucanases (PR's 2, N, and O) and one basic 1,3-β-glucanase have been identified in addition to two acidic chitinases (PR's P and Q) and two basic chitinases. The acidic hydrolases are found in the intracellular space of the leaf but the basic counterparts are not. The specific activities of the tobacco enzymes *in vitro* and their abundances in TMV-infected plants have been compared (Kauffmann *et al*. 1987, Legrand *et al*. 1987). The acidic PR proteins account for approximately one-third of the total 1,3-β-glucanase activity and one-third of the total chitinase activity in infected plants. It has been speculated that the extracellular acidic hydrolases and the intracellular basic hydrolases act as first and second lines of defence against invading pathogens, respectively (Bol *et al*. 1988).

The acidic tobacco chitinases ($M_r \sim$ 29/30 kDa) show in the region of 65 per cent amino acid sequence identity to the basic tobacco chitinases ($M_r \sim$ 32/34 kDa) and a similar degree of homology is found between the acidic ($M_r \sim$ 40 kDa) and basic ($M_r \sim$ 33 kDa) tobacco 1,3-β-glucanases (Hooft van Huijsduijnen *et al*. 1987, Linthorst *et al*. 1990. Moreover, the tobacco chitinases show a close amino acid sequence homology to the basic chitinases from bean and potato but no apparent homology is found with the acidic chitinase of cucumber. However, this cucumber chitinase shows a high degree of homology to a bifunctional lysozyme/chitinase isolated from *Parthenocissus quinquifolia* (Metraux *et al*. 1989). Where the analysis has been performed all acidic and basic hydrolases are synthesized as precursors with an N-terminal signal peptide that is probably cleaved off during transport of the protein through the endoplasmatic reticulum.

Following this N-terminal processing the basic tobacco glucanase is glyco-sylated at a C-terminal sequence though it is subsequently lost to give the mature protein (Shinshi *et al.* 1988). The acidic tobacco glucanases appear to be synthesized without this C-terminal extension (Linthorst *et al.* 1990).

Thaumatin-like proteins

A cDNA clone of a TMV-induced mRNA from tobacco was found to encode an acidic protein with a striking homology to the sweet-tasting basic protein thaumatin that accumulates in the fruits of a shrub from the African rain forests (Cornelissen *et al.* 1986b). The tobacco protein was identified as PR-S in the nomenclature of Van Loon (1982) (PR-R in the nomenclature of Pierpoint 1986). N-terminal amino acid sequence determination revealed that the PR-S preparation contained two closely related proteins in an approximate 3:2 ratio (Pierpoint *et al.* 1987). In salt-adapted tobacco cells a cationic protein is induced. Called osmotin, it appeared to be the basic counterpart of PR-S (Singh *et al.* 1987). While PR-S is excreted into the apoplast, osmotin accumulates in dense inclusion bodies within the vacuole. Neither of the two proteins tastes sweet. A possible function for the tobacco thaumatin-like proteins is suggested by their homology to a bifunctional inhibitor of α-amylase and protease of insects (Richardson *et al.* 1987) isolated from maize. It could also be that the tobacco proteins are involved in defence against insects. A protein similar to osmotin has been identified in salt-stressed tomato cells (King *et al.* 1988).

Other virus-induced proteins

In addition to PR proteins, plants stressed by pathogens may accumulate a variety of other proteins, for example enzymes from the phenylpropanoid pathway (Van Loon 1982), hydroxyproline rich glycoproteins (Lawton and Lamb 1987), proteinase inhibitors (Thornburg *et al.* 1987), peroxidases (Van Loon 1982, Lagrimini 1987), and manganese superoxide dismutase (Bowler *et al.* 1989). Using tobacco plants we have cloned cDNA to two TMV-induced tobacco mRNAs which were initially called mRNAs 'A' and 'C' (Hooft van Huijsduijnen *et al.* 1986a). mRNA 'C' was found to encode a glycine-rich protein (GRP) of 109 amino acids with a putative N-terminal signal peptide of 26 amino acids (Van Kan *et al.* 1988). By analogy to other glycine-rich proteins the tobacco GRP may be a cell wall component. Like the PR-1 mRNA, the GRP mRNA is induced strongly by spraying plants with 5mM salicylate solution (Hooft van Huijsduijnen *et al.* 1986a).

Isolation of PR genes

We have reported the isolation and sequencing of genes from tobacco Samsun NN that encode PR-1, GRP and PR-S proteins (Cornelissen *et al.* 1987, Van Kan *et al.* 1988, Van Kan *et al.* 1989). Southern blot analysis showed that PR-1 and GRP are each encoded by a family of approximately eight genes while the complexity of PR-S genes is much lower. We have reported the sequence of the PR-1a gene and the structure of two putative pseudo genes (Cornelissen *et al.* 1987). Subsequently, the sequence of the PR-1a gene has been determined by Oshima *et al.* (1987), Pfitzner *et al.* (1988), and Payne *et al.* (1988a). Some of these authors also reported the isolation of PR-1 pseudo genes.

Four genomic GRP clones have been isolated, two of which were sequenced (Van Kan *et al.* 1988). S1-nuclease mapping studies indicated that the gene in one clone (GRP-8) containing an intron of 555 bp, is expressed after TMV-infection of tobacco, whereas the gene in the other clone (GRP-4), containing an intron of 1954 bp, is not expressed. As is the case with the PR-1 genes, the upstream sequences of the GRP genes contain several direct and inverted repeats that may have some regulatory function. Most prominent in the GRP promoter region is a 64 bp inverted repeat that occurs in a similar position in the tobacco ribulose bisphosphate carboxylase small subunit gene from tobacco.

Screening of a tobacco genomic library for PR-S genes yielded two classes of clones corresponding to two PR-S genes which were called E2 and E22 (Van Kan *et al.* 1989). Figure 6.1 shows a dot-blot comparison of E2 and E22 and their flanking sequences. cDNA clones corresponding to both genes have been identified (Cornelissen *et al.* 1986b, Payne *et al.* 1988b). Genes E2 and E22 correspond to the minor and major forms of PR-S, respectively.

Isolation of cDNA clones corresponding to the mRNAs of 1,3-β-glucanases and chitinases from a number of plant species has been reported by several groups (Mohnen *et al.* 1985, Broglie *et al.* 1986, Shinshi *et al.* 1987, Hooft van Huijsduijnen *et al.* 1987, Shinshi *et al.* 1988, Hedrick *et al.* 1988, Vögeli *et al.* 1988, Metraux *et al.* 1989). Isolation of three chitinase genes from bean has been reported (Broglie *et al.* 1986) but no genomic sequences of chitinases or 1,3-β-glucanases have been published as yet.

Constitutive expression of PR genes

Insight into the putative role of PR proteins in defence mechanisms could be obtained by analysis of the susceptibility to pathogens of those healthy plants which constitutively express one or more PR proteins. A hybrid of *Nicotiana glutinosa* × *Nicotiana debneyi* showed a constitutive expression

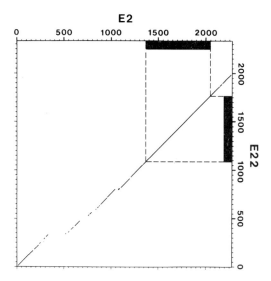

Fig. 6.1. Dot-blot comparison of the PR-S genes (boxes) and flanking sequences in clones E2 and E22. The sequences were compared with a window of 21 nucleotides and stringency of 16 nucleotides.

of a PR-1 protein (b1") that is induced in either of the two parent plants only after the hypersensitive reaction to virus infection. This hybrid was highly resistant to TMV infection, suggesting an antiviral role of the PR-1 protein (Ah1 and Gianinazzi 1982). Moreover, treatment of tobacco with salicylate results in the induction of PR-1 and GRP, and inhibits virus multiplication by over 90 per cent without a concomitant inhibition of host metabolism (Hooft van Huijsduijnen *et al.* 1986b). To test PR proteins for a possible role in the inhibition of virus multiplication or other defence mechanisms, we have constitutively expressed PR-1a, GRP and PR-S in separate transgenic tobacco plants (Linthorst *et al.* 1989a). The coding sequences of these genes were fused to the CaMV 35S promoter and introduced into the genome of Samsun NN tobacco by an *Agrobacterium tumefaciens* mediated transformation procedure. Figure 6.2 shows two Western blots performed with proteins from a PR-1 transformed plant (plant B6) and a PR-S transformed plant (plant E9), together with samples from non-transformed control plants that were either healthy (H) or TMV-infected (Tx and Tt). The left panel (B) was incubated with a PR-1 antiserum, the right panel (B+E) was incubated with a mixture of antisera to PR-1 and PR-S. It is clear that plants B6 and E9 selectively express PR-1a and PR-S, respectively. For biological assays another PR-1 transformed plant that expressed PR-1a to a higher level, comparable with that of TMV-infected

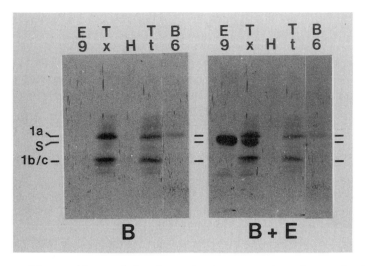

Fig. 6.2. Western blot analysis of proteins in transgenic and control tobacco plants. B6: PR-la transformed plant; E9: PR-S transformed plant; H: healthy non-transformed plant; Tx: TMV-infected plant, extracellular fluid; Tt: TMV-infected plant, total protein extract. In panel B the blot was incubated with PR-1a anti-serum; in panel (B+E) the blot was incubated with a mixture of antisera against proteins PR-la and PR-S.

plants was used. GRP-transformed plants showed a high level of GRP-mRNA accumulation but could not be analysed at the protein level due to the lack of a suitable antiserum. Compared to control plants, the transgenic plants expressing PR-1a, GRP, or PR-S showed no significant difference in susceptibility to infection with TMV or alfalfa mosaic virus (see Fig. 6.3). Leaves of these transgenic plants were as readily digested by insects (larvae of *Spodoptera exigua*) as were leaves of control plants (Linthorst *et al.* 1989b). This may suggest that PR-1a, GRP, or PR-S alone are not effective in protecting plants from virus infection or insect attack. Recently, it has been shown that engineering of a cowpea trypsin inhibitor into tobacco resulted in a resistance to insect pests (Hilder *et al.* 1987).

Conclusions

The molecular cloning of PR cDNAs and genes, and the demonstration that a number of PR proteins are hydrolytic enzymes have considerably increased our insight into the nature and interrelationships of this group of virus-induced proteins. The present data indicate that the hypersensitive response of plants to virus infection triggers the activation of a number of

Fig. 6.3.   Infection of transgenic tobacco with TMV. Symptoms of TMV-infection on non-transformed (control) and transgenic tobacco plants expressing PR-la, GRP, or PR-S as indicated. Reprinted with permission of H.J.M. Linthorst *et al.* 1989a.

general defence mechanisms directed towards different classes of pathogens, and support the notion that at least some of the PR proteins play a
role in these defences. Although the PR-1 type proteins were detected
first, are highly conserved in the plant kingdom and accumulate to high
levels under various stress conditions, their biological function is still enig-
matic. In addition to further research on the function of PR proteins, it will
also be of interest to identify and to compare the cis-acting elements and
trans-acting factors that are involved in the co-ordinate expression of the
many plant genes that are induced by the incompatible interaction of plants
with pathogens.

References

Ahl, P. and Gianinazzi, S. (1982). *b*-Protein as a constitutive component in highly
(TMV) resistant interspecific hybrids of *Nicotiana glutinosa* × *Nicotiana debneyi*. *Plant Science Letters* **26**, 173–81.

Antoniw, J.F. and White, R.F. (1986). Changes with time in the distribution of
virus and PR protein around single local lesions of TMV infected tobacco. *Plant
Molecular Biology* **6**, 145–9.

Asselin, A., Grenier, J., and Cote, F. (1985). Light-influenced extra cellular
accumulation of *b* (pathogenesis-related) proteins in *Nicotiana* green tissue
induced by various chemicals or prolonged flotation on water. *Canadian Journal
of Botany* **63**, 1276–83.

Bol, J.F. (1988). Structure and expression of plant genes encoding pathogenesis-
related proteins. In *Plant gene research; temporal and spatial regulation of plant
genes* (ed. D.P.S. Verma and R.B. Goldberg), p. 201–21. Springer Verlag,
Wien.

Bol, J.F. and Van Kan, J.A.L. (1988). The synthesis and possible functions of
virus-induced proteins in plants. *Microbiological Sciences* **5**, 47–52.

Bol, J.F., Van Kan, J.A.L., and Cornelissen, B.J.C. (1988). Plant defense genes
induced by virus infection. In *Molecular Biology of Plant–Pathogen Interactions*
(ed. B. Staskawicz, P. Ahlquist, and O. Yoder). Alan R. Liss Inc., New York,
in press.

Bol, J.F., Linthorst, H.J.M., and Cornelissen, B.J.C. (1990). Plant pathogenesis-
related proteins induced by virus infection. *Annual Review of Phytopathology*
28, 113–38.

Boller, T. (1988). Ethylene and the regulation of antifungal hydrolases in plants. In
Oxford surveys of plant molecular and cell biology, Vol. 5, (ed. B.J. Miflin).
Oxford University Press.

Bowler, C., Alliotte, T., De Loose, M., Van Montagu, M., and Inze, D. (1989).
The induction of manganese superoxide dismutase in response to stress in *Nicotiana plumbaginifolia*. *EMBO Journal* **8**, 31–8.

Broglie, K.E., Gaynor, J.J., and Broglie, R.M. (1986). Ethylene-regulated gene
expression: molecular cloning of the genes encoding an endochitinase from *Phaseolus vulgaris*. *Proceedings of the National Academy of Sciences USA* **83**, 6820–4.

Carr, J.P., Dixon, D.C., and Klessig, D.F. (1985). Synthesis of pathogenesis-related proteins in tobacco is regulated at the level of mRNA accumulation and occurs on membrane-bound polysomes. *Proceedings of the National Academy of Sciences USA* **82**, 7999–8003.

Carr, J.P., Dixon, D.C., Nikolau, B.J., Voelkerding, K.V., and Klessig, D.F. (1987). Synthesis and localization of pathogenesis-related proteins in tobacco. *Molecular and Cellular Biology* **7**, 1580–3.

Cornelissen, B.J.C., Hooft van Huijsduijnen, R.A.M., Van Loon, L.C., and Bol, J.F. (1986a). Molecular characterization of messenger RNAs for 'pathogenesis-related' proteins 1a, 1b and 1c, induced by TMV infection of tobacco. *EMBO Journal* **5**, 37–40.

Cornelissen, B.J.C., Hooft van Huijsduijnen, R.A.M., and Bol, J.F. (1986b). A tobacco mosaic virus-induced tobacco protein is homologous to the sweet-tasting protein thaumatin. *Nature* **321**, 531–2.

Cornelissen, B.J.C., Horowitz, J., Van Kan, J.A.L., Goldberg, R.B., and Bol, J.F. (1987). Structure of tobacco genes encoding pathogenesis-related proteins from the PR-1 group. *Nucleic Acids Research* **15**, 6799–811.

Cutt, J.R., Dixon, D.C., Carr, J.P., and Klessig, D.F. (1988). Isolation and nucleotide sequence of cDNA clones for the pathogenesis-related proteins PR1a, PR1b and PR1c of *Nicotiana tabacum* cv. Xanthi nc induced by TMV infection. *Nucleic Acids Research* **16**, 9861.

Dumas, E., Gianinazzi, S., and Cornu, A. (1987). Genetic aspects of polyacrylic-acid induced resistance to tobacco mosaic virus and tobacco necrosis virus in *Nicotiana* plants. *Plant Pathology* **36**, 544–50.

Dumas, E., Lherminier, J., Gianinazzi, S., White, R.F., and Antoniw, J.F. (1988). Immunocytochemical location of pathogenesis-related b1 protein induced in tobacco mosaic virus-infected or polyacrylic acid-treated tobacco plants. *Journal of General Virology* **69**, 2687–94.

Fang, K.S.Y., Vitale, M., Fehlner, P., and King, T.P. (1988). cDNA cloning and primary structure of a white face hornet venom allergen, antigen 5. *Proceedings of the National Academy of Sciences USA* **85**, 895–9.

Gaynor, J.J. (1988). Primary structure of an endochitinase mRNA from *Solanum tuberosum*. *Nucleic Acids Research* **16**, 5210.

Gianinazzi, S. (1983). Genetic and molecular aspects of resistance induced by infection or chemicals. In *Plant–microbe interactions; molecular and genetic perspectives*, Vol. 1, (ed. E.W. Nester and T. Kosuge), pp. 321–342. Macmillan, New York.

Hedrick, S.A., Bell, J.N., Boller, T., and Lamb, C.J. (1988). Chitinase cDNA cloning and mRNA induction by fungal elicitor, wounding and infection. *Plant Physiology*, **86**, 182–6.

Hilder, V.A., Gatehouse, A.M.R., Sheermann, S.E., Barker, R.F., and Boulter, D. (1987). A novel mechanism of insect resistance engineered into tobacco. *Nature* **300**, 160–3.

Hooft van Huijsduijnen, R.A.M., Cornelissen, B.J.C., Van Loon, L.C., Van Boom, J.H., Tromp, M., and Bol, J.F. (1985). Virus-induced synthesis of messenger RNAs for precursors of pathogenesis-related proteins in tobacco. *EMBO Journal* **4**, 2167–71.

Hooft van Huijsduijnen, R.A.M., Van Loon, L.C., and Bol, J.F. (1986a). cDNA cloning of six mRNAs induced by TMV infection of tobacco and a characterization of their translation products. *EMBO Journal* **5**, 2057–61.

Hooft van Huijsduijnen, R.A.M., Alblas, S.W., De Rijk, R.H., and Bol, J.F. (1986b). Induction by salicylic acid of pathogenesis-related proteins and resistance to alfalfa mosaic virus infection in various plant species. *Journal of General Virology* **67**, 2135–43.

Hooft van Huijsduijnen, R.A.M., Kauffmann, S., Brederode, F.Th., Cornelissen, B.J.C., Legrand, M., Fritig, B., and Bol, J.F. (1987). Homology between chitinases that are induced by TMV infection of tobacco. *Plant Molecular Biology* **9**, 411–420.

Hosokawa, D. and Ohashi, Y. (1988). Immuno-chemical localization of pathogenesis-related proteins secreted into the intercellular spaces of salicylate-treated tobacco leaves. *Plant Cell Physiology* **29**, 88–128.

Jamet, E. and Fritig, B. (1986). Purification and characterization of 8 of the pathogenesis-related proteins in tobacco leaves reacting hypersensitively to tobacco mosaic virus. *Plant Molecular Biology* **6**, 69–80.

Kauffmann, S., Legrand, M., Geoffroy, P., and Fritig, B. (1987). Biological function of 'pathogenesis-related' proteins. Four PR-proteins of tobacco have 1,3-β-glucanase activity. *EMBO Journal* **6**, 3209–12.

King, G.J., Turner, V.A., Hussey, V.E., Wurtele, E.S., and Lee, S.M. (1988). Isolation and characterization of a tomato cDNA clone which codes for a salt-induced protein. *Plant Molecular Biology* **10**, 401–12.

Koike, S., Hibi, T., and Yora, K. (1976). Infection of cowpea mesophyll protoplasts by TMV. *Annals of the Phytopathological Society of Japan* **42**, 105.

Kombrink, E., Schröder, M., and Hahlbrock, K. (1988). Several 'pathogenesis-related' proteins in potato are 1,3-β-glucanases and chitinases. *Proceedings of the National Academy of Sciences USA* **85**, 782–6.

Lagrimini, L.M., Burkhart, W., Moyer, M., and Rothstein, S. (1987). Molecular cloning of complementary DNA encoding the lignin-forming peroxidase from tobacco: molecular analysis and tissue-specific expression. *Proceedings of the National Academy of Sciences USA* **84**, 7542–6.

Lawton, M.A. and Lamb, C.J. (1987). Transcriptional activation of plant defence genes by fungal elicitor, wounding and infection. *Molecular and Cellular Biology* **7**, 335–41.

Legrand, M., Kauffmann, S., Geoffroy, P., and Fritig, B. (1987). Biological function of 'pathogenesis-related' proteins: four tobacco PR-proteins are chitinases. *Proceedings of the National Academy of Sciences USA* **84**, 6750–4.

Linthorst, H.J.M., Meuwissen, R.L.J., Kauffmann, S., and Bol, J.F. (1989a). Constitutive expression of pathogenesis-related proteins PR-1, GRP and PR-S in tobacco has no effect on virus infection. The *Plant Cell* **1**, 285–91.

Linthorst, H.J.M., Cornelissen, B.J.C., Van Kan, J.A.L., Van de Rhee, M.D., Meuwissen, R.L.J., Gonzalez Jaen, M.T., and Bol, J.F. (1989b). Induction of plant genes by compatible and incompatible virus-plant interactions. In *Recognition and response in plant–virus interactions*, (ed. R.S.S. Fraser), pp. 361–73. Springer Verlag, Wien.

Linthorst, H.J.M., Melchers, L.S., Mayer, A., Van Roekel, J.S.C., Cornelissen, B.J.C., and Bol, J.F. (1990). Analysis of gene families encoding acidic and basic 1,3-β-glucanases of tobacco. *Proceedings of the National Academy of Sciences USA* (in press).

Lucas, J., Camacho Henriquez, A., Lottspeich, F., Henschen, A., and Sanger, H.L. (1985). Amino acid sequence of the 'pathogenesis-related' leaf protein p14 from viroid-infected tomato reveals a new type of structurally unfamiliar proteins. *EMBO Journal* **4**, 2745–9.

Matsuoka, M., Yamamoto, N., Kano-Murakami, Y., Tanaka, Y., Ozeki, Y., Hirano, H., Kagawa, H., Oshima, M., and Ohashi, Y. (1987). Classification and structural comparison of full-length cDNAs for pathogenesis-related proteins. *Plant Physiology* **85**, 942–6.

Mauch, F. and Staehelin, L.A. (1988). Subcellar localization of chitinase and β-1, 3-glucanase in bean leaves. Functional implications for their involvement in plant pathogen interactions. *Journal of Cellular Biochemistry* **12C**, 269.

Mauch, F., Mauch-Mani, B., and Boller, T. (1988). Antifungal hydrolases in pea tissue. II. Inhibition of fungal growth by combinations of chitinase and β-1,3-glucanase. *Plant Physiology* **88**, 936–42.

Memelink, J., Hoge, J.H.C., and Schilperoort, R.A. (1987). Cytokinin stress changes the developmental regulation of several defence-related genes in tobacco. *EMBO Journal* **6**, 3579–83.

Metraux, J.P., Burkhart, W., Moyer, M., Dincher, S., Middlesteadt, W., Williams, S., Payne, G., Carnes, M., and Ryals, J. (1989). Isolation of a complementary DNA encoding a chitinase with structural homology to a bifunctional lysozyme/chitinase. *Proceedings of the National Academy of Sciences USA* **86**, 896–900.

Mohnen, D., Shinshi, H., Felix, G., and Meins, F. (1985). Hormonal regulation of β-1,3-glucanase messenger RNA levels in cultured tobacco tissues. *EMBO Journal* **4**, 1631–5.

Nassuth, A. and Sänger, H.L. (1986). Immunological relationships between 'pathogenesis-related' leaf proteins from tomato, tobacco and cowpea. *Virus Research* **4**, 229–42.

Ohashi, Y. and Matsuoka, M. (1987). Localization of pathogenesis-related proteins in the epidermis and intercellular spaces of tobacco leaves after their induction by potassium salicylate or tobacco mosaic virus infection. *Plant Cell Physiology* **28**, 1227–35.

Oshima, M., Matsuoka, M., Ymamoto, N., Tanaka, Y., Kano-Murakami, Y., Ozeki, Y., Kato, A., Harada, N., and Ohashi, Y. (1987). Nucleotide sequence of the PR-1 gene of *Nicotiana tabacum*. *FEBS Letters* **225**, 243–6.

Parent, J.G. and Asselin, A. (1984). Detection of pathogenesis-related proteins (PR or b) and of other proteins in the intercellular fluid of hypersensitive plants infected with tobacco mosaic virus. *Canadian Journal of Botany* **62**, 564–9.

Payne, G., Parks, T.D., Burkhart, W., Dincher, S., Ahl, P., Metraux, J.P., and Ryals, J. (1988a). Isolation of the genomic clone for pathogenesis-related protein 1a from *Nicotiana tabacum* cv. Xanthinc. *Plant Molecular Biology* **11**, 89–94.

Payne, G., Middlesteadt, W., Williams, S., Desai, N., Parks, T.D., Dincher, S., Carnes, M., and Ryals, J. (1988b). Isolation and nucleotide sequence of a novel cDNA clone encoding the major form of pathogenesis-related protein R. *Plant Molecular Biology* **11**, 223–4.

Pierpoint, W.S. (1986). The pathogenesis-related proteins of tobacco leaves. *Phytochemistry* **25**, 1595–601.

Pierpoint, W.S., Tatham, A.S., and Pappin, D.J.C. (1987). Identification of the virus-induced protein of tobacco leaves that resembles the sweet protein thaumatin. *Physiological and Molecular Plant Pathology* **31**, 291–8.

Pfitzner, U.M. and Goodman, H.M. (1987). Isolation and characterization of cDNA clones encoding pathogenesis-related proteins from tobacco mosaic virus infected tobacco plants. *Nucleic Acids Research* **15**, 4449–65.

Pfitzner, U.M., Pfitzer, A.J.P., and Goodman, H.M. (1988). DNA sequence analysis of a PR-1a gene from tobacco: molecular relationship of heat shock and pathogen responses in plants. *Molecular and General Genetics* **211**, 290–5.

Ponz, F., Glascock, C.B., and Bruening, G. (1988). An inhibitor of polyprotein processing with the characteristics of a natural virus resistance factor. *Molecular Plant–Microbe Interactions* **1**, 25–31.

Richardson, M., Valdes-Rodriguez, S., and Blanci-Labra, A. (1987). A possible function for thaumatin and a TMV-induced protein suggested by homology to a maize inhibitor. *Nature* **327**, 432–4.

Shinshi, H., Mohnen, D., and Meins, F. (1987). Regulation of a plant pathogenesis-related enzyme: inhibition of chitinase and chitnase mRNA accumulation in cultured tobacco tissues by auxin and cytokinin. *Proceedings of the National Academy of Sciences USA* **84**, 89–93.

Shinshi, H., Wenzler, H., Neuhaus, J.-M., Felix, G., Hofsteenge, J., and Meins, F. (1988). Evidence for N- and C-terminal processing of a plant defense-related enzyme: primary structure of tobacco prepro-β-1,3-glucanase. *Proceeding of the National Academy of Sciences USA* **85**, 5541–5.

Singh, N.K., Bracker, C.A., Hasegawa, P.M., Handa, A.K., Buckel, S., Hermodson, M.A., Pfankoch, E., Reguicr, F.E., and Bressan, R.A. (1987). Characterization of osmotin. A thaumatin-like protein associated with osmotic adaptation in plant cells. *Plant Physiology* **85**, 529–36.

Thornburg, R.W., An, G., Cleveland, T.E., Johnson, R., and Ryan, C. (1987). Wound-inducible expression of a potato inhibitor II-chloramphenicol acetyltransferase gene fusion in transgenic tobacco plants. *Proceedings of the National Academy of Sciences USA* **84**, 744–8.

Van Kan, J.A.L., Cornelissen, B.J.C., and Bol, J.F. (1988). A virus-inducible tobacco gene encoding a glycine-rich protein shares putative regulatory elements with the ribulose bisphosphate carboxylase small subunit gene. *Molecular Plant–Microbe Interactions* **1**, 107–12.

Van Kan, J.A.L., Van de Rhee, M.D., Zuidema, D., Cornelissen, B.J.C., and Bol, J.F. (1989). Structure of tobacco genes encoding thaumatin-like proteins. *Plant Molecular Biology* **12**, 153–5.

Van Loon, L.C. (1982). Regulation of changes in proteins and enzymes associated with active defence against virus infection. In *Active defence mechanisms in plants*, (ed. R.K.S. Wood), pp. 247–273. Plenum press, New York.

Van Loon, L.C., Gerritsen, Y.A.M., and Ritter, C.E. (1987). Identification, purification and characterization of pathogenesis-related proteins from virus-infected Samsun NN tobacco leaves. *Plant Molecular Biology* **9**, 593–609.

Vera, P., Hernandez Yago, J., and Conejero, V. (1988). Immuno-cytochemical localization of the major 'pathogenesis-related' (PR) protein of tomato plants. *Plant Science* **55**, 223–30.

Vögeli, U., Meins, F., and Boller, T. (1988). Co-ordinated regulation of chitinase and β-1,3-glucanase in bean leaves. *Planta* **174**, 364–72.

White, R.F., Rybicki, E.P., Von Wechmar, M.B., Dekker, J.L., and Antoniw, J.F. (1987). Detection of PR-1 type proteins in *Amaranthaceae, Chenopodiaceae, Gramineae, and Solanaceae* by immuno-electroblotting. *Journal of General Virology* **68**, 2043–8.

7 Biochemical and molecular events in the hypersensitive response of bean to *Pseudomonas syringae* pv. *phaseolicola*

ALAN J. SLUSARENKO

Institut für Pflanzenbiologie, Zollikerstrasse 107, CH-8008 Zürich

KEVAN P. CROFT AND CHRISTINE R. VOISEY

Department of Applied Biology, University of Hull, Hull HU6 7RX, UK

Introduction

Interactions between plants and bacteria can be classified into three categories (Klement 1971).

1. Plants exposed to saprophytic bacteria;
2. The compatible or susceptible reaction;
3. The incompatible or resistant reaction.

In leaves the general characteristics of interactions in the first category are little or no bacterial multiplication and no apparent damage to the leaf cells (Lyon (neé O'Brien) and Wood 1976).

In the second category bacteria multiply rapidly to reach high levels, for example *Pseudomonas syringae* pv. *phaseolicola* inoculated into leaves of susceptible French bean plants (*Phaseolus vulgaris*) increased to 10^6 times the original population after 5–6 days (Omer and Wood 1969). Typical symptoms of bacterial leaf-spot diseases are the production of greasy-looking, water-soaked lesions which increase in size gradually and eventually become necrotic in the centre. Often they are associated with toxin-induced systemic and/or localized chlorosis. These responses, along with growth distortions of newly forming leaves and a general stunting of the whole plant, are characteristic symptoms of halo blight of French bean caused by *P.s.* pv. *phaseolicola*.

In the third category, bacteria initially multiply at rates similar to those in the compatible combination, but after a few hours numbers stop increasing, or fall a little (Lyon (neé O'Brien) and Wood 1976). This pattern is often accompanied by necrosis of plant cells at the site of inoculation—the characteristic hypersensitive reaction (HR). However, resistance is not always associated with an HR in plant/bacteria interactions. For example, resistance of Soybean cv. Clark 63 to pathogenic strains of *Xanthomonas campestris* pv. *glycines*, conditioned by the recessive *rpx* gene pair, appears unusual in that bacteria multiply at the same rate as in the susceptible combination, but symptoms of infection are not produced (Fett 1984). In general, however, an HR occurs when cells of a phytopathogenic bacterium are introduced into tissues of a non-host plant, e.g. *P.s.* pv. *pisi* in tobacco, or when cells of an avirulent strain of a pathogen are introduced into a resistant host, e.g. race 1 isolates of *P.s.* pv. *phaseolicola* in *Phaseolus vulgaris* cv. Red Mexican.

The hypersensitive response

The HR of plants to phytopathogenic bacteria can be divided into three phases (Klement 1971).

1. The INDUCTION PHASE, which requires the presence of living bacteria.
2. The LATENT PHASE, during which living bacteria are no longer required. No macroscopic symptoms occur during this phase, but the permeability of the plant cell membranes increases and ions leak out into the intercellular spaces (Cook and Stall 1968; Goodman 1968). Ultrastructural changes can also be detected (Roebuck *et al.* 1978).
3. The PRESENTATION or COLLAPSE PHASE, during which host cells in the inoculated region collapse and desiccate, taking on first a silvered, then a bronzed appearance. Accumulation of antibacterial phytoalexins associated with the HR may help to restrict the growth of bacteria in the lesion (Gnanamanickam and Patil 1977; Holliday *et al.* 1981; Lyon and Wood 1975; Keen and Kennedy 1974). 'However, reducing the HR to a consideration of only tissue necrosis and phytoalexin production is almost certainly an oversimplification (Klement 1982).

The exact timing of the phases in the HR varies depending upon the host/pathogen combination and the environmental conditions. Tobacco reacts very quickly, host cell collapse occurring as early as 6–8 h after inoculation. In contrast in French bean inoculated with an avirulent race of *P.s.* pv. *phaseolicola*, cell collapse occurs from 18–24 h after inoculation. The requirement for live bacteria during the induction period was defined originally by infiltration of prokaryotic-specific antibiotics into leaves already inoculated with bacteria. At some point after inoculation killing the bacteria no longer prevented host cell collapse occurring at the presentation phase. In the French bean/*P.s.* pv. *phaseolicola* interaction, there is a requirement for live bacteria to be present for about 4 h after inoculation to set in motion the events which lead to host cell collapse some 14–20 hours later (Roebuck *et al.* 1978).

Seemingly rather high concentrations of inoculum are required to induce the confluent necrosis associated with the macroscopically visible HR: about 5×10^6 cells cm^{-3} of *P.s.* pv. *syringae* in tobacco (Klement and Goodman 1967), and about 2×10^6 cells cm^{-3} *P. mors-prunorum* or *P.s.* pv. *phaseolicola* in *P. vulgaris* cv. Red Mexican (Lyon (neé O'Brien) and Wood 1976). In the latter case this corresponds to about 2 bacterial cells for each necrotic leaf cell. Viewed in these terms the initial concentration of inoculum does not seem so unrealistic compared to natural infections in the field. At inoculum concentrations below the threshold necessary to induce confluent necrosis, *P.s.* pv. *pisi* introduced into tobacco gave rise to dead leaf cells at a ratio of 1 *per* bacterium (Turner and Novacky 1974). Similarly in bean, individual, dead plant cells can be observed microscopically in the absence of a confluent HR.

The ability of bacteria to induce HR is conditioned both by avirulence (*avr*) genes and hypersensitive reaction and pathogenicity (*hrp*) genes (Gabriel 1986; Lindgren *et al.* 1986). Avirulence genes confer the race-specific ability of the bacterium to induce an HR in a given resistant cultivar of the host (Staskawicz *et al.* 1984). However, the products of those avirulence genes characterized to date do not appear to be membrane localized or secreted, and do not induce an HR in their own right (Tamaki *et al.* 1988). The *hrp* gene cluster was identified in *P.s.* pv. *phaseolicola* as a chromosomal region where mutations render the bacteria unable to cause disease in bean (path$^-$ phenotype) and unable to cause the HR in the non-host, tobacco (HR$^-$ phenotype). The *hrp* genes have homology with DNA sequences in other *Pseudomonas syringae* pathovars (Lindgren *et al.* 1988). The transcriptional organization of the *hrp* gene cluster was studied using an ice-nucleation reporter-gene-construct with transposon Tn3 (Lindgren *et al.* 1989). It was found that some members of the *hrp* gene cluster were induced *in planta* within two hours of inoculation, and that some of the *hrp* genes appear to have a regulatory function (Ibid, Rahme *et al.* 1988). Thus, the requirement for live bacteria for induction of HR, probably depends upon expression of *hrp* and *avr* genes. The mechanism by which expression of these genes induces processes in the host which lead to HR cell collapse is unknown, however.

Race-specific resistance to *P.s.* pv. *phaseolicola*, which is associated with an HR, is conditioned by single dominant genes inherited in a simple Mendelian fashion (Taylor *et al.* 1978). The products of such resistance genes may function as part of a recognition system (Day 1984) which, when triggered by the pathogen, sets in motion, either directly, or indirectly via second messengers, the various primary host responses which lead ultimately to resistance (Slusarenko 1987). Race-specific avirulence determinants from the pathogen have been cloned (Hitchin *et al.* 1989), and so the interaction between *P.s.* pv. *phaseolicola* and French bean appears to fall into the gene-for-gene category described by Flor (1971).

The idea that the factors responsible for HR induction and for the establishment of host cell collapse are different is not new (Sequeira 1976), and it is supported by the observation that HR cell collapse depends upon host protein synthesis (Keen *et al.* 1981; Lyon and Wood 1977). Slusarenko and Longland (1986) observed a co-ordinated sequence of changes in several host mRNA activities in bean leaves in the early stages of the incompatible interaction with *P.s.* pv. *phaseolicola*. It seemed probable that at least some of the mRNA species that accumulated early on might produce proteins whose activity could have brought about cell collapse in the HR. Indeed, the HR can be viewed as a form of programmed cell death (Collinge and Slusarenko 1987; Slusarenko *et al.* 1986).

The earliest physiological changes which can be detected in cells that will undergo HR collapse indicate the occurrence of early, irreversible membrane damage. Thus, electrolyte leakage (Cook and Stall 1968; Goodman 1968; Lyon (neé O'Brien) and Wood 1976) and failure of cells to plasmolyse properly (Wood *et al.* 1988) have both been interpreted as indicating early membrane dysfunction. Pavlovkin *et al.* (1986) examined changes in the electrical membrane potential (E_m) in cotyledons of cotton (*Gossypium hirsutum*) inoculated with *P.s.* pv. *tabaci*. The E_m is made up of components contributed both by the electrogenic pump (E_p—resulting from the plasmalemma H^+-ATPase) and by the diffusion potential (E_D—resulting from passive diffusion of ions). In the HR, in the first two hours after inoculation, the E_p dropped rapidly to a level between 40 and 60 per cent of the control value, whereas the E_D dropped only 9 per cent. From 2h upto 10 h after inoculation E_p remained more or less constant, whereas E_D continued its steady decline to a level of 45 per cent of the control value. In contrast, inoculation with *Xanthomonas campestris* pv. *malvacearum* (susceptible reaction), led to total loss of E_p but the value of E_D remained unchanged. The authors suggested that the primary alteration in membranes at the onset of the HR was not in the activity of the proton extrusion pump (E_p), but rather in the function of the sites of passive ion diffusion, i.e. the lipid matrix and/or protein channels (diffusion potential E_D). Increase in the level of lipid peroxidation was observed in cucumber cotyledons during HR and it was suggested that peroxidative membrane damage might lead to HR cell collapse (Keppler and Novacky 1986).

The decrease in E_m was parallelled by loss of electrolytes from the tissue during the HR. In control tissue 59 per cent of the electrolyte loss, which was predominantly the result of non-specific electrolyte efflux from the cut surface, was represented by K^+. In HR tissues 82 per cent of the loss was K^+, suggesting an increased permeability to K^+ occurred during the HR. After complete HR cell collapse (24 h), the relative amount of K^+ present in the tissue decreased to 30 per cent of the total electrolytes. The authors suggested that changes in K^+ concentration resulted initially from electrogenic ion flux and that the mass of electroneutral leakage occurred later. Others (Atkinson *et al.* 1985; Atkinson and Baker 1987) have suggested that HR cell collapse was brought about by activation of a passive plasmalemma K^+/H^+ exchange mechanism. Changes in the membrane lipid phase, or in protein channels were proposed to explain the K^+/H^+ exchange (Keppler *et al.* 1988). The authors monitored uptake of fluorescein diacetate into plant cells during the HR, and concluded that the decrease in membrane fluidity and permeability to fluorescein diacetate might result from lipid peroxidation, or phospholipase activity. More recently Atkinson and Baker (1989) proposed that the plasmalemma H^+-ATPase activity is required for the H^+/K^+ exchange response observed in the HR. Using a variety of ATPase inhibitors, inhibitors of respiration, a protonophore and a slightly alkaline external pH, the authors showed that the K^+/H^+ exchange response could be inhibited to varying degrees.

Lipid peroxidation and HR cell collapse

Lipid peroxidation can occur by both enzymic and non-enzymic means, through the action of various active oxygen species and organic free radicals (Thompson *et al.* 1987). Lipoxygenase (EC 1.3.11.12) will oxidize unsaturated fatty acids that have a cis-1,4-pentadiene system, e.g. linoleic (18:2) and linolenic (18:3) acids (Taylor and Morris 1983), both of which are common constituents of plant membrane lipids (Leshem 1987). Lipoxygenase (LOX) activity has been shown to increase in resistant tissues of several host-pathogen combinations for example, *Cucumis sativa/ Pseudomonas syringae* pv. *pisi* (Keppler and Novacky 1987), oats/*Puccinia coronata avenae* (Yamamoto and Tani 1986), rice/*Pyricularia oryzae* (Ohta *et al.* 1988), wheat/*Puccinia coronata avenae* or *P. graminis tritici* (Ocampo *et al.* 1986), and French bean (*Phaseolus vulgaris/P.s.* pv. *phaseolicola* (Croft *et al.* 1990). Lipoxygenase has been studied extensively in soybean, where high activities are detected in the seed. In soybean three distinct isoenzyme forms, LOX 1, LOX 2 and LOX 3, have been defined on the basis of differences in pH optimum, substrate specificity and product formation (Axelrod *et al*, 1981). Corresponding isoforms have been reported in French bean (Boyer and Vanderploeg 1986).

Fatty acids that are components of membrane lipids are not good substrates for lipoxygenase, and it is usually considered that the activity of lipolytic acyl hydrolase (LAH) results in release of fatty acids which LOX can then oxidize (Thompson *et al.* 1987). LOX activity produces singlet oxygen ($^1O_2$) and superoxide anions ($O_2^{\cdot-}$) as by-products of its activity (Lynch and Thompson 1984, Thompson *et al.* 1987). The singlet oxygen produced is sufficiently reactive to oxidize lipids in membranes, as are hydroxyl radicals (OH$^{\cdot}$) which can be derived from $O_2^{\cdot-}$ by the action of superoxide dismutase and the iron-catalysed Haber–Weiss process (Epperlein *et al.* 1986). However, Salzwedel *et al.* (1988) showed that quenchers of singlet oxygen, supplied to tobacco cell suspension cultures which had been inoculated with *P.s.* pv. *pisi*, did not suppress the HR. Superoxide anion production has been linked with membrane damage and *in vivo* toxicity in numerous instances (Wolff *et al.* 1986). There have been several recent reports that $O_2^{\cdot-}$-initiated lipid peroxidation is involved in the bacterially-induced hypersensitive reaction in tobacco cell-suspension-cultures (Keppler and Baker 1989, Keppler *et al.* 1989), tobacco leaves (Adam *et al.* 1989) and cucumber cotyledons (Keppler and Novacky 1989). Because $O_2^{\cdot-}$ is not indiscriminately reactive, it has often been postulated that its deleterious effects resulted from its conversion to OH$^{\cdot}$ radicals by the mechanism described above. However, protonation of $O_2^{\cdot-}$ to give its conjugate acid, the perhydroxyl radical (HO$_2^{\cdot}$), can occur in specific cellular microenvironments where there are local concentrations of protons, for example at the negatively charged surfaces of membranes (Bielski *et al.* 1983, Fridovich 1988). The perhydroxyl radical is a stronger oxidizing agent than $O_2^{\cdot-}$ and will react with polyunsaturated fatty acids directly (Bielski *et al.* 1983, Halliwell 1988). Thus, LOX activity can lead to the production of various active oxygen species capable of initiating enzyme-independent lipid peroxidation.

Once membrane lipid peroxidation has begun, the lipid peroxy radicals (ROO$^{\cdot}$) that are produced will propagate a chain reaction among the tightly stacked fatty acids in the lipid bilayer of the membrane:

ROO$^{\cdot}$	+	RH ------>	ROOH	+	R$^{\cdot}$
LIPID PEROXY		LIPID	LIPID		LIPID
RADICAL			HYDROPEROXIDE		RADICAL

R$^{\cdot}$	+	O_2	-------------->	ROO$^{\cdot}$

It has been suggested that active oxygen species play an important role in causing cell-damage and phytoalexin accumulation in several host pathogen interactions (Doke 1985; Doke and Chai 1985; Sekizawa *et al.* 1987), and in the biotic and abiotic elicitation of phytoalexins (Epperlein *et al.* 1986; Rogers *et al.* 1988).

In our experiments with French bean cv. Red Mexican, intact seedlings were inoculated with either *P.s.* pv. *phaseolicola* race 1 (incompatible combination, HR), or race 2 or 3 (compatible combinations, susceptible). Bacteria were infiltrated into the leaves under reduced pressure through the stomata (Fig. 7.1a). Thus, we were able to infiltrate large areas of leaf tissue, in a relatively uniform manner, without wounding. Leaves inoculated with avirulent race 1 cells responded with an HR where visible necrotic flecks were interspersed with apparently healthy tissue (Fig. 7.1b). HR at the single cell level was apparent in the green areas of the leaf after trypan blue staining (Fig. 7.1c).

The temporal relationships of some early events in the HR are shown in Fig. 7.2. Increases in LOX 1 and LOX 3 activities were apparent within 4 h of inoculation and such increases appeared to depend on *de novo* protein synthesis because cycloheximide prevented the effect. Evidence that membrane-lipid peroxidation occurred *in vivo* during the HR was provided by a comparison of ethane evolution in plants inoculated either with an avirulent or a virulent isolate of *P.s.* pv. *phaseolicola*. Ethane is produced as one of the final breakdown products of the fatty acid hydroperoxides that are derived from linolenic acid (Dumelin and Tappel 1977; Gutteridge 1988), and it is a specific indicator of membrane-lipid peroxidation (Konze and Elstner 1978; Riely *et al.* 1974). Ethane evolution increased steadily from about 6 hours after inoculation in the incompatible combination Fig. 7.2. Production of ethane correlated well with other indicators of plasma membrane dysfunction occurring in the HR, such as electrolyte leakage and increase in the extracellular pH. A steady increase in lipolytic acyl hydrolase activity was observed during progression of the HR.

Our data are consistent with the hypothesis that membrane damage in the HR is caused by lipid peroxidation that may be **initiated** by the action of LOX isozymes, and propagated autoxidatively in membranes *in situ*. Active oxygen species produced as a consequence of LOX activity might contribute to further initiation of peroxidation and might damage other cell constituents. Some of the possible interrelationships between events in the HR are summarized in Fig. 7.3. Lipid hydroperoxides and their breakdown products are very damaging to cellular constituents, especially proteins, and damage to membrane transport-proteins has been suggested as a primary cause of radical-induced cell death in animal cells (Wolff *et al.* 1986).

Superoxide dismutase (SOD), peroxidase (POX), and catalase can be regarded as members of a defensive team. Catalase activity was found to show a similar pattern in both the compatible and incompatible combinations and increased only marginally over the period of the experiment. However, SOD and POX activities increased approximately 8 h after the increase in LOX activity occurred. It is possible that they are induced in cells surrounding those undergoing a hypersensitive reaction and that they help to limit the spread of necrosis. We hope to test this hypothesis by histochemical and *in situ* hybridization studies. We also hope to characterize

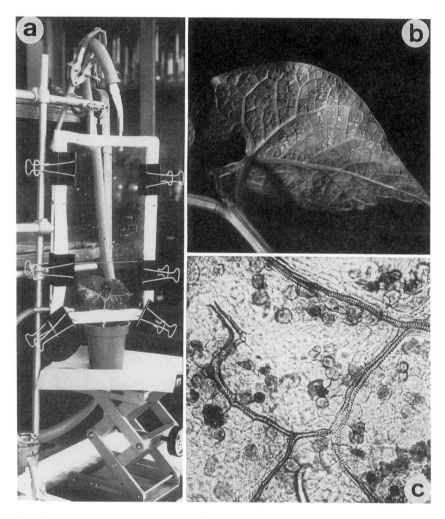

Fig. 7.1. (a) Apparatus used to infiltrate leaves with bacterial suspensions. The apparatus consists of two perspex plates separated by a rubber gasket and is based on a design by G. Wolf (Göttingen). A series of valves allows inoculum to be drawn in under reduced pressure and entry of air through a bleed-valve facilitated infiltration of inoculum into leaves. (b) 24 h after inoculation of Red Mexican with avirulent race 1 cells of *Pseudomonas syringae* pv. *phaseolicola*, local areas of collapsed, necrotic HR tissue can be seen. (c) After trypan blue staining, the green areas of the leaf in (b) could be seen to show HR at the single cell level.

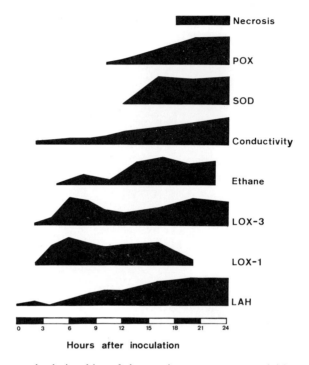

Fig. 7.2.   The temporal relationships of changes in some enzyme activities, on an arbitrary scale (LAH = lipolytic acyl hydrolase, LOX1 and LOX3 lipoxygenase isoforms, SOD = superoxide dismutase, POX = peroxidase); signs of membrane damage (ethane evolution, increase in conductivity of the bathing solution for leaf discs); and necrosis in the HR of Red Mexican to race 1 cells of *P.s.* pv. *phaseolicola*.

the genes encoding lipoxygenase in French bean and to study their regulation during the HR.

Timing of other defence responses in relation to HR

Accumulation of antibacterial isoflavonoid phytoalexins is associated with the HR in bean (Lyon and Wood 1975). In our system probing of RNA dot blots and Northern blots with cDNA probes showed co-ordinated regulation of the genes for phenylalanine ammonia lyase (PAL) and chalcone synthase (CHS), enzymes from the phenylpropanoid pathway leading to isoflavonoid biosynthesis. Maximum accumulation of transcripts occurred approximately 12 h after inoculation in the incompatible combination, but in contrast no accumulation of these transcripts was detected in the compatible combination over the same period. In the incompatible combination PAL enzyme activity increased following mRNA induction and reached maximum level at around 20 h after inoculation, (Fig. 7.4). Phytoalexins were detectable by 24 h after inoculation.

The hydrolytic enzyme chitinase is induced differentially in bean leaves inoculated with either avirulent or virulent isolates of *P.s.* pv. *phaleolicola* (Voisey and Slusarenko 1989). Chitinase mRNA concentration and activity began to increase in the HR by 6 h after inoculation, whereas in the compatible combination this increase occurred between 20 and 24 h after inoculation. The increases in chitinase enzyme activity were delayed with respect to changes in mRNA activity by 3–6 h. Interestingly, heat-killed or UV-killed cells of both virulent and avirulent races of *P.s.* pv. *phaseolicola* induced chitinase enzyme activity in bean leaves. This demonstrated that chitinase was induced separately from the hypersensitive reaction and phytoalexin synthesis; both of which require the bacteria to be live and metabolically active. However, the role of chitinase in the resistance of bean to bacterial infection is unclear.

Conclusions and prospects

The relationship of the hypersensitive response to other plant defence responses is only poorly understood. Bailey (1982) observed that host cell injury, often leading to host cell death, was a common denominator in the action of biotic and abiotic elicitors, and pathogen challenge which preceded accumulation of phytoalexins. In compatible interactions the rapid accumulation of phytoalexins was not observed, at least in the initial biotrophic phase where gross cell injury was avoided. Bailey (1982) postulated that injured cells released constitutive elicitors into the surrounding healthy cells, which responded by synthesizing phytoalexins. The dead cells also act as a sink for phytoalexin accumulation (Hargreaves and Bailey 1978). In some model systems, for example bean cell-suspension-cultures treated with elicitor, transcriptional activation of the genes involved in phytoalexin biosynthesis was very rapid and occurred within 5–10 min of treatment (Templeton and Lamb 1988). Maximum transcript levels occurred at 3–4 h after treatment with elicitor. This would seem to argue against Bailey's hypothesis. However, in more natural infections involving whole plant tissues, events seem to proceed at a more leisurely pace. Thus, when French bean hypocotyls are inoculated with conidia of *Colletotrichum lindemuthianum*, a 30–40 h period is required for spore germination, infection peg production, and penetration of the host to occur. Accumulation of PAL and CHS transcripts can only be detected 39 h after inoculation, with maximum concentration of mRNA occurring between 70 and 75 h after inoculation (Cramer *et al.* 1985; Lamb *et al.* 1986).

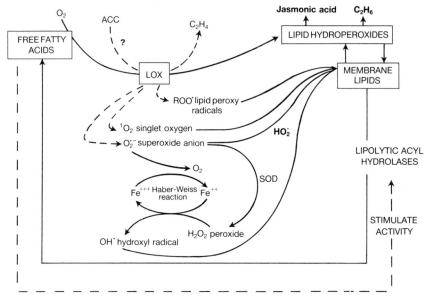

Fig. 7.3.   Possible interrelationships, as described in the text, between events contributing to membrane damage in HR cell collapse.

Obviously, in those cells stimulated directly by the pathogen, signal transduction and defence gene activation may occur very rapidly. The role of the HR might be to amplify the response and to stimulate surrounding healthy cells where much of the phytoalexin synthesis occurs. Wyman and Vanetten (1982) inoculated beans with *X.c.* pv. *phaseoli* using concentrations just above and just below that required to induce confluent necrosis (i.e. HR visible to the naked eye, rather than at the single cell level). They found no accumulation of phytoalexin in the absence of a visible HR. The transition to large scale host-cell-damage seemed to amplify the response, and to bring phytoalexin accumulation to levels that were easily detectable.

Cell death is a visible marker that the HR has occurred. That the cell is dead may be of little consequence in itself; how the cell died may be much more important. Thus, membrane lipid peroxidation may have a role to play in a signal transduction process that is important for some, even if not all of the plant's defence responses. Certainly, some fatty acid hydroperoxides are calcium ionophores (Serhan *et al.* 1981), and jasmonic acid, a compound produced in the linolenic acid cascade as a metabolite of the 13-hydroperoxide, has been shown to alter gene expression in barley

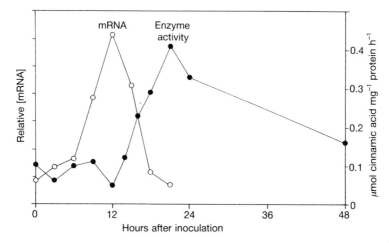

Fig. 7.4. Timing of induction of phenylalanine ammonia lyase (PAL) mRNA and enzyme activity in the HR of Red Mexican to *P.s.* pv. *phaseolicola* race 1. Relative mRNA concentration was determined by probing dot blots and Northern blots with a cDNA probe for bean PAL. Chalcone synthase mRNA showed a similar pattern of induction. PAL enzyme activity increased in parallel with mRNA concentration but about 6 h later. Antibacterial isoflavonoids were detectable by 24 h after inoculation.

(Mueller-Uri *et al.* 1988). Interestingly, one of the precursors of jasmonic acid, 12-oxo-phytodienoic acid (12-oxo-PDA), has a structure similar to that of prostaglandin A_1 of animals. Another product of the lipoxygenase pathway is 12-oxo-trans-10-dodecenoic acid, which has been proposed to be the active component of the wound hormone traumatin (Vick and Zimmerman 1987).

Potential antimicrobial effects of active oxygen species, lipid hydroperoxides, and their breakdown products may also play a role in limiting pathogen spread from the HR lesion.

Hopefully, some of the speculative hypotheses outlined in this review will be tested critically in the next few years and lead to a greater understanding of this interesting phenomenon.

Acknowledgements

We are grateful for financial support from the AFRC (A.J.S.), Gatsby Foundation (A.J.S.) and SERC (K.P.C. & C.R.V.). Thanks are due to Chris Lamb (Salk Institute) for kind gifts of bean cDNA clones. The data for PAL enzyme activity in Fig. 7.4 formed part of an undergratuate research project (Clare Nunn, University of Hull). M. Pröschel (Zürich) prepared the figures.

References

Adam, A., Farkas, T., Somlyai, G., Hevesi, M., and Kiraly, Z. (1989). Conse-quence of O_2^- generation during a bacterially induced hypersensitive reaction in Tobacco: deterioration of membrane lipids. *Physiological and Molecular Plant Pathology* **34**, 13–26.

Atkinson, M.M. and Baker, C.J. (1985). Hypersensitivity of suspension-cultured tobacco cells to pathogenic bacteria. *Phytopathology* **75**, 1270–4.

Atkinson, M.M. and Baker, C.J. (1987). Association of host plasma membrane K^+/H^+ exchange with multiplication of *Pseudomonas syringae* pv. *syringae* in *Phaseolus vulgaris*. *Phytopathology* **77**, 1273–9.

Atkinson, M.M. and Baker, C.J. (1989). Role of the plasmalemma H^+-ATPase in *Pseudomonas syringae*-induced K^+/H^+ exchange in suspension-cultured tobacco cells. *Plant Physiology* **91**, 298–303.

Atkinson, M.M., Huang, J.S., and Knopp, J.A. (1985). The hypersensitive reac-tion of tobacco to *Pseudomonas syringae* pv. *pisi*. *Plant Physiology* **79**, 843–7.

Axelrod, B., Cheesbrough, T.M., and Laakso, S. (1981). Lipoxygenase from Soy-beans. *Methods in Enzymology* **71**, 441–51.

Bailey, J.A. (1982). Physiological and biochemical events associated with the expression of resistance to disease. In *Active defence mechanisms in plants*, (ed R.K.S. Wood), pp. 39–65. NATO ASI Series. Plenum Press, New York.

Bielski, B.H.J., Arudi, R.L., and Sutherland, M.W. (1983). A study of the reacti-vity of HO_2/O_2^- with unsaturated fatty acids. *Journal of Biological Chemistry* **258**, 4759–61.

Boyer, R.F. and Vanderploeg, J.R. (1986). The influence of nutrient iron on bean leaf lipoxygenase. *Journal of Plant Nutrition* **9**, 1585–600.

Collinge, D.B. and Slusarenko, A.J. (1987). Plant gene expression in response to pathogens. *Plant Molecular Biology* **9**, 389–410.

Cook, A.A. and Stall, R.E. (1968). Effect of *Xanthomonas vesicatoria* on loss of electrolytes from leaves of *Capsicum annuum*. *Phytopathology* **58**, 617–19.

Cramer, C.L., Bell, J.N., Ryder, T.B., Bailey, J.A., Schuch, W., Bolwell, G.P., Robbins, M.P., Dixon, R.A., and Lamb, C.J. (1985). Co-ordinated synthesis of Phytoalexin Biosynthetic enzymes in biologically-stressed cells of bean (*Phaseo-lus vulgaris* L.). *EMBO Journal* **4**, 285–9.

Croft, K.P.C., Voisey, C.R., and Slusarenko, A.J. (1990). Mechanisms of hyper-sensitive cell collapse: correlation of increased lipoxygenase activity with mem-brane damage in leaves of *Phaseolus vulgaris* (L.) cv. Red Mexican inoculated with avirulent race 1 cells of *Pseudomonas syringae* pv. *phaseolicola*. *Physiologi-cal and Molecular Plant Pathology* **36**, 49–62.

Day, P.R. (1984). Genetics of recognition systems in host-parasite interactions. Chapter 8, pp. 134–147. In *Cellular interactions* (ed. H.F. Linskens and J. Heslop-Harrison), p. 743. Springer Verlag, Berlin.

Doke, N. (1985). NADPH-dependent O_2^- generation in membrane fractions isolated from wounded potato tubers inoculated with *Phytophthora infestans*. *Physiological Plant Pathology* **27**, 311–22.

Doke, N. and Chai, H.B. (1985). Activation of superoxide generation and enhancement of resistance against compatible races of *Phytophthora infestans* in potato plants treated with digitonin. *Physiological Plant Pathology* **27**, 323–34.

Dumelin, E.E. and Tappel, A.L. (1977). Hydrocarbon gases produced during *in vitro* peroxidation of polyunsaturated fatty acids and decomposition of preformed hydroperoxides. *Lipids* **12**, 894–900.

Epperlein, M.M., Noronha-Duta, A.A., and Strange, R.N. (1986). Involvement of the hydroxyl radical in the abiotic elicitation of phytoalexin in legumes. *Physiological and Molecular Plant Pathology* **28**, 67–77.

Fett, W.F. (1984). Accumulation of isoflavonoids and isoflavone glucosides after inoculation of soybean leaves with *Xanthomonas campestris* pv. *glycinea* and pv. *campestris* and a study of their role in resistance. *Physiological and Molecular Plant Pathology* **28**, 67–77.

Flor, H.H. (1971). Current status of the gene-for-gene concept. *Annual Review of Phytopathology* **9**, 275–96.

Fridovich, I. (1988). The biology of oxygen radicals. In *Oxygen radicals and tissue injury*, (ed. B. Halliwell), pp. 1–5. Proceedings of a Brook Lodge Symposium, Upjohn.

Gabriel, D.W. (1986). Specificity and gene function in plant-pathogen interactions. *American Society for Microbiology News* **52**, 19–25.

Gnanamanickam, S.S. and Patil, S.S. (1977) Accumulation of antibacterial isoflavonoids in hypersensitively responding bean leaf tissues inoculated with *Pseudomonas phaseolicola*. *Physiological Plant Pathology* **10**, 159–68.

Goodman, R.N. (1968). The hypersensitive reaction in tobacco: a reflection of changes in host cell permeability. *Phytopathology* **58**, 872–3.

Gutteridge, J.M.C. (1988). Lipid peroxidation: some problems and concepts. In *Oxygen Radicals and Tissue Injury*, (ed. B. Halliwell), pp. 9–19 Proceedings of a Brook Lodge Symposium, Upjohn.

Halliwell, B. (1988). *Oxygen Radicals and Tissue Injury*. Proceedings of a Brook Lodge Symposium, Upjohn.

Hargreaves, J.A. and Bailey, J.A. (1978). Phytoalexin production by hypocotyls of *Phaseolus vulgaris* in response to constitutive metabolites released by damaged bean cells. *Physiological Plant Pathology* **13**, 89–100.

Hitchin, F.E., Jenner, C.E., Harper, S., Mansfield, J.W., Barber, C.E., and Daniels, M.J. (1989). Determinant of cultivar specific avirulence cloned from *Pseudomonas syringae* pv. *phaseolicola* race 3. *Physiological and Molecular plant pathology* **34**, 309–22.

Holliday, M.J. Keen, N.T., and Long, M. (1981). Cell death patterns and accumulation of fluorescent material in the hypersensitive response of soybean leaves to *Pseudomonas syringae* pv. *glycinea*. *Physiological Plant Pathology* **18**, 279–87.

Keen, N.T. and Kennedy, B.W. (1974). Hydroxyphaseollin and related isoflavonoids in the hypersensitive resistance reaction of soybeans to *Pseudomonas glycinea*. *Physiological Plant Pathology* **4**, 173–85.

Keen, N.T., Ersek, T., Long, M., Bruegger, R., and Holliday, M. (1981).

Inhibition of the hypersensitive reaction of soybean leaves to incompatible *Pseudomonas* spp. by blasticidin S, streptomycin or elevated temperature. *Physiological Plant Pathology* **18**, 325–37.

Keppler, L.D. and Baker, C.J. (1989). O_2-Initiated lipid peroxidation in a bacteria-induced hypersensitive reaction in tobacco cell suspensions. *Phytopathology* **79**, 555–62.

Keppler, L.D. and Novacky, A. (1986). Involvement of lipid peroxidation in the development of a bacterially induced hypersensitive reaction. *Phytopathology* **76**, 104–108.

Keppler, L.D. and Novacky, A. (1987). The initiation of membrane lipid peroxidation during bacteria-induced hypersensitive reaction. *Physiological and Molecular Plant Pathology* **30**, 233–45.

Keppler, L.D. and Novacky, A. (1989). Changes in cucumber cotyledon membrane lipid fatty acids during paraquat treatment and a bacteria-induced hypersensitive reaction. *Phytopathology* **79**, 705–8.

Keppler, L.D., Atkinson, M.M., and Baker, C.J. (1988). Plasma membrane alteration during bacteria-induced hypersensitive reaction in tobacco suspension cells as monitored by intracellular accumulation of fluorescein. *Physiological and Molecular Plant Pathology* **32**, 209–19.

Keppler, L.D., Baker, C.J., and Atkinson, M.M. (1989). Active oxygen production during a bacteria-induced hypersensitive reaction in tobacco suspension cells. *Phytopathology* **79**, 974–9.

Klement, Z. (1971). Development of the hypersensitivity reaction induced by plant pathogenic bacteria. *Proceedings of the IIIrd International Conference on Plant Pathogenic Bacteria, Wageningen*, pp. 157–164.

Klement, Z. (1982). Hypersensitivity. In *Phytopathoenic Prokaryotes*, Vol. 2 (ed. M.S. Mount and G.H. Lacy), pp. 149–77. Academic Press, New York.

Klement, Z. and Goodman, R.N. (1967). The hypersensitive reaction to infection by bacterial pathogens. *Annual Review of Phytopathology* **5**, 17–44.

Konze, J.R. and Elstner, E.F. (1978). Ethane and ethylene formation by mitochondria as indication of aerobic lipid degradation in response to wounding of plant tissue. *Biochimica et Biophysica Acta* **25**, 213–21.

Lamb, C.J., Corbin, D.R., Lawton, M.A., Sauer, N., and Wingate, V.P.M. (1986). Recognition and response in plant: pathogen interactions. In *Recognition in microbe–plant symbiotic and pathogenic interactions* (ed. B. Lugtenberg), pp. 333–44. NATO ASI Series, Springer Verlag, Berlin.

Lesham, Y.Y. (1987). Membrane phospholipid catabolism and Ca^{2+} activity in control of senescence. *Physiologia Plantarum* **69**, 551–9.

Lindgren, P.B., Peet, R.C., and Panopoulos, N.J. (1986). Gene cluster of *Pseudomonas syringae* pv. *phaseolicola* controls pathogenicity on bean plants and hypersensitivity on nonhost plants. *Journal of Bacteriology* **168**, 512–22.

Lindgren, P.B., Panopoulos, N.J., Staskawicz, B.J., and Dahlbeck, D. (1988). Genes Required for pathogenicity and hypersensitivity are conserved and interchangeable among pathovars of *Pseudomonas syringae*. *Molecular and General Genetics* **211**, 499–506.

Lindgren. P.B., Frederick, R., Govindarajan, A.G., Panopoulos, N.J., Staskaw-icz, B.J., and Lindow, S.W. (1989). An ice nucleation reporter gene system: Identification of inducible pathogenicity genes in *Pseudomonas syringae* pv. *phaseolicola. EMBO Journal* **8**, 1291–301.

Lynch, D.V. and Thompson, J.E. (1984). Lipoxygenase-mediated production of superoxide anion in senescing plant tissue. *FEBS* **173**, 251–4.

Lyon, F.M. and Wood, R.K.S. (1975). Production of Phaseollin, coumestrol and related compounds in bean leaves inoculated with *Pseudomonas* spp. *Physiological Plant Pathology* **6**, 117–24.

Lyon, F. (neé O'Brien) and Wood, R.K.S. (1976). The hypersensitive reaction and other responses of bean leaves to bacteria. *Annals of Botany* **40**, 479–91.

Lyon, F. and Wood, R.K.S. (1977). Alteration of response of bean leaves to com-patible and incompatible bacteria. *Annals of Botany* **41**, 359–67.

Mueller-Uri, F., Pathier, B., and Nover, L. (1988). Jasmonate-induced alteration of gene expression in barley leaf segments analysed by *in-vivo* and *in-vitro* pro-tein synthesis. *Planta* **176**, 241–47.

Ocampo, C.A., Moerschbacher, B., and Grambow, H.J. (1986). Increased lipoxy-genase activity is involved in the hypersensitive response of wheat leaf cells infected with avirulent rust fungi or treated with fungal elicitor. *Zeitschrift für Naturforschung* **41c**, 559–63.

Ohta, H., Shida, K., Morita, Y., Reng, Y.L., Furusawa, I., and Shishiyama, J. (1988). Increase in the activities of lipoxygenase and lipid hydroperoxide decomposing-enzyme in rice leaves infected with an incompatible race of *Pyri-cularia oryzae* (Abstract). *5th International Congress of Plant Pathology, Kyoto.*

Omer, M.E.H. and Wood, R.K.S. (1969). Growth of *Pseudomonas phaseolicola* in susceptible and in resistant bean plants. *Annals of Applied Biology* **63**, 103–16.

Pavlovkin, J., Novacky, A., and Ullrich-Eberius, C.I. (1986). Membrane potential changes during bacteria-induced hypersensitive reaction. *Physiological and Molecular Plant Pathology* **28**, 125–35.

Rahme, L., Frederick, R.T., Grim, C., Minderinos, M., and Panopoulos, N.J. (1988). Transcriptional organization of pathogenicity/hypersensitivity control-ling genes (*hrp*) in *Pseudomonas syringae* pv. *phaseolicola* (Abstract). *Fifth International Congress of Plant Pathology, Japan.*

Riely, C.A., Cohen, G., and Lieberman, M. (1974). Ethane evolution: a new index of lipid peroxidation. *Science* **183**, 208–10.

Roebuck, P., Sextion, R., and Mansfield, J.W. (1978). Ultrastructural obser-vations on the development of the hypersensitive reaction in leaves of *Phaseolus vulgaris* cv. red Mexican inoculated with *Pseudomonas phaseolicola* (Race 1). *Physiological Plant Pathology* **12**, 151–7.

Rogers, K.R., Albert, F., and Anderson, A.J. (1988). Lipid peroxidation is a con-sequence of elicitor activity. *Plant Physiology* **86**, 547–53.

Salzwedel, J.L., Daub, M.E., and Huang, J.S. (1988). Effects of singlet oxygen quenchers and pH on the bacterially induced hypersensitive reaction in tobacco suspension cell cultures. *Plant Physiology* **90**, 25–8.

Sekizawa, Y., Haga, M., Hirabayashi, E., Takeuchi, N., and Takino, Y. (1987). Dynamic behaviour of superoxide generation in rice leaf tissue infected with blast fungus and its regulation by some substances. *Agricultural and Biological Chemistry* **51**, 763–70.

Sequeira, L. (1976). Induction and supression of the hypersensitive reaction caused by phytopathogenic bacteria: specific and non-specific components. In *Specificity in plant diseases* (ed. R.K.S. Wood and A. Graniti), pp. 289–309. NATO ASI Series. Plenum Press, London.

Serhan, C., Anderson, P., Goodman, E., Dunham, P., and Weissmann, G. (1981). Phosphatidate and oxidized fatty acids are calcium ionophores. *Journal of Biological Chemistry* **256**, 2736–41.

Slusarenko, A.J. (1987). Gene expression and resistance of french bean to *Pseudomonas phaseolicola*. In *Genetics and plant pathogenesis* (ed. P.R. Day and G.J. Jellis), pp. 55–64. Blackwell, Oxford.

Slusarenko, A.J. and Longland, A. (1986). Changes in gene activity during expression of the hypersensitive response in *Phaseolus vulgaris* cv. Red Mexican to an avirulent race 1 isolate of *Pseudomonas syringae* pv. *phaseolicola*. *Physiological and Molecular Plant Pathology* **29**, 79–94.

Slusarenko, A.J., Longland, A., and Friend, J. (1986). Expression of plant genes in the hypersensitive reaction of french bean (*Phaseolus vulgaris*) to the plant pathogenic bacterium *Pseudomonas syringae* pv. *phaseolicola*. In *Recognition in microbe–plant symbiotic and pathogenic interactions*, (ed. B. Lugtenberg), pp. 367–76. NATO ASI Series. Springer-Verlag, Berlin.

Staskawicz, B.J., Dahlbeck, D., and Keen N.T. (1984). Cloned avirulence gene of *Pseudomonas syringae* pv. *glycinea* determines race-specific incompatibility on *Glycine max* (L). *Merr. Proceedings of the National Academy of Sciences USA* **81**, 6024–8.

Tamaki, S., Dahlbeck, D., Staskanwicz, B., and Keen, N.T. (1988). Characterization and expression of two avirulence genes cloned from *Pseudomonas syringae* pv. *glycinea*. *Journal of Bacteriology* **170**, 4846–54.

Taylor, G.W. and Morris, H.R. (1983). Lipoxygenase pathways. *British Medical Bulletin* **39**, 219–22.

Templeton, M.D. and Lamb, C.J. (1988). Elicitors and defence gene activation. *Plant Cell and Environment* **11**, 395–401.

Thompson, J.E., Legge, R.L., and Barber, R.F. (1987). The role of free radicals in senescence and wounding. *New Phytologist* **105**, 317–44.

Turner, J.G. and Novacky, A. (1974). The quantitative relation between plant and bacterial cells involved in the hypersensitive reaction. *Phytopathology* **64**, 885–90.

Vick, B.A. and Zimmerman, D.C. (1987). The lipoxygenase pathway. In *The metabolism, structure and function of plant lipids*, (ed. P. K. Stumpf, J. B. Mudd, and W. D. Nes), pp. 383–90. Plenum Press, New York.

Voisey, C.R. and Slusarenko, A.J. (1989). Chitinase mRNA and enzyme activity in *Phaseolus vulgaris* (L.) show faster increase in response to avirulent than to virulent cells of *Pseudomonas syringae* pv. *phaseolicola*. *Physiological and Molecular Plant Pathology* **35**, 403–12.

Wolff, S.P., Garner, A., and Dean, R.T. (1986). Free radicals, lipids and protein degradation. *Trends in Biochemical Sciences* **11**, 27–31.

Woods, A.M., Fagg, J., and Mansfield, J. (1988). Fungal development and irreversible membrane damage in cells of *Lactuca sativa* undergoing the hypersensitive reaction to the downy mildew fungus *Bremia lactucae*. *Physiological and Molecular Plant Pathology* **32**, 483–97.

Wyman, J.G. and Vanetten, H.D. (1982). Isoflavonoid phytoalexins and nonhypersensitive resistance of beans to *Xanthomonas campestris* pv. *phaseoli*. *Phytopathology* **72**, 1419–24.

Yamamoto, H. and Tani, T. (1986). Possible involvement of lipoxygenase in the mechanisms of resistance of oats to *Puccinia coronata avenae*. *Journal of Phytopathology* **116**, 329–37.

8 Molecular genetics of bacterial blight of peas

ALAN VIVIAN

Science Department, Bristol Polytechnic, Coldharbour Lane, Frenchay, Bristol, BS16 1QY, UK

Introduction

Bacterial blight of peas

Bacterial blight of peas is a seed-borne disease of world-wide occurrence. First described by Sackett (1916) in the United States, it has been found subsequently in many countries. It infects all parts of the plant which are above ground, and can occur at any time throughout the growing season (Young and Dye 1970). However, it did not occur in the United Kingdom until its accidental introduction occurred via infected seed in 1985, resulting in major outbreaks in several areas of the country (Stead and Pemberton 1987). Spread of the disease is favoured by cool, wet conditions, both of which were a feature of the summers of 1985 and 1986.

Economic aspects

The current pea crop in the United Kingdom is some 100 000 ha, and is projected to rise to about 300 000 ha over the next few years. Peas are an alternative to soybean for animal feed, and also they can be grown on land that may otherwise have been used for cereal production. Thus as a crop the pea is advantageous both as a substitute for imports and as an alternative, permitting reduction in the surplus of cereals. In 1985, the pea crop was some 80 000 ha, valued at £40 million. Taylor (1986) has estimated that for a 10 per cent loss occurring every four years, the monetary loss would be £1 million per annum. Following further outbreaks in 1986, control of the disease was achieved by strict hygiene and quarantine measures combined with rigorous testing of seed (Stead and Pemberton 1987).

Pea blight as a model system

A considerable number of genes, which are involved in the interaction of host and pathogen have been identified among bacteria pathogenic to plants. These may be categorized in three main groups: *avr* or avirulence genes, responsible for the recognition between host and pathogen which results in the induction of a resistant reaction in the host; *hrp* or harp genes, involved in production of a hypersensitive response (HR) in non-host plants, and pathogenicity in compatible combinations of host and pathogen; and *dsp* or disease-specific genes (Boucher *et al.* 1987). Both *avr* and *hrp* genes have been studied in the plant-pathogenic Pseudomonads. Here I shall describe the investigation of *avr* genes in *Pseudomonas syringae* pathovar *pisi* (hereafter *P.s. pisi*) in relation to cultivar specificity in the host (*Pisum sativum*).

Bacterial blight of peas has been chosen as a model host/pathogen system to investigate the genetic basis of race structure in bacterial plant pathogens for the following reasons: the host is an annual which is readily-cultivated in glasshouses or growth chambers; it is naturally-selfing, facilitating the production of genetically-uniform cultivar stocks; and it is genetically well-characterized. In 1972 Taylor identified two races of the pathogen and since that time, a further five races have been characterized (J.D. Taylor, personal communication), so that the pathogen exhibits a well-defined race-structure. Strains can be readily cultured in the laboratory on simple defined media. Vivian *et al.* (1989) have developed a rapid, reliable, and simple test involving stem-inoculation of seedlings to characterize the interaction of the pathogen with its host.

Gene-for-gene hypothesis

Flor (1956), studied the relationship between genes segregating in the fungal pathogen, *Melampsora lini*, the cause of flax rust, and genes for resistance that segregated in the host, flax (*Linum usitatissimum*). These crosses in both host and pathogen demonstrated that single dominant genes, with one exception, were responsible for the specificity observed between certain cultivars of the host and races of the pathogen. The exception has since been shown to involve interaction of two genes in the pathogen (an avirulence gene and an epistatic inhibitor of the avirulence gene) with the M_1 resistance gene in cv. Williston Brown (Lawrence *et al.* 1981).

An incompatible interaction between host and pathogen resulted from the presence in the host of a dominant resistance gene, and the presence in the pathogen of a corresponding gene for avirulence in that particular race of the pathogen. The observed phenotypic outcome of the incompatibility was, in many cases, a hypersensitive response in the host, which resulted in the absence of disease. Both resistance and avirulence genes were postulated to have alleles for susceptibility and virulence, respectively. All combinations of alleles other than the specific avirulence/resistance allele combination result in compatibility, and disease. This gene-for-gene

Table 8.1 Current race structure in *Pseudomonas syringae pv. pisi*.

Host cultivar	Genotype	1	2	3	4	5	6	7
		Pathogen race+						
	(R)*							
Kelvedon Wonder (KW)	nil	+	+	+	+	+	+	+
Early Onward (EO)	2, 5	+	−	+	+	−	+	−
Puget (P)/Shasta (Sh)	1, 3	−	+	−	+	+	+	−
Hursts Greenshaft (HG)	1, 3, 4, 5	−	+	+	−	−	+	−
1404	1, 3, 5	−	+	−	+	−	+	−
Partridge (Pa)	1, 3, 4, 5	−	+	−	−	−	+	−
Sleaford Triumph (ST)	1, 2, 4, 5	−	−	+	−	−	+	−
Vinco (V)	1, 2, 3, 5	−	−	−	+	−	+	−
Fortune (F)	1, 2, 3, 4, 5	−	−	−	−	−	+	−

− incompatible, + compatible.
* Dominant R genes postulated to be present, based on the observed interactions with the seven races.
+ Race 6 is postulated to lack all avirulence genes, while races 2, 3, and 4 have single avirulence genes 2, 3 and 4, respectively; races 1 and 5 have avirulence genes 1 and 5, respectively, and each may harbour two additional avirulence genes; race 7 is postulated to possess all five avirulence genes (based on Taylor *et al.* 1989).

hypothesis was proposed to account for the pattern of interactions observed between host and pathogen (Flor 1956).

The pattern of interaction observed between *P.s. pisi* and a series of differential pea cultivars enabled Taylor *et al.* (1989) to propose a gene-for-gene relationship based on five pairs of matching dominant genes governing resistance in the host and avirulence in the pathogen. In this scheme, while some races harbour single avirulence genes, others have at least two and possibly three *avr* genes (Table 8.1). Seven distinct races of *P.s. pisi* are currently recognised on the basis of their differential interaction with a range of pea cultivars (Table 8.1; Taylor *et al.* 1989; J.D. Taylor personal communication). The pattern of interaction can be accounted for in terms of five matching gene-pairs in host and pathogen (Table 8.1; Taylor *et al.* 1989).

Cloning of race 2 specificity

The cosmid vector, pLAFR3 (Staskawicz *et al.* 1987) was used to construct a partial gene library for the race 2 strain, 203 (Atherton 1987). The library cosmids were mobilized into a rifampicin-resistant derivative of the *P.s. pisi* race 1 strain, 299A, using the helper plasmid, pRK2013 and a replica-plating procedure (Vivian *et al.* 1989). Transconjugants bearing individual cosmid clones were stem-inoculated into pea cultivar Early Onward. The race 1 strain is compatible with this cultivar, but race 2 is incompatible.

A single cosmid clone (pAV270) was identified, which conferred on race 1 avirulence towards cv. Early Onward. However, this transconjugant remained virulent towards pea cultivar Kelvedon Wonder (the universal suscept). Leaf inoculations, using a technique that involved wounding with pin-pricks followed by aerosol spray application of a suspension of the pathogen, confirmed a similar pattern of compatible and incompatible reactions in the leaves.

A 4. O kb *Eco*RI subclone of pAV270 (designated pAV200) was introduced into rifampicin-resistant derivatives of other races of *P.s. pisi* (Table 8.2). The results of inoculations with this transconjugant generally confirm the race 2-specificity of the *avr* gene on pAV200, but an anomalous result was observed with Shasta, a cultivar which is postulated to harbour resistance genes, R1 and R3. Introduction of pAV200 into race 3 resulted in a change from an incompatible to compatible interaction with cv. Shasta, implying that the introduction of this cosmid has affected the expression of the race 3 avirulence gene towards this particular cultivar. The reason for this result remains unclear.

In order to demonstrate a true gene-for-gene situation existed, it was necessary to show that the race 2 avirulence gene from strain 203 was interacting with a single resistance gene in the host pea. Two host crosses were performed and a sample of both the F_1 and the F_2 progeny were inoculated sequentially with three strains race 1, race 2, and race 1 plus pAV270. One

Table 8.2 Pattern of interactions observed between host and pathogen involving race 2 specific avirulence present on pAV200
(For genotypes of pea cultivars see Table 8.1)

	Host cultivar				
Race	KW	EO	Sh	V	F
1	+	+	−	−	−
1 + 2*	+	−	−	−	−
2	+	−	+	−	−
3	+	+	−	−	−
3 + 2*	+	−	+**	−	−
4	+	+	+	+	−
4 + 2*	+	−	+	−	−
6	+	+	+	+	+
6 + 2*	+	−	+	−	−

* race 2-specific avirulence introduced into the appropriate race on cosmid pAV200.
** anomalous result; predicted race 3 specificity not observed.

Table 8.3 Co-segregation of host resistance gene, R2 and race 2-specific avirulence in a cross betwen cvs. Fortune and Kelvedon Wonder

	Phenotypes			
Expected ratio:	9:	3:	3:	1
race 1	−	−	+	+
race 2	−	+	−	+
race 1 + 2 (pAV270)	−	−	−	+
Observed totals	86	32	30	7*

1 A total of 155 F$_2$ progeny were sequentially inoculated with strains of races 1, 2, and race 1 harbouring pAV270. Since cv. Fortune harbours R genes postulated to match avirulence genes for specificity to races 1 and 2, segregation of each gene should be detected in the F$_2$.
2 For race 1 specificity, the ratio of resistant to susceptible progeny is 118:37 (3:1) and for race 2 specificity 116:39 (3:1). The ratio of susceptible to resistant plants to both races 1 and 2 is 7:148 (1:15), in agreement with the independent segregation ratio expected for two unlinked dominant genes.
* includes one anomalous plant which was compatible with races 1 and 2, but resistant to race 1 harbouring pAV270.

cross, between cvs. Kelvedon Wonder (lacking all known race-specific resistance genes) and Early Onward (postulated to harbour R2 and R5) produced the expected F$_1$ progeny that were resistant to races 1, 2, and the race 1 transconjugant harbouring pAV270. The race 2-resistant progeny F$_2$ were also resistant to race 1 harbouring pAV270, with the numbers of susceptible (39) and resistant (116) progeny observed being in the predicted 1:3 ratio for a single dominant host resistance gene segregating in a Mendelian fashion. In a further cross between cvs. Kelvedon Wonder and Fortune (postulated to harbour five resistance genes, R1 to R5) where resistance genes R1 and R2 could be observed to segregate in the F$_2$, the results confirmed the gene-for-gene specificity of the race 2 cloned avirulence gene (Table 8.3; Vivian *et al.* 1989). The results demonstrate clearly the co-segregation of resistance to race 2 and the race 1 pAV270 transconjugant, and show that the avirulence gene located on the cosmid interacts with a single dominant host resistance gene, designated R2.

Cloning of race 3 specificity

In 1975, a new race of *P.s. pisi* was detected in seed imported from the USA; this strain, designated 870A remains a unique representative of race 3 (Taylor *et al.* 1989). It was shown by Malik (1985) to harbour three cryptic plasmids, designated pAV230, pAV231, and pAV232 of approximate sizes 54, 42, and 106 kb, respectively. A cryptic plasmid (pAV212), similar in size to pAV230, was also detected in a number of race 1 strains, including 299A, together with a smaller plasmid pAV213 (Malik 1985). Both race

1 and race 3 were incompatible with a group of pea cultivars which included Puget, and which were deemed to harbour resistance genes R1 and R3 (Taylor *et al.* 1989).

Malik *et al.* (1987) using the transposon vector, pSUP2011 (Simon *et al.* 1983) in 299A, isolated a mutant, designated PF24. This was one of a class of mutants (designated class III), which were found to exhibit reduced pathogenicity towards the cv. Early Onward. Subsequent to its isolation, PF 24 became sensitive to kanamycin, the antibiotic resistance mediated by the transposon, but retained its class III phenotype. It was also found to lack (at least as an autonomous replicon) the cryptic plasmid pAV212.

Testing of PF24 on a range of host cultivar groups, revealed a hitherto undetected change of phenotype with respect to its progenitor, 299A; namely acquisition by PF24 of virulence towards cultivars of the group typified by Puget (A.D. Bavage and A. Vivian, unpublished results). Preliminary results from crosses involving cultivars similar to Puget, have cast doubt upon their genotype; R1 and R3 resistance was never observed to segregate among the F_2 progeny from crosses with cv. Kelvedon Wonder. This indicates either very close linkage of R1 and R3, or the possibility that this group has only a single R gene conditioning resistance to races 1 and 3 (J.R. Bevan, personal communication).

A partial gene library of the race 3 strain, 870A was mobilized into a rifampicin-resistant derivative of PF24 (designated PF303), and the transconjugants tested on cv. Puget. Five cosmid clones, which were shown to share some common insert fragments when digested with *Eco*RI (data not shown), restored avirulence toward cv. Puget in PF303. This confirmed the presence in race 3, of a novel avirulence gene which is capable of restoring avirulence in the race 1 mutant (A.D. Bavage and A. Vivian, unpublished results).

It remains necessary to confirm the association of *avrB* with plasmids in races 1 and 3, but the relatively high representation of restoring clones in the race 3 library might result from a copy number effect due to a plasmid-borne gene.

Conclusions

There has been a rapid advance in our knowledge of the genetic basis of host specificity in bacterial plant pathogens since the application of recombinant DNA technology. The creation of gene libraries for races of the pathogen has enabled precise location and characterization of the genes responsible for avirulence (Vivian 1989). In *P.s. glycinea*, the initial cloning of an avirulence gene (Staskawicz *et al.* 1984) provided a clear indication that the Flor hypothesis, involving single dominant avirulence genes in the pathogen, was likely to be correct for the leaf-spotting Pseudomonads. Subsequent work with *P.s. glycinea* (Staskawicz *et al.* 1987) and in

Xanthomonas campestris pv. *vesicatoria* (Stall *et al.* 1986) has clearly established the existence of further race-specific avirulence genes, and the precise gene-for-gene relationships of some of these genes with single dominant resistance genes in their respective hosts, soybean and pepper (Swanson *et al.* 1988; Kearney *et al.* 1988). The work described here adds *P.s. pisi* to the list of proven gene-for-gene relationships in terms of race 2 specificity, and also indicates the likely involvement of plasmids in the location of some of these avirulence genes.

The next few years should see the elucidation of much of the genetic basis of race structure in the plant pathogenic Pseudomonads. However, the precise nature of the products of avirulence and resistance genes and their mode of operation remain unclear, in spite of DNA sequencing and attempts at *in vitro* expression of cloned *avr* genes.

References

Atherton, G.T. (1987). The genetics of pathogenicity and host specificity of *Pseudomonas syringae* pathovar *pisi*. Ph.D. thesis, CNAA, Bristol Polytechnic.

Boucher, C., Van Gijsegem, F., Barberis, P.A., Arlat, M., and Zischek, C. (1987). *Pseudomonas solanacearum* genes controlling both pathogenicity on tomato and hypersensitivity on tobacco are clustered. *Journal of Bacteriology* **169**, 5626–32.

Flor, H.H. (1956). The complementary genetic systems in flax and flax rust. *Advances in Genetics* **8**, 29–54.

Kearney, B., Ronald, P.C., Dahlbeck, D., and Staskawicz, B.J. (1988). Molecular basis for evasion of plant host defence in bacterial spot disease of pepper. *Nature* **332**, 541–43.

Lawrence, G.J., Mayo, G.M.E., and Shepherd, K.W. (1981). Interactions between genes controlling pathogenicity in the flax rust fungus. *Phytopathology* **71**, 12–19.

Malik, A.N. (1985). Genetic studies with *Pseudomonas syringae* pathovar *pisi*. Ph.D. thesis, CNAA, Thames Polytechnic.

Malik, A.N., Vivian, A., and Taylor, J.D. (1987). Isolation and partial characterization of three classes of mutant in *Pseudomonas syringae* pathovar *pisi* with altered behaviour towards their host, *Pisum sativum*. *Journal of General Microbiology* **133**, 2393–9.

Sackett, W.G. (1916). A bacterial stem blight of field and garden peas. *Colorado Agricultural Experimental Station Bulletin* 218.

Simon, R., Priefer, U., and Puhler, A. (1983). A broad host range mobilization system for *in vivo* genetic engineering: transposon mutagenesis in Gram-negative bacteria. *Bio/Technology* **1**, 784–9.

Stall, R.E., Loschke, D.C., and Jones, J.B. (1986). Linkage of copper resistance and avirulence loci on a self-transmissible plasmid in *Xanthomonas campestris* pv. *vesicatoria*. *Phytopathology* **76**, 240–3.

Staskawicz, B.J., Dahlbeck, D., and Keen, N.T. (1984). Cloned avirulence gene of *Pseudomonas syringae* pv. *glycinea* determines race-specific incompatibility on *Glycine max*(L) Merr. *Proceedings of the National Academy of Sciences USA* **81**, 6024–8.

Staskawicz, B., Dahlbeck, D., Keen, N., and Napoli, C. (1987). Molecular characterization of cloned avirulence genes from race 0 and race 1 of *Pseudomonas syringae* pv. *glycinea*. *Journal of Bacteriology* **169**, 5789–94.

Stead, D.E. and Pemberton, A.W. (1987). Recent problems with *Pseudomonas syringae* pv. *pisi* in UK. *EPPO Bulletin* **17**, 291–4.

Swanson, J., Kearney, B., Dahlbeck, D., and Staskawicz, B. (1988). Cloned avirulence gene of *Xanthomonas campestris* pv. *vesicatoria* complements spontaneous race-change mutants. *Molecular Plant–Microbe Interactions*. **1**, 5–9.

Taylor, J.D. (1972). Races of *Pseudomonas pisi* and sources of resistance in field and garden peas. *New Zealand Journal of Agricultural Research* **15**, 441–7.

Taylor, J.D. (1986). Bacterial blight of compounding peas. In *1986 British Crop Protection Conference – Pests and Diseases*, Vol. 2, pp. 733–6. BCPC Brighton, UK.

Taylor, J.D., Bevan, J.R., Crute, I.R., and Reader, S.L. (1989). Genetic relationship between races of *Pseudomonas syringae* pathovar *pisi* and cultivars of *Pisum sativum*. *Plant Pathology* **38**, 364–75.

Vivian, A. (1989). Recognition in resistance to bacteria. In *Recognition and response in plant–virus interactions*, (ed. R.S.S. Fraser), pp. 17–29 NATO ASI Series. Springer-Verlag, Berlin.

Vivian, A., Atherton, G.T., Bevan, J.R., Crute, I.R., Mur, L.A.J., and Taylor, J.D. (1989). Isolation and characterization of cloned DNA conferring specific avirulence in *Pseudomonas syringae* pv. *pisi* to pea (*Pisum sativum*) cultivars, which possess the resistance allele, R2. *Physiological and Molecular Plant Pathology* **34**, 335–44.

Young, J.M. and Dye, D.W. (1970). Bacterial blight of peas caused by *Pseudomonas pisi* Sackett, 1916 in New Zealand. *New Zealand Journal of Agricultural Research* **13**, 315–24.

9 Molecular genetic dissection of pathogenicity of *Xanthomonas*

M.J. DANIELS, C.E. BARBER, J.M. DOW, C.L. GOUGH, A.E. OSBOURN, and J.L. TANG

The Sainsbury Laboratory, John Innes Institute, Colney Lane, Norwich NR4 7UH, UK

Introduction

Xanthomonas is a genus of yellow-pigmented, Gram-negative bacteria, almost all members of which are plant pathogens. At present the genus contains five species, *X. albilineans*, *X. ampelina*, *X. axonopodis*, *X. campestris*, and *X. fragariae* (Bradbury 1986), but it is likely that further groups, at present distinguished by subspecific epithets, will soon be elevated to species rank. Of particular interest in the present context is the species *X. campestris*, in which more than 120 pathovars, defined as groups which can be distinguished from one another with certainty only by their plant host range, have been recognized. Pathovars generally take the name of the plant from which they were first isolated, e.g. *X. campestris* pathovar *oryzae* is a pathogen of rice (*Oryza*), *X.c.* pathovar *phaseoli* a pathogen of beans (*Phaseolus*). Although any pathovar usually infects only a limited range of related plant species, exhaustive plant tests are obviously impracticable, and it is possible that some pathovars which have been described are identical. Recent molecular genetic data indicate a high degree of relatedness between pathovars (*Sawczyc et al.* 1989; Lazo *et al.* 1987). A further degree of host specificity is signified by division of certain pathovars into races. A race interacts in a characteristic differential manner with cultivars of a particular plant species. The pathovar with which this paper is principally concerned is *X.c.* pathovar *campestris*, which is the causal agent of black rot of crucifers, one of the major worldwide diseases of cruciferous crops (Williams 1980). *X.c. campestris* (the word pathovar is usually omitted in shorter designations) is not known to possess a race structure, although resistant varieties of certain *Brassica* crops are known.

It has been recognized for many years that a genetic approach to understanding host–pathogen interactions is likely to be a fruitful way of gaining an insight into a very complex biological phenomenon (Day 1974). Until relatively recently the only pathogens which could be studied genetically were fungi. Systems for gene exchange between plant pathogenic bacteria were developed in the mid 1970s (Lacy and Leary 1979), but before these techniques could be used to study pathogenicity itself conventional microbial genetics had been partially eclipsed by molecular genetics (i.e. recombinant DNA technology). The development of gene-cloning vectors with broad-host range was crucial, because it meant that genes from a pathogen could not only be cloned and propagated in *Escherichia coli*, but could also be transferred into any Gram-negative bacterium (such as *Xanthomonas*) and thus studied in their natural genetic environment.

This paper summarizes work carried out in our laboratory and elsewhere on *X.c. campestris*. The choice of this pathogen as a model was empirical. It was apparent that the ideal experimental organism would have to possess certain attributes, such as ease of testing for pathogenicity, ability to harbour broad-host range vectors, etc., and *X.c. campestris* appeared to offer many advantages. With the passage of time other properties have become important, but fortunately *X.c. campestris* still remains well suited for our purposes. It must be emphasized that pathogenicity involves an interaction between two distinct organisms, the host and the pathogen, and a full understanding can only be attained through complementary studies of both partners. Flor's work on flax rust represents a milestone in the genetics of fungal diseases because for the first time the genetic variation in both the pathogen and the host, and the way in which the genes interact was studied (Flor 1956). Thus he was led to enunciate the 'Gene-for-gene hypothesis' which, although often misunderstood, has been important in influencing recent work on host–pathogen specificity (Keen and Staskawicz 1988). This paper is concerned only with the pathogen, because another chapter in this volume deals with work on the 'plant' aspects of the *X.c. campestris–Brassica* interaction (Dow *et al.*; this volume Chapter 10).

Molecular cloning of pathogenicity genes

Since molecular cloning of genes is a fundamental step towards their analysis and the subsequent acquisition of useful biological information, this section will deal with strategies which have been used to clone *X.c. campestris* pathogenicity genes. The phrase 'pathogenicity genes' is a working term to indicate genes which are required for pathogenicity to plants, but which are not apparently required for growth *in vitro*. It is almost certainly unrealistic to classify genes in this way, because plants are the only known niche occupied by *X.c. campestris*, and the whole genome will have evolved to exploit the habitat. Nevertheless the concept of a subset of genes that is required for pathogenicity is useful (probably essential) in designing experiments.

The most obvious manifestations of plant–pathogen interactions are the disease symptoms, which in the case of black rot include extensive tissue destruction. *X.c. campestris* produces a range of extracellular enzymes, including protease, amylase, pectinases (pectinmethylesterase and polygalacturonate lyase), and endoglucanase, which have the potential capacity for degrading components of plant cells. It also produces copious quantities of the extracellular polysaccharide (EPS), known as xanthan gum, which is an important commercial product. Because of the avirulence of EPS-deficient variants of many plant pathogenic bacteria (e.g. Kelman 1954), it is believed that EPS is a pathogenicity determinant, but proof of this assertion is lacking.

It is possible to clone genes encoding extracellular enzymes, and genes involved in EPS biosynthesis, because simple visual tests of the relevant phenotype are available. The starting point for gene cloning is a library of wild type *X.c. campestris* DNA cloned in a broad-host range vector such as pLAFR1, pLAFR3, or pIJ3200 (Daniels *et al.* 1984a; Gough *et al.* 1988; Liu *et al.* 1990), consisting of 1000–3000 clones with an average insert size of 25 kb, and which is maintained in a *recA* strain of *E. coli*.

Clones containing the *X.c. campestris* protease gene can be identified by plating the library (in *E. coli*) on medium containing skimmed milk; positive colonies are surrounded by a clear halo where the enzyme degrades the casein (Tang *et al.* 1987). However, most *X.c. campestris* genes are not expressed in *E. coli*, and an alternative host must be used for these other genes. The *X.c. campestris* libraries were transferred by conjugation from *E. coli* into other *X. campestris* pathovars, *translucens* and *vesicatoria*, which do not produce the same range of extracellular enzymes as *X. campestris*. Clones carrying amylase, or endoglucanase or polygalacturonate lyase genes were identified by enzyme production on indicator media containing starch or carboxymethyl-cellulose or sodium polygalacturonate, respectively (Dow *et al.* 1989a; Gough *et al.* 1988; and unpublished data). EPS-deficient mutants can be recognized by colony appearance on sugar-containing plates (Whitfield *et al.* 1981), and the genes wherein mutation gave the EPS⁻ phenotype were cloned by complementation (Barrère *et al.* 1986).

The list of potential pathogenicity genes which can be cloned by such a direct strategy is short, and as it is likely that many other genes will be necessary for the normal interaction with the host, alternative strategies are necessary to clone these other genes. One of the most commonly-used techniques is to isolate mutants altered in pathogenicity, and then to complement them with cloned DNA and restore pathogenicity (Daniels *et al.* 1984a, b). Since, by definition, the only phenotypic difference between mutants and wild-type is altered pathogenicity, the mutants can only be isolated by individually testing survivors of mutagenesis on plants. Consequently pathogenicity assays suitable for testing hundreds or thousands of strains are necessary. Having built up a collection of pathogenicity mutants, the genes in question can be cloned by transferring the wild type DNA library into each mutant in turn. The resulting individual transconjugants are then screened on plants to detect clones which complement the mutations and so restore pathogenicity. Again, this requires the performance of hundreds to thousands of separate plant tests.

A third general strategy is to isolate genes from a library by means of hybridization of the insert DNA to a nucleic acid or oligonucleotide probe that is of interest. The probe might be a gene, perhaps from another organism, which is believed to be homologous to the gene which is to be isolated. For example, Boucher *et al.* (1987) showed that *hrp* genes in *Pseudomonas solanacearum*, which are required both for pathogenicity in susceptible host plants and for induction of hypersensitive resistance in non-hosts, hybridize with sequences present in *X. campestris* pathovars. The *P. solanacearum hrp* genes are being used as probes to isolate clones containing putative *X.c. campestris hrp* genes from genomic libraries, so that the role of these genes in pathogenicity can be studied. Two other examples of this hybridization strategy for gene cloning may be cited. (1) Osbourn *et al.* (1987) described a technique for identifying plant-inducible genes in *X.c. campestris*, using a promoter-cloning vector. In view of the close relationship between plants and pathogens, it is likely that pathogen gene expression will be regulated by the plant environment, and consequently it was supposed that the set of genes transcribed from plant-inducible promoters might include pathogenicity genes. Plant-inducible promoters were cloned and the promoter fragments were used as probes to isolate surrounding genomic fragments from a library, and a novel pathogenicity gene was thereby found (Osbourn *et al.* 1990a). (2) As will be described below, a cluster of genes functioning in positive regulation of extracellular enzyme synthesis is necessary for pathogenicity (Daniels *et al.* 1984a). Sequencing showed that some of these genes encode products which are members of the family of prokaryotic two-component regulatory systems (Tang 1989). These proteins, which are generally involved in regulating gene expression in response to environmental changes, have highly conserved domains (Ronson *et al.* 1987). In order to determine whether *X.c. campestris* contains more genes of this type that could be involved in reacting to changes in the (plant) environment, and hence be important for pathogenicity, we used a 24-mer oligonucleotide, which would code for a conserved octapeptide sequence highly conserved among two-component systems, as a probe to isolate genomic clones containing such genes. Several new putative regulatory genes have been identified, but it is not yet known if they have a role in interactions with plants (Osbourn *et al.* 1990b).

Biological information from cloned pathogenicity genes

The purpose of cloning pathogenicity genes is to help us to understand the molecular basis of host–pathogen interactions, and this section will outline some of our findings with *X.c. campestris*. A detailed review discusses all plant pathogenic bacteria (Daniels *et al.* 1988), and a brief overview may be found in the proceedings of a recent conference (Daniels 1988).

The result of applying the gene cloning strategies described above will be isolation (from a genomic library) of a set of recombinant cosmids, each of which is presumed to carry a pathogenicity gene (or genes). Since commonly used vectors such as pLAFR1 can accommodate inserts of foreign DNA up to 30 kb in size, it is apparent that the pathogenicity gene(s) may account for only a small proportion of the genetic information in the clone. The first priority is to define the limits of the DNA of interest. Three general techniques which have been applied are subcloning, deletion analysis, and transposon mutagenesis. Subcloning following complete or partial digestion of the original cosmid clone with restriction enzymes, is most efficient if a restriction map is available, so that specific fragments whose relative positions are known can be tested. The construction of a detailed map of a large DNA fragment is tedious and time consuming, however. Deletion of sequences that are bounded by restriction sites can be achieved by partial or complete digestion of the cosmid clone, depending on the number of sites for the enzyme in the insert and vector, followed by religation of the mixture. Since intramolecular ligation will usually be favoured, the result of the experiment will be generation of a preponderance of product that consists of the vector plus insert sequences adjacent to the original cloning site, and lacking one or more internal restriction fragments. Alternatively exonuclease III can be used to generate deletions that are not dependent on the presence of convenient restriction sites. A versatile broad-host range vector, pIJ3200, has been constructed which enables deletions to be made and then either tested for biological activity or subjected to further physical analysis (Liu *et al.* 1990).

Transposon mutagenesis has been widely used in many laboratories. Although the general strategies used are common to all systems, the detailed methodology usually has to be adapted to suit the specific application, because much inter-strain (not to mention inter-pathovar and inter-species) variation is found in properties such as plasmid stability. For *X.c. campestris* the protocol described by Turner *et al.* (1985) has been used routinely to analyse numerous clones. The recombinant cosmid is introduced by transformation into a strain of *E. coli* carrying a chromosomal Tn5 insertion, and from a pool of transformants a sample of Tn5-containing plasmids is isolated, (these particular transformants are identified by the ability to confer resistance to tetracycline, from the cloning vector and kanamycin, from Tn5). The transposon insertion in each plasmid is mapped in relation to restriction sites; this usually also helps to supplement the restriction mapping data of the original plasmid. A subset of plasmids is chosen in which Tn5 mutations are distributed throughout the insert DNA, ideally no more than 1 kb apart. Each is then tested for loss of the biological function of the original plasmid (e.g. encoding production of an enzyme, or complementation of a mutation). In this way the gene in question is roughly localized. Finally all the Tn*5* insertions are transferred by marker-exchange into the homologous position in the wild-type *X.c. campestris* genome. Plasmid incompatibility phenomena are used to select for mutants (Turner *et al.* 1985), and the phenotype of each mutant is tested. It is possible by this method to detect genes linked to the originally identified gene, and there are now several examples of clustered pathogenicity genes which have been found in this way (Daniels 1988). A refinement of this kind of analysis is the use of modified transposons such as Tn*51ac* or Tn*4431* (Kroos and Kaiser 1984; Shaw *et al.* 1988) which generate fusions of reporter genes to the target gene, thereby giving data on orientation, organization, and regulation of transcription units.

Extracellular enzymes

A cosmid clone pIJ3070, which contains the structural gene for the major *X.c. campestris* extracellular protease was isolated (Tang *et al.* 1987). By subcloning fragments into pIJ3200, followed by exonuclease III deletion analysis and sequencing, an open reading frame (ORF) encoding a protein of 57 kDa was found. The amino acid sequence showed homology to several serine proteases (Liu *et al.* 1990). Biochemical data from experiments with inhibitors had previously suggested that the enzyme belonged to this class of protease. Marker-exchange of Tn*5* insertions from the cloned DNA to the *X.c. campestris* produced mutants lacking the protease. Surprisingly the mutants retained near-normal virulence when they were inoculated into the hypocotyls of turnip seedlings or infiltrated into intercellular spaces of mature turnip leaves, and both the growth rate and final titre of mutant bacteria in seedlings was indistinguishable from the wild type. These tests for pathogenicity were devised as convenient laboratory procedures suitable for screening numerous strains, and bear little relation to the natural disease. In view of this the mutants were inoculated into veins at the edges of mature leaves, simulating natural entry through hydathodes (Williams 1980), and in this test the protease-deficient strain showed very low virulence (Daniels *et al.* in preparation). This observation emphasizes the need to select pathogenicity tests carefully and to exercise caution in interpreting results. More interestingly it appears that in certain tissues or anatomical regions of the leaf protease may be necessary in the early stages of infection, but not thereafter. Shaw and Kado (1987) have also pointed out the need to use complementary tests for pathogenicity.

The endoglucanase ('cellulase') gene of *X.c. campestris* has been similarly analysed (Gough *et al.* 1988, 1990). The nucleotide sequence predicts a mature protein of 53 kDa, in close agreement with the estimated Mr of the purified protein. Amino acid sequencing of the N-terminus of the protein confirmed the DNA sequence, and showed that a 25 amino acid signal peptide is cleaved from the N-terminus, presumably during secretion of the protein through the inner membrane. The protein has a 'hinge' region consisting of twelve thr-pro direct repeats, a common feature of cellulases (Knowles *et al.* 1987). Cellulase-deficient mutants were produced by marker-exchange from the Tn5-containing cloned gene, and like the protease-deficient mutants, these retained pathogenicity in seedlings and infiltrated leaves. However, even when inoculation was carried out at the leaf margins or by soaking seeds in bacterial suspensions, their virulence was similar to that of the wild type (Gough *et al.* 1988). Thus the role of the cellulase (the major extracellular protein produced by *X.c. campestris*) in host–pathogen interaction remains obscure.

Mutants deficient in the production of extracellular polysaccharide also behave almost normally in seedling or leaf infiltration tests, but like protease mutants they are less virulent when inoculated at leaf margins (Daniels *et al.* in preparation).

X.c. campestris produces at least three isoforms of polygalacturonate lyase (Dow *et al.* 1987). The structural gene of one of these has been cloned and used to generate mutants deficient in the one form (but still producing the others). Despite the relatively small decrease in total enzyme activity, the mutants were significantly less virulent than the wild type (Dow *et al.* 1989a).

Export of proteins

Turner *et al.* (1985) showed that a cosmid clone pIJ3000, which complemented a pathogenicity mutant (Daniels *et al.* 1984a), contains a cluster of genes spanning approximately 11 kb of DNA, all of which were required for pathogenicity, although the flanking regions (totalling 15 kb) were not necessary. Dow *et al.* (1987) showed that the genes were necessary for export of protease, cellulase and polygalacturonate lyase, from the periplasm into the medium. Mutations in any one of the genes caused the enzymes to accumulate in the periplasm, although the total activity was scarcely altered. Passage from the periplasm into the medium is not apparently accompanied by cleavage of a signal peptide (Dow *et al.* 1989b), and the mechanism is obscure. The *X.c. campestris* export genes are analogous to the *out* genes of *Erwinia*, described elsewhere in this volume. Their role in pathogenicity is of course indirect, because the enzymes which they export are themselves pathogenicity determinants that must be exported in order to attack plant components.

Regulatory genes

A non-pathogenic mutant of *X.c. campestris*, isolated by screening mutagenized bacteria on plants, was found to be defective in production of protease, cellulase, polygalacturonate lyase, amylase, and EPS, and a recombinant plasmid, pIJ3020, concomitantly restored synthesis of all these factors, and pathogenicity (Daniels *et al.* 1984a). Tn*5* mutagenesis of the genomic region cloned in pIJ3020 revealed the presence of five separate regions in which mutations resulted in phenotypes which were indistinguishable from the original mutant. Subsequent analysis by subcloning, Tn*51ac* mutagenesis, promoter probing and DNA sequencing has indicated that at least nine genes lie in these regions, although it has not yet been proved that all of them are involved in regulation of enzyme and EPS biosynthesis. Three of the genes are of particular interest: the deduced gene C product contains domains showing strong homology (at the level of amino acid sequence) with consensus sensor and regulator domains of prokaryotic 'two component' regulatory systems, as well as containing regions that are probably membrane-spanning. In other systems, with one exception, the sensor and regulator elements are encoded by separate genes (Ronson *et al.* 1987; Stibitz *et al.* 1989). The gene G product also shows homology with regulatory proteins, while the gene F product is related to the phoM protein (Tang 1989). Complementation experiments show that the regulatory genes in the pIJ3020 cluster interact in a complex fashion. It is not yet known how the presumed 'cascade' functions in order to regulate synthesis of the diverse enzymes and EPS, and there is also no information on the nature of the primary triggering substance.

These genes act as positive regulators, that is activators of biosynthesis. We considered the possibility that a balancing negative regulatory gene might exist. A candidate gene has been identified which, when introduced into the wild-type strain (thereby increasing the 'dosage' of the gene) severely depresses synthesis of the extracellular enzymes and EPS. However the growth rate of the bacteria is not affected and hence there is presumably no inhibition of general protein synthesis. Mutation of the gene causes an increased production of enzymes and EPS compared with the wild type, and the mutants are pathogenic. Nothing is yet known about the role of this gene; it does however, appear to be independent of the positive regulating genes (Tang *et al.* 1990).

Finally, it should be noted that some of the extracellular enzymes are subject to individual regulation. Protease is induced by proteins or peptides, and repressed by amino acids. Polygalacturonate lyase is induced by degradation products of pectin. Thus regulation of synthesis of these pathogenicity factors is complex, and points to a subtle adaption of *X.c. campestris* to its habitat of rotting plant tissues.

Other pathogenicity genes

Much of this discussion has focused on genes involved in extracellular enzyme production because these have been studied in greatest detail.

However, other types of pathogenicity gene are known. For example, a gene was detected adjacent to a plant-inducible promoter. Tn*5* mutants in this gene showed greatly reduced virulence in all pathogenicity tests. The bacteria were still prototrophic, motile, grew at the normal rate both *in vitro* and *in planta*, and produced the normal complement of enzymes and EPS. Sequencing of the gene has given no clues about the function of the gene product (Osbourn *et al.* 1990a). So far about 25 pathogenicity genes have been cloned from *X.c. campestris*. It has been suggested that between 20 and 100 genes may be required for pathogenicity (Daniels 1988), and a consideration of all the plant bacteria which have been studied indicates that a total of 50–70 types of genes have been cloned. Of course not all classes of gene will be represented in a particular pathogen. *X. campestris* pathovars do not elaborate toxins, and *P. syringae* pathovars do not usually produce extracellular enzymes. Nevertheless it is clear that pathogenicity is a complex process and application of molecular genetic techniques is making an important contribution to unravelling this complexity.

Acknowledgements

The Sainsbury Laboratory is supported by a grant from the Gatsby Charitable Foundation. Research described in this paper has also been supported by the Agricultural and Food Research Council, the John Innes Foundation and the Commission of the European Communities.

References

Barrère, G.C., Barber, C.E., and Daniels, M.J., (1986). Molecular cloning of genes involved in the production of the extracellular polysaccharide xanthan by *Xanthomonas campestris* pv *campestris*. *International Journal of Biological Macromolecules* **8**, 372–4.

Boucher, C.A., van Gijsegem, F., Barberis, P.A., Arlat, M., and Zischek, C. (1987). *Pseudomonas solanacearum* genes controlling both pathogenicity on tomato and hypersensitivity on tobacco are clustered. *Journal of Bacteriology* **169**, 5626–32.

Bradbury, J.F. (1986). *Guide to plant pathogenic bacteria*. CAB International, Slough.

Daniels, M.J. (1988). Molecular genetics of host–pathogen interactions. In *Molecular genetics of plant–microbe interactions 1988* (ed. R. Palacios and D.P.S. Verma), p. 229. APS Press, St Paul.

Daniels, M.J., Barber, C.E., Turner, P.C., Sawczyc, M.K., Byrde, R.J.W., and Fielding, A.H. (1984a). Cloning of genes involved in pathogenicity of *Xanthomonas campestris* pv *campestris* using the broad host range cosmid pLAFR1. *EMBO Journal* **3**, 3323–8.

Daniels, M.J., Barber, C.E., Turner, P.C., Cleary, W.G., and Sawczyc, M.K. (1984b). Isolation of mutants of *Xanthomonas campestris* pv *campestris* showing altered pathogenicity. *Journal of General Microbiology* **130**, 2447–55.

Daniels, M.J., Dow, J.M., and Osbourn, A.E. (1988). Molecular genetics of pathogenicity in phytopathogenic bacteria. *Annual Review of Phytopathology* **26**, 285–312.

Day, P.R. (1974). *Genetics of host–parasite interaction.* W.H. Freeman, San Francisco.

Dow, J.M., Scofield, G., Trafford, K., Turner, P.C., and Daniels, M.J. (1987). A gene cluster in *Xanthomonas campestris* pv *campestris* required for pathogenicity controls the excretion of polygalacturonate lyase and other enzymes. *Physiological and Molecular Plant Pathology* **31**, 261–71.

Dow, J.M., Milligan, D.E., Jamieson, L., Barber, C.E., and Daniels, M.J. (1989a). Molecular cloning of a polygalacturonate lyase gene from *Xanthomonas campestris* pv *campestris* and role of the gene product in pathogenicity. *Physiological and Molecular Plant Pathology* **35**, 113–20.

Dow, J.M., Daniels, M.J., Dums, F., Turner, P.C., and Gough, C.L. (1989b). Genetic and biochemical analysis of protein export from *Xanthomonas campestris*. Journal of *Cell Science Supplement* **11**, 59–72.

Flor, H.H. (1956). The complementary genic systems in flax and flax rust. *Advances in Genetics* **8**, 29–54.

Gough, C.L., Dow, J.M., Barber, C.E., and Daniels, M.J. (1988). Cloning of two endoglucanase genes of *Xanthomonas campestris* pv *campestris*: analysis of the role of the major endoglucanase in pathogenesis. *Molecular Plant–Microbe Interactions* **1**, 275–81.

Gough, C.L., Dow, J.M., Keen, J., Henrissat, B., and Daniels, M.J. (1990). Nucleotide sequence of the *engXCA* gene encoding the major endoglucanase of *Xanthomonas campestris* pv. *campestris*. *Gene* **89**, 53–9.

Keen, N.T. and Staskawicz, B.J. (1988). Host range determinants in plant pathogens and symbionts. *Annual Review of Microbiology* **42**, 421–40.

Kelman, A. (1954). The relationship of pathogenicity in *Pseudomonas solanacearum* to colony appearance on a tetrazolium medium. *Phytopathology* **44**, 693–5.

Knowles, J., Lehtovaara, P., and Teeri, T. (1987). Cellulase families and their genes. *Trends in Biotechnology* **5**, 225–61.

Kroos, L. and Kaiser, D. (1984). Construction of Tn*5lac*, a transposon that fuses *lacZ* expression to exogenous promoters, and its introduction into *Myxococcus xanthus*. *Proceedings of the National Academy of Sciences USA* **81**, 5816–20.

Lacy, G.H. and Leary, J.V. (1979). Genetic systems in phytopathogenic bacteria. *Annual Review of Phytopathology* **17**, 181–202.

Lazo, G.R., Roffey, R., and Gabriel, D.W. (1987). Pathovars of *Xanthomonas campestris* are distinguishable by restriction fragment length polymorphisms. *International Journal of Systematic Bacteriology* **37**, 214–21.

Liu, Y.-N., Tang, J.-L., Clarke, B.R., Dow, J.M., and Daniels, M.J. (1990). A multipurpose broad host range cloning vector and its use to characterise an extracellular protease gene of *Xanthomonas campestris* pathovar *campestris*. *Molecular and General Genetics* **220**, 433–40.

Osbourn, A.E., Barber, C.E., and Daniels, M.J. (1987). Identification of plant-induced genes of the bacterial pathogen *Xanthomonas campestris* pathovar *campestris* using a promoter-probe plasmid. *EMBO Journal* **6**, 23–8.

Osbourn, A.E., Clarke, B.R., and Daniels, M.J. (1990). Identification and DNA sequence of a pathogenicity gene of *Xanthomonas campestris* pv. *campestris*. *Molecular Plant–Microbe Interactions* **3**, 280–5.

Osbourn, A.E., Clarke, B.R., Stevens, B.J.H., and Daniels, M.J. (1990). Use of oligonucleotide probes to identify members of two-component regulatory systems in *Xanthomonas campestris* pathovar *campestris*. *Molecular and General Genetics* **222**, 145–51.

Ronson, C.W., Nixon, B.T., and Ausubel, F.M. (1987). Conserved domains in bacterial regulatory proteins that respond to environmental stimuli. *Cell* **49**, 579–81.

Sawczyc, M.K., Barber, C.E., and Daniels, M.J. (1989). The role in pathogenicity of some related genes in *Xanthomonas campestris* pathovars *campestris* and *translucens*: a shuttle strategy for cloning genes required for pathogenicity. *Molecular Plant–Microbe Interactions* **2**, 249–55.

Shaw, J.J. and Kado, C.I. (1987). Whole plant wound inoculation for consistent reproduction of black rot of crucifers. *Phytopathology* **78**, 981–6.

Shaw, J.J., Settles, L.G., and Kado, C.I. (1988). Transposon Tn*4431* mutagenesis of *Xanthomonas campestris* pv. *campestris*: characterisation of a non-pathogenic mutant and cloning of a locus for pathogenicity. *Molecular Plant–Microbe Interactions* **1**, 39–45.

Stibitz, S., Aaronson, W., Monack, D., and Falkow, S. (1989). Phase variation in *Bordetella pertussis* by frameshift mutation in a gene for a novel two-component system. *Nature* **338**, 266–9.

Tang, J–L. (1989). *Aspects of extracellular enzyme production by Xanthomonas campestris*. PhD Thesis, *University of East Anglia, Norwich*.

Tang, J.–L., Gough, C.L., and Daniels, M.J. (1990). Cloning of genes involved in negative regulation of production of extracellular enzymes and polysaccharide of *Xanthomonas campestris* pathovar *campestris*. *Molecular and General Genetics* **222**, 157–60.

Tang, J–L., Gough, C.L., Barber, C.E., Dow, J.M., and Daniels, M.J. (1987). Molecular cloning of protease gene(s) from *Xanthomonas campestris* pv *campestris*: expression in *Escherichia coli* and role in pathogenicity. *Molecular and General Genetics* **210**, 443–8.

Turner, P., Barber, C., and Daniels, M. (1985). Evidence for clustered pathogenicity genes in *Xanthomonas campestris* pv *campestris*. *Molecular and General Genetics* **199**, 338–43.

Whitfield, C., Sutherland, I.W., and Cripps, R. (1981). Surface polysaccharides in mutants of *Xanthomonas campestris*. *Journal of General Microbiology* **124**, 385–92.

Williams, P.H. (1980). Black rot: a continuing threat to world crucifers. *Plant Disease* **64**, 736–42.

10 Interaction of *Xanthomonas campestris* and *Brassica*: genetic and biochemical aspects of the plant response

JOHN MAXWELL DOW, DAVID COLLINGE,
DAWN E. MILLIGAN, ROMELIA PARRA,
JUTTA CONRADS-STRAUCH, and MICHAEL
J. DANIELS

*The Sainsbury Laboratory, John Innes Institute, Colney Lane,
Norwich NR4 7UH, UK*

Introduction

Xanthomonas campestris is a species of gram-negative phytopathogen in which over 100 pathovars can be recognized, differentiated by their pathogenicity on different plants or groups of plants (Dye *et al.* 1980). Pathovars are generally named after the host from which the bacteria were isolated, often the full host range of a particular isolate is not known. The known host ranges of each pathovar have been summarized by Leyns *et al.* (1984). Accordingly, *Xanthomonas campestris* pv.*campestris* (hereafter referred to as *X.c. campestris*) is an important worldwide pathogen of almost all cultivated brassicas and non-crop crucifers (Williams 1980), whereas *X.c. vitians*, a pathogen of lettuce (*Lactuca sativa*) or *X.c.translucens*, a pathogen of wheat, are incompatible with brassica and do not cause disease (Daniels *et al.* 1987). We are interested in the mechanisms of resistance of *Brassica* spp., in particular turnip (*Brassica campestris*), to incompatible pathovars of *X. campestris*. Elucidation of the molecular bases of these responses may provide a rationale for engineering resistance in *Brassica* to a wider spectrum of potential pathogens including *X.c. campestris*;

Resistance mechanisms in *Brassica*

In general resistance of plants to incompatible pathogens may depend on both preformed and induced mechanisms. Examples of preformed barriers would include the presence of constitutive antimicrobial compounds and the plant cuticle. In many cases, however, the incompatible pathogen triggers a hypersensitive response (HR) in the plant whereby cells at or

adjacent to the site of invasion collapse and die. Associated with this is an
increase in the transcriptional activity of a number of 'defence' genes.
Some of these genes encode enzymes involved in the synthesis of phytoalexins and lignin; while others are responsible for production of pathogenesis-related (PR) proteins, including hydrolytic enzymes such as
chitinase and 1,3-β-glucanase; cell wall hydroxyproline rich glycoproteins
(HRGP)' and inhibitors of enzymes such as protease from the pathogen
(Collinge and Slusarenko 1987; Ebel and Grisebach 1988; Hedrick *et al.*
1988; Kombrink *et al.* 1988; Lamb *et al.* 1987; Somssich *et al.* 1988; Walter
et al. 1988). Such changes in gene expression may take place not only in
healthy cells surrounding the collapsed tissue, but also in cells which are in
direct contact with the pathogen early-on, prior to any hypersensitive collapse. (Lamb *et al.* 1987; Somssich *et al.* 1988). The resulting biochemical
changes serve to contain the pathogen through changes in the plant cell
wall that produce a physical barrier (lignification or deposition of HRGP),
and/or by direct effects on the pathogen such as the antibiotic action of
phytoalexins or through hydrolysis of pathogen structures by enzymes such
as chitinase and 1,3-β-glucanase. Similar changes in plant gene expression
may also take place at later stages of compatible reactions, and in some
cases may result in lesion limitation (Bailey 1987).

Currently, resistance mechanisms in *Brassica* are poorly understood.
Brassica spp. do contain constitutive compounds, called glucosinolates, a
group of sulphur-containing glycosides. Following tissue damage these are
hydrolysed to indole or isothiocyanate derivatives, compounds that are
toxic to fungi, by the endogenous enzyme myrosinase (Mithen *et al.* 1986;
1987), and there is some correlation between levels of glucosinolates and
expression of resistance to the fungal pathogen *Leptosphaeria maculans*
(Mithen *et al.* 1986). However, as most *Brassica* species are susceptible to
X.c. campestris, we would infer that these preformed antimicrobial compounds, or indeed any preformed barriers are not particularly effective. As
to induced mechanisms, there are reports of elicitation of phytoalexins in
Brassica spp. challenged with either bacterial (Takasugi *et al.* 1986) or fungal (Conn *et al.* 1988) pathogens and the difference in susceptibility of different crucifers to *Alternaria brassicae* may be due in part to qualitative
and/or quantitative differences in phytoalexin production (Conn *et al.*
1988). The biosynthetic pathways for these phytoalexins are as yet
unknown.

To study potential resistance mechanisms in *Brassica campestris* we have
used two approaches:

1. Investigation of changes in gene expression in turnip in response both to
   the compatible pathovar *X.c. campestris* and the incompatible pathovar
   X.c. vitians. This approach may help to identify *Brassica* defence genes,

the transcription of which is specifically increased in the resistance response.

2. Investigation of biochemical changes in turnip plants responding to the incompatible and to the compatible pathovar. An increasing body of work in other plant-pathogen systems has suggested likely changes to monitor and some of these have been alluded to earlier. In particular we have studied the hydrolytic enzymes 1,3-β-glucanase and chitinase. The latter is especially interesting in the context of bacterial diseases since almost all plant chitinases so far described have some lysozyme activity (Boller 1987; 1988 see also chapter 15), and are thus potentially effective against both bacterial and fungal pathogens.

Firstly we shall describe the characteristics of the host–pathogen system.

X.c. vitians induces a hypersensitive response in *Brassica campestris*

Inoculation of mature leaves of turnip cv. Just Right with *X.c.vitians* induces collapse of the leaf tissue in the area of inoculation, accompanied by vein-blackening (Collinge *et al.* 1987). These effects are relatively rapid occurring within 18 hours after inoculation. By contrast, the compatible pathovar *X.c.campestris* produces no visible effects on the leaves within the same time period, but after 48 hours causes an initial yellowing of the inoculated area which subsequently spreads (Fig. 10.1). This is followed by tissue rotting. Loss of cellular integrity, as indicated by the leakage of electrolytes from the plant cells, is much more rapid in leaves inoculated with *X.c. vitians* than those inoculated with *X.c. campestris*.

Development of visible symptoms does not occur when the inoculum of *X.c. vitians* contains fewer than 10^6 cells/ml and requires the bacteria to be metabolically active; bacteria treated with antibiotics or autoclaved bacteria fail to produce symptoms of collapse. The numbers of viable *X.c. vitians* cells decrease rapidly in *B. campestris* following inoculation, whereas those of *X.c. campestris* increase. Inoculations with a mixture of marked strains of the two pathovars suggested that the response induced by *X.c. vitians* was capable of eliminating *X.c. campestris* (Collinge *et al.* 1987). Rapid cellular collapse, electrolyte leakage, a requirement for metabolically active bacteria and limitation of bacterial spread, all imply that the plant response to *X.c. vitians* has the characteristics of the HR according to the definition of Klement (1982). However, not all incompatible pathovars induce the same response in turnip. The collapse observed with pathovars *malvacearum*, *phaseoli*, and *pisi* was less pronounced than that with *vitians*, and moreover, pathovar *translucens*, the wheat pathogen, does not grow in turnip and gives a null response. This suggests that factors

166

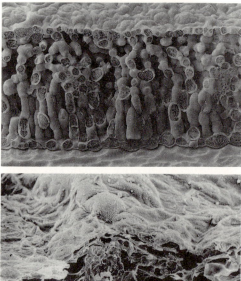

Fig. 10.1. Response of mature leaves of turnip (*Brassica campestris*) 24 h after inoculation with *X.c. campestris* (compatible) or *X.c. vitians* (incompatible) at 10^8 cells/ml. Upper panel: vein blackening associated with the resistance response (right hand side of leaf) with only inoculation damage in the compatible response (left hand side of leaf). Middle panel: Freeze-fractured cross section through healthy leaf as viewed with the scanning electron microscope. Leaves undergoing the susceptible response are identical at this time. Lower panel: Cross section through a leaf undergoing the resistance response.

other than triggering of the HR (nutritional?) also serve to restrict the host range of pathovars.

Analysis of *de novo* gene expression following inoculation of *Brassica* leaves

The transcriptional activity of the plant genes was followed by preparation and translation of both total and poly A^+ RNA from mature leaves at different time points following inoculation with *X.c. campestris*, *X.c. vitians*, or sterile medium control (Collinge *et al.* 1987). RNA preparations were translated *in vitro* in a rabbit reticulocyte lysate or wheat germ translation system in the presence of ^{35}S methionine, and radio-labelled translation products were analysed by two-dimensional polyacrylamide gel electrophoresis followed by fluorography.

Major changes in gene expression were observed solely as a consequence of the inoculation technique. In addition, however, after inoculation with *X.c. vitians*, up to fifteen additional major polypeptides had appeared or were greatly increased in concentration within 4 h. Some of these had subsequently disappeared by nine hours while several more had appeared. No major polypeptides disappeared or even decreased greatly in intensity following inoculation with *X.c. vitians*. Both the changes in polypeptide population attributable to inoculation stress and those changes which correlated with the development of HR were scattered over the major area of the gels. This implies that the polypeptides involved are heterogeneous with respect to size and isoelectric point, and therefore do not fall into a single empirically defined group such as the PR proteins. No changes were recorded in response to *X.c. campestris* within 9 h of inoculation which were not also observed in response to sterile medium. In addition, no major differences could be detected in the polypeptide products obtained from *in vitro* translation of RNA enriched for polyadenylated sequences compared to those translated from total RNA. Total *Xanthomonas* RNA did not give detectable polypeptide translation products using rabbit reticulocyte lysate.

These results demonstrate that, as in other systems, development of HR in *Brassica* is associated with specific changes in gene expression, some of which may be transient. They provide a basis for the cloning and characterization of potential defence genes by cross-screening of cDNA libraries complementary to the appropriate mRNA preparations.

Hydrolytic enzymes in the response of turnip to *Xanthomonas campestris* pathovars

Induction of the hydrolytic enzymes 1,3-β-glucanase and chitinase has been studied in a number of plants in response to wounding, infection and treatment with heat-killed pathogens, pathogen cell walls or chitosan (reviewed by Boller 1988). Some of the pathogenesis-related proteins induced in several different plants have now been identified as 1,3-β-gluca-nases and chitinases (De Wit *et al*. 1988; Kombrink *et al*. 1988; Metraux *et al*. 1988). These hydrolases are of interest because of their ability to attack directly specific structures of pathogens. Plant chitinases are potent inhibi-tors of fungal growth (Schlumbaum *et al*. 1986), and in combination with 1,3-β-glucanase are able to attack a number of fungi *in vitro* (Boller 1988; Mauch *et al*. 1988b). This antifungal action is probably due to hydrolysis of the non-crystalline chitin present at the apex of the growing fungal hypha (Wessels 1986). 1,3-β-glucanase and chitinase are co-ordinately induced in a number of plant tissues both by pathogen attack and by elicitors (Vogeli *et al*. 1988). Bean and pea chitinases also possess lysozymal activity that is capable of hydrolysing the peptidoglycan of bacterial cell walls (Boller *et al*. 1983; Mauch *et al*. 1988a). Indeed, all plant lysozymes that have been described so far have been shown to act as endochitinases (Boller 1988). Fleming (1932) described lysozymal activity in turnip, and the enzyme has been partially characterized in turnip roots (Bernier *et al*. 1971) and caulif-lower (Ereifej and Markakis 1980). In this section, we describe recent experiments on the induction of chitinase/lysozyme and 1,3-β-glucanase in response to *X.c. campestris*, *X.c. vitians* and *Escherichia coli* (a non-patho-gen). These observations are extended to characterization of isoforms of chitinase/lysozyme that show differential regulation in response to these different bacteria.

Time course of induction of hydrolytic-enzyme activity in turnip leaves

The specific activities of chitinase and 1,3-β-glucanase were determined in leaves which had been inoculated in two slightly different ways. In both cases bacterial suspensions were introduced through the stomata on the underside of the leaves using a syringe (with no needle). Inoculation was performed either in a patchwise fashion or over the whole leaf area. The patchwise method of inoculation with *X.c. vitians* lead to collapse of the tissue only within the area of inoculation, so that each collapsed area was surrounded by 'healthy' tissue which comprised at least 50 per cent of the leaf area. This was done in order to mimic 'natural' HR processes in which tissues both directly in contact with, and distant to, the triggering pathogen may contribute to the overall response. In contrast, leaves inoculated over the whole surface might be expected to show largely the responses of tissue directly in contact with the bacteria.

Differences were observed in both the kinetics and magnitude of induction of 1,3-β-glucanase in response to *X.c. campestris* and *X.c. vitians* (Fig. 10.2); *X.c. vitians* caused a larger and more rapid change in

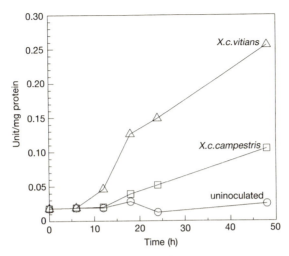

Fig. 10.2. Time course of the induction of β-glucanase in leaves inoculated in a patchwise fashion with *X.c. campestris* (compatible) or *X.c. vitians* (incompatible) at 10^8 cells/ml.

glucanase activity than did *X.c. campestris*. This effect was detected with both methods of inoculation and at all concentrations of bacteria used, although the magnitude of the response decreased with decreasing bacterial numbers. Inoculation with *Escherichia coli*, or with heat- or antibiotic-killed *X.c. vitians*, or with sterile water caused no change from the basal level of glucanase that was found in uninoculated leaves. Similar increases in specific activity of glucanase were detected in both the inoculated areas and in the tissues surrounding them in patchwise inoculated plants.

Differences between treatment with one pathovar and treatment with the other, in respect of the kinetics of chitinase induction were less pronounced than those observed with β-glucanase (Fig. 10.3). With the patchwise method of inoculation, no distinct differences in chitinase levels were seen within the first 24 h, between leaves inoculated with one pathovar and those inoculated with the other, although chitinase levels were higher than in uninoculated plants. After 24 h, the level of chitinase in leaves inoculated with *X.c. vitians* was higher than that in leaves inoculated with *X.c. campestris*. With the full leaf inoculations, both bacteria induced increases in the level of chitinase that were equally rapid, although again the level was slightly higher in leaves inoculated with *X.c. vitians*. Although inoculation with sterile water had no effect on the level of chitinase, inoculation with heat- or antibiotic-killed *X.c. vitians*, or with *E. coli* caused an increase in chitinase activity. This is in distinct contrast to the results with induction of β-glucanase. In plants inoculated with *X.c. vitians* in a

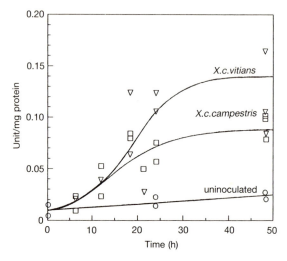

Fig. 10.3. Time course of the induction of lysozyme/chitinase in leaves inoculated in a patchwise fashion with *X.c. campestris* (compatible) or *X. c. vitians* (incompatible). Points from a number of different experiments are superimposed.

patchwise fashion, increases in chitinase activity were detected in inoculated and collapsed areas, and in the surrounding apparently healthy tissue. In contrast, changes in chitinase activity in plants inoculated with *E. coli* were only observed in the inoculated areas; the specific activity of chitinase in the surrounding tissue was the same as that of uninoculated leaves. These changes in hydrolytic enzyme activity occurred relatively slowly compared to those of other defence-related products; in leaves inoculated with *X.c. vitians* or *X.c. campestris* using the patchwise method, the activity of phenylalanine ammonia lyase increased within 2 h after inoculation and reached a maximum after 6 h. At this time, no changes in hydrolytic enzyme activity were detected.

Characterization of 1,3-β-glucanase and isozymes of chitinase, and their differential induction in response to pathogens and non-pathogens

Multiple isoforms of chitinase and β-glucanase have been described in a number of plants, and in some cases these isoforms show differential regulation in response to pathogens or with ageing (Mauch *et al.* 1988a). Accordingly we wished to determine the number of isoforms of β-glucanase and chitinase present in healthy plants, and whether different isoforms were induced specifically by the different treatments employed.

Basic chitinases and β-glucanases were purified from turnip leaves by a

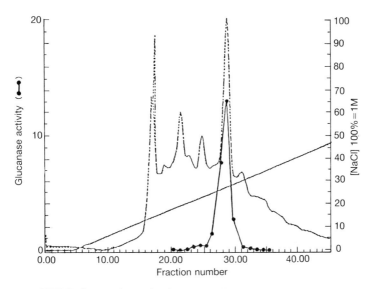

Fig. 10.4. HPLC (ion-exchange) of extracts from leaves inoculated with *X.c. vitians* (prepared as described in the text) showing a single peak of β-glucanase activity.

modification of the method of Boller *et al.* (1988) for pea tissue. Briefly, chitinases were separated from a crude dialysed extract of leaves, by affinity chromatography on a column of regenerated chitin (Molano *et al.* 1977) at pH 8. Following their elution from the column at pH 3, individual chitinase isozymes were separated from each other by HPLC on a cation exchange column (SP-5PW), at pH 4.3. β-glucanases were purified by the same HPLC method after passage of the dialysed extract through a column of DEAE-Sephadex followed by adjustment of the pH of the eluate from 8 to 4.3 and removal of the precipitated material by centrifugation.

One major peak of glucanase activity was detected following ion-exchange chromatography of extracts from both uninoculated plants and those inoculated with *X.c. vitians* (Fig. 10.4). The active fractions were pooled, diluted, and re-chromatographed at the same pH using a shallower gradient of salt concentration. Again only one peak of activity was evident. SDS-polyacrylamide gel electrophoresis of the active fractions showed a single species, Mr 36.5 kDa, from both uninoculated and *X.c. vitians* inoculated plants. Isoelectric focusing confirmed the presence of a single species with a pI of about 8.5 from both treatments. Although these results suggest that a single gene product is present in both healthy plants and those showing an HR, the presence of very closely related isoforms cannot be eliminated by this data. The purified enzyme degraded laminarin (a 1,3-β-glucan) by an endo-mechanism; analysis, by thin layer chromatography, of the products of digestion showed that even after long incubation times, monomeric glucose was not a product; the trimer (laminaritriose) predominated, with smaller amounts of the dimer (laminaribiose) and higher oligomers.

In contrast to the situation with glucanase, several isozymes of chitinase were resolved by ion-exchange HPLC (Fig. 10.5). In addition, the pattern of chitinase isozymes induced by treatment with *X.c. vitians* was different from that induced by heat- or antibiotic-killed *X.c. vitians*, *X.c. campestris*, or *E. coli*. Treatment with live *X.c. vitians* induced largely an isoform (CH 1) which eluted early but with significant amounts of a later eluting form (CH 2). All the other treatments induced largely CH 2. Both CH 1 and CH 2 had very similar molecular weights as revealed by SDS-polyacrylamide gel electrophoresis, (ca. 30 000), and both had pI's above 8. A number of minor isozymes were observed in uninoculated plants, but their activity did not increase following inoculation. Analysis of the distribution of isozymes in HPLC fractions was performed by using their lysozyme activity which can be assayed by a relatively rapid spectrophotometric method. However, thin layer chromatography revealed that CH 1 and CH 2 were also both capable of degrading regenerated chitin to produce chitobiose, chitotriose, and higher oligomers as soluble reaction products, although the ratio of lysozyme to chitinase activity was much higher for CH 2 than for CH 1. Both isozymes were immunologically cross-reactive with the bean chitinase; antiserum against bean chitinase (kindly provided by Thomas Boller) detected both isoforms on Western blots although not with high sensitivity.

Contribution of the hydrolytic enymes to resistance

The potential antifungal activities of the purified chitinases and 1,3-β-glucanase were tested against a number of fungi, including the *Brassica* pathogen *Leptosphaeria maculans*. Combinations of chitinase and β-glucanase were more effective than the individual enzymes at inhibiting hyphal extension, as determined by microscopic examination. These are similar results to those described by Mauch *et al.* (1988a) and Boller (1988) and like them suggest that the hydrolases may contribute to resistance to fungal pathogens.

The potential antibacterial activities of the enzymes were tested against *X.c. vitians*. Incubation of the bacteria with chitinases or β-glucanase either alone or in combination, under a variety of conditions of pH and nutrient status, did not lead to a loss of viability as revealed by subsequent plating on nutrient agar medium. Although we do not yet know whether the turnip chitinase can degrade the peptidoglycan of *Xanthomonas*, we

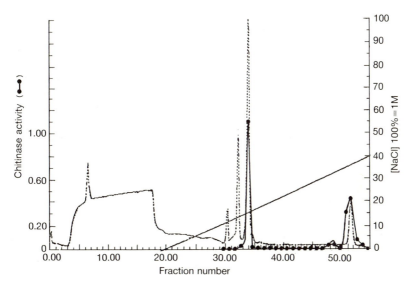

Fig. 10.5. Separation by HPLC (ion-exchange) of chitinases isolated from leaves inoculated with *X.c. vitians*, after affinity chromatography on a column of regenerated chitin.

could speculate that other enzymes, whose function it is to destroy or permeabilize the bacterial outer membrane, may be required to potentiate its action.

Concluding remarks

The results presented here suggest that the induction of 1,3-β-glucanase and of one chitinase isoform (CH 1) are specifically associated with the HR induced in turnip leaves by live, incompatible bacterial pathogens, induction occurring both in directly inoculated tissue and surrounding areas. Another isoform (CH2) is induced by a range of living and dead bacteria including non-pathogens such as *E. coli*, but in a localized fashion. Differential activation, under different stress conditions, of members of small multigene families that encode defence related products has been described for HRGP and chalcone synthase in *Phaseolus vulgaris* (Corbin *et al*. 1987; Ryder *et al*, 1987). In addition, transcripts of some defence genes (HRGP, PR 1) have been shown to accumulate in areas distant to the site of infection, whereas others may be more localized (Sommsich *et al*. 1988; Templeton and Lamb 1988; Kombrink *et al*. Chapter 15).

The localized chitinase induction in *Brassica* may be elicited by the peptidoglycan component of the bacterial envelope although we have as yet no evidence for this. The complete bacterial envelope of *Pseudomonas syringae* pv. *pisi* is capable of enhancing disease-resistance-response mRNA levels in pea, and of eliciting pisatin production although the lipo-polysaccharide, lipoprotein-peptidoglycan or outer membrane prep-arations were unable to elicit pisatin (Daniels *et al.* 1988). The induction of β-glucanase and CH 1 in cells distant from the site of inoculation suggests the participation of some diffusible signal, perhaps produced by cells collapsing in the hypersensitive response, as envisaged by Collinge and Slusarenko (1987). By analogy with other systems, we might expect that this regulation operates at the level of transcriptional activation of the genes. Construction of probes using the regulatory regions of these genes fused to structural genes for reporter proteins such as GUS, may provide assays for such intercellular signals in plants.

Acknowledgements

The work described in this paper was supported by the Gatsby Foundation and by the Agricultural and Food Research Council (via a grant-in-aid to the John Innes Institute).

References

Bailey, J.A. (1987). Phytoalexins: a genetic view of their significance. In *Genetics and plant pathogenesis*, (ed. P.R. Day and G.J. Jellis), pp. 233–44. Blackwell Scientific Publications, Oxford.

Bernier, I. Van Leemputten, E., Horisberger, M., Bush, D.A., and Jolles, P. (1971). The turnip lysozyme. *FEBS letters* **14**, 101–4.

Boller, T. (1987). Hydrolytic enzymes in plant disease resistance. *Plant–microbe interactions, molecular and genetic perspectives*. Vol. 2 (ed. T. Kosuge and E. W. Nester), pp. 385–413. Macmillan, New York.

Boller, T. (1988). Ethylene and the regulation of antifungal hydrolases in plants. *Oxford surveys of plant molecular and cell biology*. Vol. 5, 145–74 (ed. B.J. Miflin), Oxford University Press.

Boller, T., Gehri, A., Mauch, F., and Vogeli, U. (1983). Chitinase in bean leaves: induction by ethylene, purification, properties, and possible function. *Planta* **157**, 22–31.

Collinge, D.B. and Slusarenko, A.J. (1987). Plant gene expression in response to pathogens. *Plant Molecular Biology* **9**, 389–410.

Collinge, D.B., Milligan, D.E., Dow, J.M., Scofield, G., and Daniels, M.J. (1987). Gene expression in *Brassica campestris* showing a hypersensitive response to the incompatible pathogen *Xanthomonas campestris* pv. *vitians*. *Plant Molecular Biology* **8**, 405–14.

Conn, K.L., Tewari, J.P., and Dahiya, J.S. (1988). Resistance to *Alternaria brassi-*

cae and phytoalexin-elicitation in rapeseed and other crucifers. *Plant Science* **56**, 21–5.

Corbin, D.R., Sauer, N., and Lamb, C.J. (1987). Differential regulation of a hydroxyproline-rich glycoprotein gene family in wounded and infected plants. *Molecular and Cellular Biology* **7**, 4337–44.

Daniels, M.J., Collinge, D.B., Dow, J.M., Osbourn, A.E., and Roberts, I.N. (1987). Molecular biology of the interaction of *Xanthomonas campestris* with plants. *Plant Physiology and Biochemistry* **25**, 353–9.

Daniels, C., Cody, Y.S., and Hadwiger, L. (1988). Host responses in peas to challenge by wall components of *Pseudomonas syringae* pv. *pisi* Races 1, 2 and 3 *Phytopathology* **78**, 1451–3.

De Wit, P.J.G.M., Toma, I.M.J., and Joosten, M.H.A.J. (1988). Race-specific elicitors and pathogenicity factors in the *Cladosporium fulvum*-tomato interaction. *Physiology and biochemistry of plant–microbe interactions*, (ed. N.T. Keen, T. Kosuge, and L.L. Walling). The American Society of Plant Physiologists.

Dye, D.W., Bradbury, J.F., Goto, M., Hayward, A.C., Lelliott, R.A., and Schroth, M.N. (1980). International standards for naming pathovars of phytopathogenic bacteria and a list of pathovar names and pathotype strains. *Review of Plant Pathology* **59**, 153–68.

Ebel, J. and Grisebach, H. (1988). Defense strategies of soybean against the fungus *Phytophthora megasperma* f.sp. *glycinea*: a molecular analysis. *Trends in Biochemical Science* **13**, 23–7.

Ereifej, K.I. and Markakis, P. (1980). Cauliflower lysozme. *Journal of Food Science* **45**, 1781–2.

Fleming, A. (1932). Lysozyme. *Proceedings of the Royal Society for Medicine* **26**, 71.

Hedrick, S.A., Bell, J.N., Boller, T., and Lamb, C J. (1988). Chitinase cDNA cloning and mRNA induction by fungal elicitor, wounding, and infection. *Plant Physiology* **86**, 182–6.

Klement, Z. (1982). Hypersensitivity. In *Phytopathogenic prokaryotes*, (ed. M.S. and G.H. Lacy), pp. 149–77. Academic Press, New York.

Kombrink, E., Schroder, M., and Hahlbrock, K. (1988). Several 'pathogenesis-related' proteins in potato are 1, 3-β-glucanases and chitinases. *Proceedings of the National Academy of Sciences USA* **85**, 782–6.

Lamb, C.J., Bell, J.N., Corbin, D.R., Lawton, M.A., Mendy, M.C., Ryder, T.B., Sauer, N., and Walter, M.H. (1987). Activation of defense genes in response to elicitor and infection. *Molecular strategies for crop protection*, pp. 49–58. Alan R. Liss, New York.

Leyns, F., de Clene, M., Swings, J-G., and de Ley, J. (1984). The host range of the genus *Xanthomonas*. *Botanical Reviews* **50**, 308–56.

Mauch, F., Hadwiger, L.A., and Boller, T. (1988a). Antifungal hydrolases in pea tissue. Purification and characterization of two chitinases and two β-1,3-glucanases differentially regulated during development and in response to fungal infection. *Plant Physiology* **87**, 325–33.

Mauch, F., Mauch-Mani, B., and Boller, T. (1988b). Antifungal hydrolases in pea tissue. II Inhibition of fungal growth by combinations of chitinase and β-1,3-glucanase. *Plant Physiology* **88**, 936–42.

Metraux, J.P., Streit, L., and Staub, T. (1988). A pathogenesis-related protein in cucumber is a chitinase. *Physiological and Molecular Plant Pathology* **33**, 1–9.

Mithen, R.F., Lewis, B.G., and Fenwick, G.R. (1986). In vitro activity of gluco-sinolates and their products against *Leptosphaeria maculans. Transactions of the British Mycological Society* **87**, 433–40.

Mithen, R.F., Lewis, B.G., Heaney, R.K., and Fenwick, G.R. (1987). Resistance of leaves of Brassica species to *Leptosphaeria maculans. Transactions of the British Mycological Society* **88**, 525–31.

Molano, J., Duran, A., and Cabib, E. (1977). A rapid and sensitive assay for chitinase using tritiated chitin. *Analytical Biochemstry* **83**, 648–56.

Ryder, T.B., Hedrick, S.A., Bell, J.N., Liang, X., Clouse, S.D., and Lamb, C.J. (1987). Organization and differential activation of a gene family encoding the plant defense enzyme chalcone synthase in *Phaseolus vulgaris. Molecular and General Genetics* **210**, 219–33.

Schlumbaum, A., Mauch, F., Vogeli, U., and Boller, T. (1986). Plant chitinases are potent inhibitors of fungal growth. *Nature* **324**, 365–7.

Somssich, I.E., Schmelzer, E., Kawalleck, P., and Hahlbrock, K. (1988). Gene structure and in situ transcript localization of pathogenesis-related protein 1 in parsley. *Molecular and General Genetics* **213**, 93–8.

Takasugi, M., Katsui, N., and Shirata, A. (1986). Isolation of three novel sulphur-containing phytoalexins from Chinese cabbage *Brassica campestris* L. ssp. *pekinensis* (Cruciferae). *Journal of the Chemical Society, Chemical Communications* **14**, 1077–8.

Templeton, M.D. and Lamb, C.J. (1988). Elicitors and defence gene activation *Plant Cell and Environment* **11**, 395–401.

Vogeli, U., Meins, F., Jr, and Boller, T. (1988). Co-ordinated regulation of chiti-nase and β-1, 3-glucanase in bean leaves. *Planta* **174**, 364–72.

Walter, M.H., Grima-Pettenati, J., Grand, C., Boudet, A.M., and Lamb, C.J. (1988). Cinnamyl-alcohol dehydrogenase, a molecular marker specific for lignin synthesis: cDNA cloning and mRNA induction by fungal elicitor. *Proceedings of the National Academy of Sciences USA* **85**, 5546–50.

Wessels, J.G.H. (1986). Cell wall synthesis in apical hyphal growth. *International Review of Cytology* **104**, 37–79.

Williams, P.H. (1980). Black rot: a continuing threat to world crucifers. *Plant Disease* **64**, 736–42.

11 Molecular genetics of regulation and export of *Erwinia* pectinases

ARUN K. CHATTERJEE, HITOSHI MURATA, and JAMES L. McEVOY

Department of Plant Pathology, 108 Waters Hall, University of Missouri, Columbia, MO 65211, USA

ALAN COLLMER

Department of Plant Pathology, Cornell University, Ithaca, New York 14853–5908, USA

Introduction

Phytopathogenic bacteria elicit an array of disease symptoms including necroses, wilts, galls, and rots. The rotting or maceration of plant tissues results from the activities of plant cell-wall-degrading enzymes produced by some of these pathogens, particularly certain species of *Erwinia*. These bacteria actively export a number of such degradative enzymes to the outside of the bacterial cells, and are capable of causing damage to plants under growing conditions as well as in storage and transit. Collectively they infect a wide-variety of plant hosts. The three soft-rotting bacteria that have been studied the most are *E. carotovora* subsp. *atroseptica* (Eca), *E. carotovora* subsp. *carotovora* (Ecc), and *E. chrysanthemi* (Echr). Their taxonomic relationships, ecology, genetics, and pathogenicity factors have been extensively reviewed (Chatterjee and Starr 1980; Chatterjee and Vidaver 1986; Collmer and Keen 1986; Kotoujansky 1987; Perombelon and Kelman 1980). Suffice it to say that in the elicitation of soft-rot diseases production of extracellular enzymes, in particular the pectinases, is crucial. In this report we summarize some of our recent findings on the regulation of two pectinases, pectate and pectin lyases, and discuss the physiological and genetical aspects of protein export.

Pectate lyases

Background

During the past several years pectolytic enzymes, particularly the pectate lyases (Pels), of several Echr strains have been the subject of intensive research in various laboratories. In most of our studies we have used the Echr Strain EC16 (Chatterjee and Starr 1977) and its derivatives. By cloning the *pel* genes and by analysing Pel production in *E. coli* clones as well as in Echr (Barras and Chatterjee 1987; Barras *et al.* 1987; Keen *et al.* 1984), it was established that the four Pel species (PelA, pI 4.2; PelB, pI 8.8; PelC, pI 9.0; PelE, pI 10.0) are encoded by four separate genes. The simultaneous cloning of *pelA* and *pelE*, and of *pelB* and *pelC* demonstrates that these genes occur in two clusters; this has subsequently been confirmed by sequence analysis (Keen and Tamaki 1986; Tamaki *et al.* 1988). *PelE* is separated from *pelA* by approximately 900 base pairs; likewise about 500 base pairs separate *pelB* and *pelC*. Southern hybridizations of *pel* DNA indicated a high homology between *pelB* and *pelC* (Schoedel and Collmer 1986) and some homology between *pelA* and *pelE* (Barras and Chatterjee, 1987). These findings again have been substantiated and extended by nucleotide sequence data (Keen and Tamaki 1986; Tamaki *et al.* 1988).

The homology between the two sets of *pel* genes, and their close linkage supported the hypothesis that each cluster arose as a result of gene duplication (Schoedel and Collmer 1986). It should be noted that strain EC16 is somewhat unusual in that it has four functional *pel* genes as opposed to the five *pel* genes observed in most other Echr strains that have been genetically characterized (see for example Collmer *et al.* 1985; Reverchon *et al.* 1986). Sequence data (Tamaki *et al.* 1988) revealed the presence of a truncated *pelD*-like gene between *pelA* and *pelE*, indicating a spontaneous *pelD* deletion in EC16.

The four *pel* genes of strain EC16 have been inactivated by *in vitro* manipulations. This entailed the insertion of an *nptI-sacB-sacR* cartridge (Ried and Collmer 1987; 1988), encoding kanamycin resistance and sucrose sensitivity, within the *pel* DNA. Subsequently, marker-exchange-recombination selecting for kanamycin resistance, followed by plasmid eviction and another cycle of marker-exchange-recombination selecting for sucrose tolerance, yielded strains of the following genotypes: Δ(*pelA-pelE*); Δ(*pelB-pelC*); Δ*pelA*Δ(*pelB-pelC*); Δ(*pelB-pelC*)Δ*pelE*; and Δ(*pelA-pelE*)Δ(*pelB-pelC*). Analysis of the constructs showed most of the maceration activity to be associated with PelE, followed by PelB and PelC. The strains containing only a functional PelA gene, and those lacking all of the four *pel* genes elicited a low pathogenic response, that is, 2 per cent of the macerating-activity of the wild-type in potato tubers, and 21 per cent in potato slices (Ried and Collmer 1988). The data showed that PelE has the highest tissue-macerating potential, PelB and PelC are intermediate in their maceration efficiency, and PelA does not contribute to the maceration-ability of the bacterium. The residual macerating-activity in strains devoid of all four *pel* genes was unexpected. It should be noted, however, that such strains produce other extracellular degradative enzymes such as exo-pel, exo-poly-α-D-galacturonosidase (exo-Peh), pectin lyase (Pnl), cellulases (Cel), protease (Prt), and phospholipase. It is possible that a combination of these activities contributes to the residual low level of maceration. This hypothesis is currently being tested. The results of studies with enzymes obtained from Pel⁺ *E. coli* clones (Barras *et al.* 1987; Payne *et al.* 1987; Thurn *et al.* 1987) paralleled findings with the mutants. PelE elicited the highest tissue maceration, electrolyte loss and cell death, followed by PelB and PelC; PelA failed to induce any of these pathological conditions in potato tuber tissue. The physiological basis for the differential response with the Pels is unknown.

The lack of tissue maceration by PelA, despite the presence of an endopectate lyase activity, is intriguing. It is doubtful that the low pI of the enzyme (4.2) is responsible for this phenotype. Tamaki *et al.* (1988) constructed chimeric plasmids carrying parts of *pelA* and *pelE* sequences. One Pel-EA hybrid protein with an altered pI (7.05) possessed high pectolytic activity but still had low macerating activity. It is possible then that the PelA activity plays a very different role in the pathogenicity of Echr and is not involved directly in tissue maceration. This hypothesis is supported by the observation that PelA-deficient mutants of the Echr strain 3937 failed to cause systemic symptoms in the normally susceptible host, *Saintpaulia ionantha* (Boccara *et al.* 1988).

Regulation

Early studies of Pel regulation revealed induction by products of polygalacturonate catabolism and susceptibility to catabolite repression (see, Chatterjee and Vidaver 1986; Kotoujansky 1987). Various types of mutants have been isolated (Chatterjee *et al.* 1985; Collmer and Bateman 1981; Condemine *et al.* 1986; Hugouviex-Cotte-Pattat *et al.* 1986) that are altered in Pel production. The use of reporter gene systems has facilitated exploration of the differential expression of individual *pel* genes and the identification of regulatory loci that act *in trans* on *pel* gene expression. Using *lacZ* transcriptional fusions, Robert-Baudouy and associates, in a series of studies (Condemine and Robert-Baudouy 1987; Hugouviex-Cotte-Pattat and Robert-Baudouy 1985; Reverchon and Robert-Baudouy 1987) have demonstrated that KdgR represses *pel*, *ogl* (oligogalacturonate lyase) and the other polygalacturonate catabolic genes in Echr strain 3937. In addition, the 5-prime end of the *pelE* gene of this bacterial strain appears to possess a putative KdgR binding motif (Reverchon *et al.* 1989). It is likely that a generalized regulation by a KdgR-like protein also occurs in strain EC16 since 3-deoxy-D-glycero-2, 5-hexodiulosonate fuctions as an inducer and activates the expression of *pel*, *ogl*, *peh*, and *kduD* (3-deoxy-D-glycero-2, 5-hexodiulosonate dehydrogenase) in this bacterium (Chatterjee *et al.* 1985). The altered expression of some of those genes in Ogl⁻ and

KduD⁻ mutants (Chatterjee *et al.* 1985; Condemine and Robert-Baudouy 1987) further indicates that the genes are subject to a common regulation. Isolation of KdgR⁻ mutants and the KdgR⁺ DNA from EC16 should facilitate additional testing of this hypothesis.

The availability of plasmids carrying *pel* genes and the EC16 derivative UM1005, which is devoid of endo-Pel activity, allowed comparison of gene expression in *E. coli* and Echr. We introduced a *pelA-pelE* plasmid (pAKC308) and a *pelB-pelC* plasmid (pAKC309) into both UM1005 and the *E. coli* strain HB101, and examined Pel production following growth of the strains in a minimal casamino acids medium (Thurn and Chatterjee 1985) either with or without 0.2 per cent sodium pectate. Basal levels of the enzymes were higher in Echr than in *E. coli*; specifically, PelB-C levels were 12-fold higher while PelA-E levels were 40-fold higher. As expected, the enzymes were inducible in Echr but not in *E. coli*; the induction ratio for PelA-E was 22 in contrast with an induction ratio of 4 with PelB-C. The response with pAKC308 was primarily due to expression of the *pelE* gene. This was determined by examination of the basal and induced levels of the enzymes in deletion mutants that retained either *pelA* (UMJ1009) or *pelE* (UMJ1008). With the PelE⁺ strain we obtained an induction ratio of 234. In contrast, with the PelA⁺ strain the corresponding value was 3, in fact the induction ratio in several experiments has ranged from 1 to 3 indicating that *pelA* expression is not stimulated, or at best is poorly stimulated, by polygalacturnate or its catabolic products. These data revealed the existence of three classes of Pels with regard to the basal levels of activity, and inducibility with products of polygalacturonate catabolism. Class I, represented by PelE, is present at a moderate basal level and is efficiently induced; Class II, represented by Pels B and C, is present at a very high basal level and is subject to weak induction; and Class III, represented by PelA, is present at a very low basal level that is not significantly altered by the polygalacturonate catabolic products.

The differential activation of *pelA* and *pelE* genes in Echr (EC16) can be explained by invoking the presence of specific transcriptional activators. The following evidence provides further indirect support for this hypothesis. In preliminary trials, we have noted some stimulation of *pelA* expression with plant extracts. The *pelA* gene of the strain 3937 also appears to be stimulated by plant extracts (Beaulieu and Van Gijsegem 1990). These observations, if confirmed, would implicate a transcriptional factor resulting from a cooperative interaction between an activator protein and a plant metabolite(s). The findings of Reverchon and Robert-Baudouy (1987) with the strain B374 are also consistent with a positive regulation of some of the *pel* genes.

For a better insight into the regulation of the *pelE* gene, we isolated by EMS mutagenesis of UMJ1008 (genotype: *pelE⁺*, *ΔpelA*, *Δ[pelB-pelC]*)

Table 11.1 Characteristics of the Pel defective mutant AC4232 of
E. chrysanthemi strain UMJ1008

Characteristics	Strain	
	UMJ1008	AC4232
Utilization of:		
Gly, Ara, Glu, Sucrose, Mtl	+	+
Pectate	+	+
Pel products [UDG; UOG][1]	+	+
Complementation by *pelE*	not applicable	−
Specific activity[2] of Pel in cultures containing:		
Glycerol	0.3	0.1
PGA	53	0.1
PGA-digest[3]	8	0.8

[1] UDG = unsaturated digalacturonate.
  UOG = unsaturated oligogalactuonate.
[2] The specific activity is expressed as units of activity per mg of protein. One unit of activity = the quantity of the enzyme that produces in one minute at 30°C a change in absorbance of 1.0 at 235 nm.
[3] Polygalacturonate digest was prepared using *E. chrysanthemi* pectate lyase as described by Chatterjee *et al.* (1985). Minimal casamino acids medium was supplemented with 650 (A_{235}) units/ml of the digest. The digest contains predominately unsaturated and saturated digalacturonates.

strains that apparently were altered in Pel production. Three classes of mutants were detected: PelE⁻, defective in the structural gene and complemented by *pelE*⁺ DNA; Out⁻, where the enzyme is localized in the periplasmic space, complemented by *out*⁺ DNA of Ecc71 (see below for the genetics of the export systems); and Pec1, which is noninducible by pectate, phenotype not restored by *out*⁺ or *pelE*⁺ DNA. The data (Table 11.1) show that the Pec1 mutant (AC4232) is not apparently defective in the *c*AMP-CRP system or in the polygalacturonate catabolic pathway. In addition, the data show that pectate does not stimulate PelE production in AC4232. However, in the presence of polygalacturonate digest some stimulation in Pel production was observed with AC4232; although the levels in the parent were much higher (10-fold) compared to that in the mutant (Table 11.1).

This limited induction may be attributed to, (i) the lack of import of the di- or oligouronide, the precursor(s) of the monomeric inducers (Chatterjee *et al.* 1985; Collmer and Bateman 1981; Condemine *et al.* 1986; Condemine and Robert-Baudouy 1987), (ii) inefficient interaction between the 'activator' and 'effector', or (iii) activation of the exo-Pel gene(s) and not of the *pelE* gene. To determine if the noninducible phenotype was exerted *in trans* we introduced a low copy number plasmid carrying *pelA* and *pelE* (pAKC308) into UMJ1008 (parent) and AC4232 (Pec1 mutant). The specific activity of Pel in AC4232/pAKC308 was less than 1.0 under both inducible and noninducible conditions. These values were 0.3 per cent of that obtained from UMJ1008 harbouring pAKC308 and grown under inducing conditions. Since it is known that *pelA* is only marginally inducible, (see above) this finding suggests that the mutation is *trans*-acting on the expression of *pelE*. While we favour the hypothesis that AC4232 has a mutation in an activator of the *pelE* gene, we cannot yet formally rule out other possibilities as stated above. The isolation of a gene that restores *pelE* induction in AC4232, however, should allow testing of the model that invokes specific activation of *pelE* expression by a positive activator.

Aside from specific regulation of *pel* genes in the Echr strains discussed above, recent evidence with both Ecc and Echr indicates that Pel production may be co-regulated along with other extracellular enzymes such as Peh, Cel, and Prt. By Tn5 mutagenesis of Ecc71 we have obtained a strain (AC5012) that is pleiotropically defective in the production of Pel, Peh, Cel, and Prt. A phenotypically similar mutant (AC4231) of Echr strain EC16 has been obtained by EMS mutagenesis. We also have isolated cosmid clones that restore enzyme production in these mutants, however, we have not detected cross complementation of Ecc and Echr mutants with the cloned DNAs. This lack of cross complementation suggests that the mutations may be in nonallelic genes. These mutants differed from the *bona fide* Out⁻ mutants (see below for the genetics of the Out system) in the following characteristics.

1. The total levels of Pel, Peh, Cel, as well as Prt were considerably reduced. In the Echr mutant, AC4231, the production of extracellular phospholipase C was also lower.
2. These pleiotropic defects were rectified by plasmids that did not restore the Out⁺ phenotype in any of the export-defective mutants.
3. Conversely, none of the Out⁺ plasmids restored enzyme production in AC4231 and AC5012.

It is possible that in these mutants there is a defect in genes that are analogous to the *sec* genes of *E. coli*; some of the *sec* genes are known to affect the production of a number of extracytoplasmic proteins (Ferro-Novick *et al.* 1984; Watanabe *et al.* 1988). Our comparative studies of these two systems should allow us to define the molecular component(s) that apparently regulates production of most of the extracellular enzymes. We should note that mutants of Ecc pleiotropically defective in extracellular enzyme production were also isolated and described by Beraha and Garber (1971).

Based upon the property of the *Erwinia* mutants and taking into consideration models of protein translocation across membranes, we propose the following hypothesis. The mutations leading to a pleiotropic defect, as noted here, may be due to a block in a step in the translocation of the exported proteins across the cytoplasmic membrane. This defect would prevent a permissive interaction between the peptide synthesizing system and such membrane components as may be necessary for peptide translocation. In the absence of a stable membrane-polysome complex, translation of the mRNAs of these proteins for export may be aborted, leading to a deficiency of the exported proteins. This hypothesis allows two predictions. First, the extracellular proteins share a common pathway, or at least some components of the pathway for their translocation across the cytoplasmic membrane. This is indirectly supported by the finding that in cells of *E. coli* carrying the genes for Pels, Peh, or Cel, the proteins are predominantly localized in the periplasm and not in the membranes or cytoplasm. This extracytoplasmic localization is probably determined by the 'Sec' system, although this remains to be confirmed. Second, the production of extracellular enzymes is controlled at two levels: specific control is exerted at the level of transcription, and a generalized control at the level of translation.

Molecular genetics of enzyme export

The soft-rotting *Erwinia* spp. are unusual among enterobacteria in producing an array of extracellular degradative enzymes. In fact, enzyme export is tightly linked with the ability of these organisms to cause tissue-macerating diseases in plants. For example, comparative studies with *Erwinia, Klebsiella*, and *Yersinia* species revealed that while plant pathogens exported most of the Pels, the enzymes remained periplasmic in the nonphytopathogenic enterobacteria (Chatterjee *et al.* 1979). More substantive evidence is the reduction of the tissue-macerating ability of the export defective (Out⁻) mutants of Echr (Andro *et al.* 1984; Chatterjee and Starr 1977; Thurn and Chatterjee 1985) and Ecc (Chatterjee *et al.* 1985; Hinton and Salmond 1987). Enzyme export is also important in pathogenicity of the soft-rotting *Pseudomonas* spp. (Liao *et al.* 1988) as well as nonsoft-rotting phytopathogens such as *P. solanacearum* (Schell *et al.* 1988; Roberts *et al.* 1988), and *Xanthomonas campestris* pv. *campestris* (Daniels *et al.* 1984; Chapter 9 this volume. Dow *et al.* 1987; Chapter 10 this volume). Aside from its contributions to microbial pathogenicity, protein export is an important physiological process with a variety of biological and industrial ramifications (Pugsley 1988; Pugsley and Schwartz 1985). For these reasons considerable effort is being directed towards understanding the genetic and biochemical aspects of protein export. In the following section we describe two different export pathways in the soft-rotting *Erwinia*, and the beginnings of an analysis of a gene cluster specifying one of the export pathways.

Characteristics of export defective mutants

Mutants of Echr and Ecc defective in extracellular enzyme production have been isolated using chemical or transposon (Tn5, Tn*phoA*, Tn*10*, Tn*10-lacZ*) mutagenesis. Most mutants are *bona fide* Out⁻ in that the mutations alter the localization of Pel, Cel, and Peh but apparently had no effect on their synthesis as determined by quantitation of the enzyme activities in cell lysates. The most straightforward interpretation of these findings is that Pel, Peh, and Cel are exported in both Ecc and Echr by a two-step pathway. In this model (Pugsley and Schwartz 1985), two distinct translocation events are postulated. The first step entails translocation across the cytoplasmic membrane by a system analogous to the 'Sec' machinery of *E. coli*. This step is not impaired in the Out⁻ strains. The second step involves translocation from the outer face of the cytoplasmic membrane (or from the periplasm) through the outer membrane by the export (Out) machinery. In the Out⁻ mutants, genetic modifications of the export machinery resulted in accumulation of the enzymes in the periplasm.

The Out⁻ mutants of Ecc and Echr described above were not affected in the export of Prt (Andro *et al*. 1984; Murata *et al*. 1988; Thurn and Chatterjee 1985). This finding implicated the operation of at least two distinct export pathways; one that mediated translocation of Pel, Peh, and Cel, and the other export of Prt. Subsequent studies have confirmed this. For example, cells of *E. coli* harbouring a plasmid-borne Echr *prt* gene exported Prt during exponential growth, and in the absence of lysis or periplasmic leakiness (Barras *et al*. 1986; Wandersman *et al*. 1987). The cells of *E. coli* that exported Prt were not proficient in Pel export since the enzyme was localized in the periplasm (Barras *et al*. 1986). These characteristics are consistent with a one-step model that postulates that the exported proteins concurrently translocate through the cytoplasmic membrane and the outer membrane. In this model, the precursors of exported proteins may accumulate in the membrane, but not in the periplasm, if the export process is interrupted. This appeared to be the case in the export of exotoxin A in *P. aeruginosa* (Lory *et al*. 1983).

Molecular cloning of *out* genes and their genetic organization

Physical analysis of the sequences flanking Tn5 in Out⁻ Echr mutants indicated that the export defective phenotype resulted from insertions of Tn5 into different chromosomal sites (Thurn and Chatterjee 1985). Subsequently, Van Gijsegem (1987) reported that *out* mutations in strain 3937 mapped in at least three different chromosomal regions. Consistent with this genetic organization is the isolation of only one *out* gene, *outJ*, as opposed to an *out* cluster, from the Echr strain 3937 (Ji *et al*. 1989). In contrast to the scattered localization of the *out* genes in Echr strain 3937 preliminary evidence in Ecc193 suggested a clustering of two *out* genes (Chatterjee *et al.* 1985). Subsequently, cosmid clones carrying several *out* genes were isolated from Ecc71 (Murata *et al.* 1988). The plasmids formed four complementation groups with regard to their ability to restore the Out phenotype in *Erwinia* mutants. One plasmid (pAKC601) restored the Out phenotype in every mutant from Eca, Ecc, and Echr derived by ethyl methanesulfonate, Tn*5*, Tn*10*, Tn*10-lacZ*, or MudI-Ap*lac* mutagenesis. We, therefore, believe this plasmid carries all or most of the *out* genes. The other plasmids probably contain only a subset of the *out* genes and, therefore, complement only certain Out mutants. The presence of a cluster of *out* genes in pAKC601 is also indicated by restoration of the Out phenotype in EC16 mutants that were shown to have Tn*5* insertions at different chromosomal sites (see above and Thurn and Chatterjee 1985), and by the complementation of mutants independently derived from several Ecc wildtype strains as well as from an Eca strain. A clustering of the *out* genes has been noted in another strain of Ecc (G.P. Salmond, personal communication). Further analysis of the 15.7 Kb DNA segment in pAKC601, the plasmid carrying the Ecc71 *out* cluster, is in progress to define clearly the limits of the *out* genes and each of the transcriptional units.

Despite functional homology in the export systems of Ecc and Echr, indicated by the restoration of enzyme export in Echr mutants by Ecc Out$^+$ DNA, the export systems appear to have diverged genetically. In Southern hybridizations there were no detectable DNA sequence homologies between *out* genes of these bacteria. Furthermore, the nonclustering of the *out* genes in some Echr strains probably reflects reassortment of the genes in the course of the evolution of the export systems. The biological significance, if any, of the two different genetic organizations awaits elucidation.

The two-step export model predicts extracytoplasmic localization of the *out* proteins where they can interact with nascent peptides emerging through the cytoplasmic membrane. Noteworthy in this context is the finding that some Out$^-$ mutants of Echr lacked periplasmic proteins (Ji *et al.* 1987; Thurn and Chatterjee 1985); the product of the *outJ* gene was also found to be a secreted (periplasmic) protein (Ji *et al.* 1989). Moreover, the PhoA$^+$ phenotype of an Ecc strain, HC131, containing an *out*$^-$ Tn*phoA* fusion (Hinton and Salmond 1987) implicates extracytoplasmic proteins in enzyme export.

The tight coupling of enzyme synthesis and enzyme export, observed in early kinetic analyses of Echr (Chatterjee *et al.* 1979), had suggested that these two processes may be co-regulated. Studies thus far, however, indicate that some of the *out* genes are expressed constitutively and, unlike some Pels, are not markedly inducible by pectate or pectin catabolic intermediates. The evidence for this is the pattern of expression of *out-lacZ* fusions of Echr (Ji *et al.* 1987) and Ecc (our unpublished results). In each instance there is a high basal level of the reporter gene product that is not stimulated further when cells are grown with either pectate or pectin. Despite this evidence, it would be premature at this juncture to surmize that none of the *out* genes respond to conditions that induce the synthesis of exo-proteins. We should note that only a few of the many *out* genes have been analysed so far. Using the cloned Ecc71 *out* cluster, we are currently defining each of the transcriptional units and testing the inducing signals from pectate, pectin, and other plant components. Such analyses may very well confirm that all of the *out* genes are expressed constitutively to produce stoichiometric amounts of each Out component, yielding a functional Out system under 'all' physiological conditions. This system would ensure export of the Pel, Cel, and Peh polypeptides synthesized under noninducing or inducing conditions. Noteworthy in this context are the findings that the endo-Peh of Ecc (Willis *et al.* 1987), and PelB and PelC of Echr strain EC16 (see above) are present at a high basal level, and these enzymes are exported out of the cell. In the absence of a functional export system these proteins could accumulate in the cell (= periplasm) at nonphysiological levels. A constitutive Out system may, therefore, have evolved to alleviate such potential physiological constraints. This hypothesis suggests that the Out function is not the limiting factor in the production of extracellular enzymes and predicts that the enzymes share a common export system. The latter view is supported by the consistent recovery of mutants pleiotropically defective in the export of Pel, Peh, and Cel, although the export of each protein is affected to a different degree. Moreover, restoration of the Out phenotype in Eca, Ecc, and Echr by a cluster of Ecc71 *out* genes provides additional evidence for interaction of the Out system with the exoproteins of these bacteria. The Pel, Peh, and Cel polypeptides of Ecc, Eca, and Echr must share similarities to allow their recognition and subsequent translocation across the cell envelope by the Ecc71 Out system.

Pectin lyase

Background

Surveys for pectolytic activities in the past had consistently revealed production of Pels by bacteria and Pnls by fungi (see for example, Fogarty and Ward 1974), thereby promoting the belief that production of Pnl was restricted to fungi or other eukaryotic organisms. However, during the 1970s and early 1980s several reports appeared (Tomizawa and Takahashi 1971; Kamimiya *et al.* 1972; Kamimiya *et al.* 1974; Itoh *et al.* 1980; Itoh *et al.* 1982) describing production of Pnls in *Erwinia* spp. in response to DNA damaging agents such as nalidixic acid, mitomycin C, and UV light. This regulation was considered an unusual response resulting from insertion of a *pnl* gene into a bacteriophage or a defective bacteriophage (= bacteriocin) genome. This notion was supported by the finding that bacteriocins (= carotovoricins) and Pnl were co-regulated in Ecc by DNA damaging agents. Subsequent work (Tsuyumu and Chatterjee 1984) revealed that (i) Pnl induction by DNA damaging agents was common to most soft-rotting *Erwinia* spp., although the extent of induction varied depending upon the strain, and (ii) Pnl was induced along with a bacteriophage or bacteriocin, depending upon the lysogenic or bacteriocinogenic state of the bacterium. Moreover, in most cases the cultures treated with DNA damaging agents generally lysed after one doubling, again depending upon the dosage of the DNA damaging agents. In the Ecc strain Ecc71, wherein various aspects of Pnl regulation have been studied, Pnl production is co-induced with culture lysis (Lss) and carotovoricin (Ctv) production. Investigations of Pnls and Pels of *Erwinia* species have revealed significant differences in their regulatory and enzymological properties, some of which are listed in Table 11.2. Pnl, like most Pel species with an endo-mode of substrate cleavage, however, can macerate plant tissues.

The hypothesis that *pnl* was inserted into a 'defective bacteriophage' genome predicted linkage between genes for *ctv*, *lss*, and *pnl*. This was tested by (i) determining the kinetics of induction of these phenotypes; (ii) Tn5 mutagenesis; and (iii) by molecular cloning of *pnl* DNA.

Kinetics of nalidixic-induced Pnl and Ctv production and culture lysis, however, suggested a noncoordinate induction in that the induced Pnl production commenced one hour prior to that of Ctv and three hours prior to cellular lysis (defined as a decrease in culture turbidity). Tn5 insertions yielded mutants of Ecc71 defective in either Pnl or Ctv production in L broth cultures and potato tuber tissue (McEvoy *et al.* 1988). The mutants were not defective for the lysis phenotype. The Tn5 insertions that produced changes in Ctv or Pnl phenotypes were in chromosomal DNA and not in the 8.2 kb indigenous plasmid of Ecc71. No polar effects due to the Tn5 insertions were noted, implicating individual transcriptional units for the coinduced genes. Moreover, hybridizations between ColE1: :Tn5, and Southern blots of *Cla*I and *Sal*I genomic digests of the Tn5 mutants revealed no linkage between *pnl* and *ctv*. This finding was further supported by our inability to simultaneously clone *pnl* and *ctv* genes by cosmid cloning (see below).

Coregulation of unlinked genes (for example, *ctv* and *pnl*) by DNA damaging agents is reminiscent of the SOS regulon in *E. coli*. In the SOS system. DNA damage activates the product of the *rec*A gene, which

Table 11.2 Properties of pectin and pectate lyases of *Erwinia* species

Characteristics	Pectin lyases	Pectate lyases
Inducer(s)	DNA damaging agents such as : UV light, mitomycin C, nalidixic acid	Deoxyketouronate(s)
Susceptibility to catabolite repression	−	+
Coinduction with bacteriocins, bacteriophages and lysis	+	−
Requirement of RecA in gene expression	+	−
Preferred substrate	Hightly esterified pectin	Polygalacturonate
Tissue macerating ability	+	+
Requirement of Ca^{++} for catalytic activity	−	+
Isoelectric point	Basic	Acidic to basic
Mode of substrate cleavage	Endo	Endo and exo
Export	Unknown	Active

cleaves LexA, the repressor of a number of unlinked genes (see Ossana *et al.* 1986; Walker 1984). In Ecc strain 71 it has been shown that an SOS-like system exists (McEvoy *et al.* 1987) and that *recA* is required for Pnl and Ctv production along with cellular lysis (Zink *et al.* 1985). Studies with the *E. coli lexA* [+] DNA indicated that although it repressed the expression of several damage inducible Ecc71 genes, *pnl* was not repressed. These findings led to the hypothesis that Pnl regulation was mediated by a molecule other than the *lexA* product and one that was susceptible to RecA processing. To analyse further the regulatory components we, at this juncture, proceeded to (i) clone the structural gene, *pnlA*, (ii) construct *pnlA-lacZ* transcriptional fusions, (iii) isolate regulator mutants by monitoring alterations in β-galactosidase activity, and (iv) clone the putative regulatory genes.

Molecular cloning of *pnlA*

Initial attempts to clone *pnl* genes entailed the construction of gene libraries in cosmid vectors pHC79 (Hohn and Collins 1980) or pSF6 (Selvaraj *et al.* 1984) and the screening of *E. coli* clones for a Pnl⁺ phenotype. We failed to obtain a Pnl⁺ *E. coli* clone using either RecA⁺ or RecA⁻ strains. This finding was in sharp contrast to the occurrence of clones carrying *pel⁺*, *peh⁺*, *cel ⁺*, *out⁺*, and other Ecc genes. Since the *E. coli recA⁺* DNA is as effective as the Ecc *recA⁺* DNA in *pnl* expression (Zink *et al.* 1985), our inability to obtain a Pnl⁺ clone led us to consider that *pnlA* expression requires a transcriptional activator, encoded by *pnlR*, that is not linked to *pnlA*, and hence cannot be simultaneously cloned. In view of this possibility we decided to screen the gene library in Ecc strain Ecc193. This strain is inducible for Pnl but produces only one-tenth the activity produced by Ecc71. We argued that a gene dosage effect (that is multiple copies of *pnlA*) or a more efficient activator (that is, *pnlR⁺* of Ecc71) might result in a higher level of Pnl activity in Ecc193. Indeed, by mobilizing the gene bank into Ecc193 and screening the transconjugants for Pnl activity, we detected two that produced higher levels of the enzyme upon induction by DNA damaging agents (McEvoy, *et al.* 1988). The plasmids also restored Pnl production in all of the Tn5 insertion mutants of Ecc71. The DNA responsible for Pnl production was localized on a 3.4 kb *Eco*RI fragment. *E. coli* strains carrying this fragment cloned into pUC derivatives, pTZ vectors or pBluescript produced low basal levels of pnl activity; the gene, however, remained noninducible by DNA damaging agents in RecA⁺ and RecA⁻ *E. coli* strains. In addition, merodiploid analysis with *pnl⁺* plasmids and with *pnl-lacZ* plasmids (see below for the construction of *lacZ* transcriptional fusions) established that the DNA segment carried the structural gene, *pnlA* and not the regulatory gene(s).

Construction of *pnlA-lacZ* transcriptional fusions and isolation of Pnl regulatory mutants

pnlA–lacZ transcriptional fusions were constructed using the transposase deficient element Mu dI1734 (Castilho *et al.* 1984). By examining the inducible β-galactosidase and Pnl levels and by mapping the insertions, we localized the *pnlA* gene to a 0.9 kb DNA stretch and determined the direction of transcription. One representative fusion was introduced into the LacZ⁻ Ecc71 derivative, AC5006, and the chromosomal copy of the *pnlA* gene was replaced by marker-exchange-recombination with the *pnlA-lacZ* DNA. The resulting strain, AC5022, was Pnl⁻ and produced inducible β-galactosidase in response to DNA damaging agents. Our date revealed that the induction of Pnl by DNA damaging agents results from transcription of *pnlA* and that RecA activity is required for this transcription. We then proceeded to look for mutants with altered regulation of β-galactosidase activity and to determine the presence of activators of *pnlA* transcription in plant tissue.

Following mutagenesis of AC5022 with ethyl methanesulfonate, we isolated a mutant wherein β-galactosidase production no longer occurred in the presence of mitomycin C or other DNA damaging agents. The phenotype was not restored in cells carrying the *recA*$^+$ plasmid. Moreover, the introduction of a *pnlA*$^+$ plasmid did not result in a Pnl$^+$ phenotype suggesting that the mutational effect occurs *in trans*. At present we are screening the gene bank for a DNA segment that would restore induction of *pnlA-lacZ* by DNA damaging agents.

In our earlier studies with RecA$^+$ and RecA$^-$ derivatives of Ecc71 we had detected production of Pnl in potato tuber tissue infected with RecA$^+$ strains but not with RecA$^-$ strains. These data suggested the presence, in infected tubers, of plant metabolites that activated *pnlA* transcription by utilizing the RecA function. With the availability of a sensitive assay for the reporter gene product (i.e., β-galactosidase) the induction of Pnl by plant metabolites was re-examined. Crude plant extracts were incorporated into a minimal salts medium. Bacterial cells grown in this medium were collected and β-galactosidase activities determined in permeabilized cells. There was no detectable activity in cells grown in minimal salts medium. When this medium was supplemented with extracts from celery, lettuce, or spinach, levels of β-galactosidase ranged from 195 to 293 Miller units (Tsuyumu *et al*. 1989). The extract from potato tuber tissue was least effective in that only 19 units of β-galactosidase activity was present. With the strain AC5006/pSK1002 (carrying *lacZ* fused with the SOS gene *umuC*, a gene whose function is required for the induction of mutations by UV) the β-galactosidase levels were consistently higher compared to the strain carrying a chromosomal copy of the *pnlA-lacZ* fusion, although the pattern of induction was the same. The higher response with the *umuC-lacZ* strain may simply reflect a gene dosage effect. The data indicate the occurrence in these plant tissues of metabolites that can activate the SOS system as well as *pnlA* transcription.

The progress that has occurred with Ecc71 has now paved the way for confirmation and clarification of several aspects of the Pnl regulatory network. For example, the occurrence of PnlR and the function of the PnlR polypeptide have to be established. It also remains to be determined whether the proteolytic activity of RecA is responsible for generation of a transcriptional activator. Another unresolved issue is the regulation of *pnlR* itself. It is conceivable that the level of PnlR may be the rate limiting step in Pnl production. In fact, the differences in the quantity of PnlR is one explanation for the variability in the extent of Pnl induction in various soft-rotting *Erwinia* (see Tsuyumu and Chatterjee 1984).

Evidence for the synthesis of DNA damaging agents by plants, either naturally or upon exposure to physical or biotic agents, has been documented (see Ames 1983; Tsuyumu *et al.* 1985; Sun *et al.* 1989). The issues that remain unresolved are (i) whether the synthesis of such components is activated upon infection by the soft-rotting bacteria, or induced by other factors at some stage during the disease cycle; and (ii) what effect these putative defence related metabolites may have upon the physiology of the pathogen, including production of the primary virulence factor, that is, the pectate lyases. This knowledge is critical in understanding the rationale for the activation of an auxiliary system that appears to augment the pathogenic potential of a pectic enzyme system which is already remarkable for its complexity and destructive potential.

Acknowledgements

Research in our laboratories is supported by grants from the National Science Foundation (grant DBM-8796262), the United States Department of Agriculture (grants 87-CRCR-1-2504 and 87-CRCR-2352), and the Food for the 21st Century program of the University of Missouri.

References

Ames, B.N. (1983). Dietary carcinogens and anticarcinogens: oxygen radicals and degenerative diseases. *Science* **221**, 1256–64.

Andro, T., Chambost, J.P., Kotoujansky, A., Cattaneo, J., Bertheau,Y., Barras, F., Van Gijsegem, F., and Coleno, A. (1984). Mutants of *Erwinia chrysanthemi* defective in secretion of pectinase and cellulase. *Journal of Bacteriology* **160**, 1199–203.

Barras, F. and Chatterjee, A.K. (1987). Genetic analysis of the *pelA-pelE* cluster encoding the acidic and basic pectate lyases in *Erwinia chrysanthemi* EC16. *Molecular and General Genetics* **209**, 615–17.

Barras, F., Thurn, K.K., and Chatterjee, A.K. (1986). Export of *Erwinia chrysanthemi* (EC16) protease by *Escherichia coli*. *FEMS Microbiology Letters* **34**, 343–8.

Barras, F., Thurn, K.K., and Chatterjee, A.K. (1987). Resolution of four pectate lyase structural genes of *Erwinia chrysanthemi* (EC16) and characterization of the enzymes produced in *Escherichia coli*. *Molecular General Genetics* **209**, 319–25.

Beaulieu, C. and Van Gijsegem, F. (1990). Identification of plant-inducible genes in *Erwinia chrysanthemi* 3937. *Journal of Bacteriology* **172**, 1569–75.

Beraha, L. and Garber, E.D. (1971). Avirulence and extracellular enzymes of *Erwinia caratovora*. *Phytopathologische Zeitschrift* **70**, 335–44.

Boccara, M., Diolez, A., Rouve, M., and Kotoujansky, A. (1988). Role of individual pectate lyases of *Erwinia chrysanthemi* strain 3937 in pathogenicity on Saintpaulia plants. *Physiological Molecular Plant Pathology* **33**, 95–104.

Castilho, B.A., Olfson, P., and Casadaban, M.J. (1984). Plasmid insertion mutagenesis and *lac* gene fusion with mini Mu bacteriophage transposons. *Journal of Bacteriology* **158**, 488–95.

Chatterjee, A.K., Buchanan, G.E., Behrens, M.K., and Starr, M. P. (1979). Synthesis and excretion of polygalacturonic acid *trans*-eliminase in *Erwinia, Yersinia*, and *Klebsiella* species. *Canadian Journal of Microbiology* **25**, 94–102.

Chatterjee, A.K., Ross, L.M., McEvoy, J.M., and Thurn, K.K. (1985a). pULB113, and RP4::mini-Mu plasmid, mediates chromosomal mobilization and R-prime formation in *Erwinia amylovora, Erwinia chrysanthemi*, and subspecies of *Erwinia carotovora*. *Applied and Environmental Microbiology* **50**, 1–9.

Chatterjee, A.K. and Starr, M.P. (1977). Donor strains of the soft-rot bacterium *Erwinia chrysanthemi* and conjugational transfer of pectolytic capacity. *Journal of Bacteriology* **132**, 862–9.

Chatterjee, A.K. and Starr, M.P. (1980). Genetics of *Erwinia* species. *Annual Review of Microbiology* **34**, 645–76.

Chatterjee, A.K., Thurn, K.K., and Tyrell, D.J. (1985b). Isolation and characterization of Tn5 insertion mutants of *Erwinia chrysanthemi* that are deficient in polygalacturonate catabolic enzymes oligogalacturonate lyase and 3-deoxy-D-glycero-2,5-hexodiulosonate dehydrogenase. *Journal of Bacteriology* **162**, 708–14.

Chatterjee, A.K. and Vidaver, A.K. (1986). Genetics of pathogenicity factors: Application to phytopathogenic bacteria. In *Advances in plant pathology* (ed. D. S. Ingram and P. H. Williams), Vol. 4, pp. 153–70. Academic Press, London.

Collmer, A. and Bateman, D.F. (1981). Impaired induction and self-catabolite repression of extracellular pectate lyase in *Erwinia chrysanthemi* mutants deficient in oligogalacturonide lyase. *Proceedings of the National Academy of Sciences USA* **78**, 3920–4.

Collmer, A. and Keen, N.T. (1986). The role of pectic enzymes in plant pathogenesis. *Annual Review of Phytopathology* **24**, 383–409.

Collmer, A., Schoedel, C., Roeder, D.L., Ried, J.L., and Rissler, J.F. (1985). Molecular cloning in *Escherichia coli* of *Erwinia chrysanthemi* genes encoding multiple forms of pectate lyase. *Journal of Bacteriology* **161**, 913–20.

Condemine, G. and Robert-Baudouy, J. (1987). Tn5 insertions in *kdgR*, a regulatory gene of the polygalacturonate pathway in *Erwinia chrysanthemi*. *FEMS Microbiology Letters* **42**, 39–46.

Condemine, G., Hugouvieux-Cotte-Pattat, N., and Robert-Baudouy, J. (1986). Isolation of *Erwinia chrysanthemi kduD* mutants altered in pectin degradation. *Journal of Bacteriology* **165**, 937–41.

Daniels, M.J., Barber, C.E., Turner, P.C., Cleary, W.G., and Sawczyc, M.K. (1984). Isolation of mutants of *Xanthomonas campestris* p.v. *campestris* showing altered pathogenicity. *Journal of General Microbiology* **130**, 2447–55.

Dow, J.M., Scofield, G., Trafford, K., Turner, P.C., and Daniels, M.J. (1987). A gene cluster in *Xanthomonas campestris* pv. *campestris* required for pathogenicity controls the excretion of polygalacturonate lyase and other enzymes. *Physiological and Molecular Plant Pathology* **31**, 261–71.

Ferro-Novick, S., Honma, M., and Beckwith, J. (1984). The product of gene *secC* is involved in the synthesis of exported protein in *E. coli*. *Cell* **38**, 211–17.

Fogarty, W.M. and Ward, O.P. (1974). Pectinases and pectic polysaccharides. *Progress in Industrial Microbiology* **13**, 61–119.

Hinton, J.C.D. and Salmond, G.P.C. (1987). Use of Tn*phoA* to enrich for extracellular enzyme mutants of *Erwinia carotovora* subspecies *carotovora*. *Molecular Microbiology* **1**, 381–6.

Hohn, B. and Collins, J. (1980). A small cosmid for efficient cloning of large DNA fragments. *Gene* **11**, 291–8.

Hugouvieux-Cotte-Pattat, N. and Robert-Baudouy, J. (1985). Isolation of *kdgK-lac* and *kdgA-lac* gene fusions in the phytopathogenic bacterium *Erwinia chrysanthemi*. *Journal of General Microbiology* **131**, 1205–11.

Hugouvieux-Cotte-Pattat, N., Reverchon, S., Condemine, G., and Robert-Baudouy, J. (1986). Regulatory mutations affecting the synthesis of pectate lyase in *Erwinia chrysanthemi*. *Journal of General Microbiology* **132**, 2099–106.

Itoh, Y., Izaki, K., and Takashi, H. (1980). Simultaneous synthesis of pectin lyase and carotovoricin induced by mitomycin C, nalidixic acid or ultraviolet light irradiation in *Erwinia carotovora*. *Agricultural and Biological Chemistry* **44**, 1135–40.

Itoh, Y., Sugiura, J., Izaki, K., and Takahashi, H. (1982). Enzymological and immunological properties of pectin lyases from bacteriocinogenic strains of *Erwinia carotovora*. *Agricultural and Biological Chemistry* **46**, 199–205.

Ji, J., Hugouvieux-Cotte-Pattat, N., and Robert-Baudouy, J. (1987). Use of Mu-*lac* insertions to study the secretion of pectate lyase by *Erwinia chrysanthemi*. *Journal of General Microbiology* **133**, 793–802.

Ji, J., Hugouvieux-Cotte-Pattat, N., and Robert-Baudouy, J. (1989). Molecular cloning of *outJ* gene involved in pectate lyase secretion by *Erwinia chrysanthemi*. *Molecular Microbiology* **3**, 285–93.

Kamimiya, S., Izaki, K., and Takahashi, H. (1972). A new pectolytic enzyme in *Erwinia aroideae* formed in the presence of nalidixic acid. *Agricultural and Biological Chemistry* **36**, 2367–72.

Kamimiya, S., Nishiya, T., Izaki, K., and Takahashi, H. (1974). Purification and properties of a pectin *trans*-eliminase in *Erwinia aroideae* formed in the presence of nalidixic acid. *Agricultural and Biological Chemistry* **38**, 1071–8.

Keen, N.T. and Tamaki, S. (1986). Structure of two pectate lyase genes from *Erwinia chrysanthemi* EC16 and their high-level expression in *Escherichia coli*. *Journal of Bacteriology* **168**, 595–606.

Keen, N.T., Dahlbeck, D., Staskawicz, B., and Belser, W. (1984). Molecular cloning of pectate lyase genes from *Erwinia chrysanthemi* and their expression in *Escherichia coli*. *Journal of Bacteriology* **159**, 825–31.

Kotoujansky, A. (1987). Molecular genetics of pathogenesis by soft-rot erwinias. *Annual Review of Phytopathology* **25**, 405–30.

Liao, C.-H., Hung, H.-Y., and Chatterjee, A.K. (1988). An extracellular pectate lyase is the pathogenicity factor of the soft-rotting bacterium *Pseudomonas viridiflava*. *Molecular Plant–Microbe Interactions* **1**, 199–206.

Lory, S., Tai, P.C., and Davis, B.D. (1983). Mechanism of protein excretion by gram-negative bacteria: *Pseudomonas aeruginosa* exotoxin A. *Journal of Bacteriology* **156**, 695–702.

McEvoy, J.L., Thurn, K.K., and Chatterjee, A.K. (1987). Expression of the *E. coli lexA⁺* gene in *Erwinia carotovora* subsp. *carotovora* and its effect on production of pectin lyase and carotovoricin. *FEMS Microbiology Letters* **42**, 205–8.

McEvoy, J.L., Murata, H., Engwall, J.K., and Chatterjee, A.K. (1988). Genetics of damage-inducible pectin lyase (PNL) of *Erwinia carotovora* subsp. *carotovora*: analysis of Tn5 insertion mutants and molecular cloning of *pnl* DNA. In *Molecular genetics of plant–microbe interactions*, (ed. R. Palacios and D. P. S. Verma). APS Press, St. Paul, pp. 257–8.

Murata, H., Fons, M., and Chatterjee, A.K. (1988). Molecular cloning of genes that specify enzyme export in *Erwina carotovora* subsp. *carotovora* and *E. chrysanthemi*. In *Molecular genetics of plant–microbe interactions*, (ed. R. Palacios and D. P. S. Verma). APS Press, St. Paul, pp. 259–60.

Ossanna, N., Peterson, K.R., and Mount, D.W. (1986). Genetics of DNA repair in bacteria *Trends in Genetics* **14**, 55–8.

Payne, J.H., Schoedel, C., Keen, N.T., and Collmer, A. (1987). Multiplication and virulence in plant tissues of *Escherichia coli* clones producing pectate lyase isozymes PLb and PLe at high levels and of an *Erwinia chrysanthemi* mutant deficient in PLe. *Applied and Environmental Microbiology* **53**, 2315–20.

Perombelon, M.C.M. and Kelman, A. (1980). Ecology of the soft rot erwinias. *Annual Review of Phytopathology* **18**, 361–87.

Pugsley, A.P. (1988). Protein secretion across the outer membrane of gram-negative bacteria. In *Protein Transfer and organelle biogenesis*, (ed. R.C. Das and P.W. Robbin), pp. 607–52. Academic Press, New York.

Pugsley, A.P. and Schwartz, M. (1985). Export and secretion of proteins by bacteria. *FEMS Microbiology Reviews* **32**, 3–38.

Reverchon, S., Huang, Y., Bourson, C., and Robert-Baudouy, J. (1989). Pectinolysis regulation in *Erwinia chrysanthemi*. In *Abstracts of the 7th International Conference on Plant Pathological Bacteria, Budapest, Hungary*. p. 197.

Reverchon, S. and Robert-Baudouy, J. (1987). Regulation of expression of pectate lyase genes *pelA*, *pelD*, and *pelE* in *Erwinia chrysanthemi*. *Journal of Bacteriology* **169**, 2417–23.

Reverchon, S., Van Gijsegem, F., Rouve, M., Kotoujansky, A., and Robert-Baudouy, J. (1986). Organization of a pectate lyase gene family in *Erwinia chrysanthemi*. *Gene* **49**, 215–24.

Ried, J.L. and Collmer, A. (1987). An *nptI-sacB-sacR* cartridge for constructing directed, unmarked mutations in Gram-negative bacteria by marker exchange-eviction mutagenesis. *Gene* **57**, 239–46.

Ried, J.L. and Collmer, A. (1988). Construction and characterization of an *Erwinia chrysanthemi* mutant with directed deletions in all of the pectate lyase structure genes. *Molecular Plant–Microbe Interactions* **1**, 32–8.

Roberts, D.P., Denny, T.P., and Schell, M.A. (1988). Cloning of the *egl* gene of *Pseudomonas solanacearum* and analysis of its role in pathogenicity. *Journal of Bacteriology* **170**, 1445–51.

Schell, M.A., Roberts, D.P., and Denny, T.P. (1988). Analysis of *Pseudomonas solanacearum* polygalacturonase encoded by *pglA* and involvement in phyto-pathogenicity. *Journal of Bacteriology* **170**, 4501–8.

Schoedel, C. and Collmer, A. (1986). Evidence of homology between pectate lyase-encoding *pelB* and *pelC* genes in *Erwinia chrysanthemi*. *Journal of Bacteriology* **167**, 117–23.

Selvaraj, G., Fong, Y.C., and Iyer, V.N. (1984). A portable DNA sequence carry-ing the cohesive site (*cos*) of bacteriophage lambda and the *mob* (mobilization) region of the broad-host-range plasmid RK2: a module for the construction of new cosmids. *Gene* **32**, 235–41.

Sun, T.J., Essenberg, M., and Melcher, U. (1989). Photoactivated DNA nicking, enzyme inactivation, and bacterial inhibition by sesquiterpenoid phytoalexins from cotton. *Molecular Plant–Microbe Interactions* **2**, 139–47.

Tamaki, S.J., Gold, S., Robeson, M., Manulis, S., and Keen, N.T. (1988). Struc-ture and organization of the *pel* genes from *Erwinia chrysanthemi* EC16. *Journal of Bacteriology* **170**,3468–78.

Tomizawa, H. and Takahashi, H. (1971). Stimulation of pectolytic enzyme form-ation of *Erwinia aroideae* by nalidixic acid, mitomycin C and bleomycin. *Agri-cultural and Biological Chemistry* **35**, 191–200.

Thurn, K.K. and Chatterjee, A.K. (1985). Single-site Tn5 chromosomal insertions affect the export of pectolytic and cellulolytic enzymes in *Erwinia chrysanthemi* EC16. *Applied and Environmental Microbiology* **50**, 894–8.

Thurn, K.K., Barras, F., Kegoya-Yoshino, Y., and Chatterjee, A.K. (1987). Pec-tate lyases of *Erwinia chrysanthemi*: PelE like polypeptides and *pelE* homolo-gous sequences in strain isolated from different plants. *Physiological and Molecular Plant Pathology* **31**, 429–439.

Tsuyumu, S. and Chatterjee, A.K. (1984). Pectin lyase production in *Erwinia chry-santhemi* and other soft-rot *Erwinia* species. *Physiological Plant Pathology* **24**, 291–302.

Tsuyumu, S., Funakubo, T., Hori, K., Takikawa, Y., and Goto, M. (1985). Pres-ence of DNA damaging agents in plants as the possible inducers of pectin lyase of soft-rot *Erwinia*. *Annals of the Phytopathological Society of Japan* **51**, 294–302.

Tsuyumu, S., Miura, M., Chatterjee, A.K., and McEvoy, J.L. (1989). Induction of pectin lyase and a SOS (*umuC*) gene in various plants. In *Abstracts of the 7th International Conference on Plant Pathological Bacteria, Budapest Hungary*. p. 195.

Van Gijsegem, F. (1987). Development of a chromosomal map of soft- rot Erwi-niae. In *Plant Pathogenic Bacteria: Proceedings of the 6th International Confer-ence on Plant Pathological Bacteria, Maryland*. (ed. E.L. Civerolo, A. Collmer, R.E. Davis, and A.G. Gillaspie), pp. 198–205. Martinus Nijhoff, Dordrecht.

Walker, G.C. (1984). Mutagenesis and inducible responses to deoxyribonucleic acid damage in *Escherichia coli*. *Microbiology Reviews* **48**, 60–93.

Wandersman, C., Delepelaire, P., Letoffe, S., and Schwartz, M. (1987). Characteri-zation of *Erwinia chrysanthemi* extracellular proteases: cloning and expression of the protease genes in *Escherichia coli*. *Journal of Bacteriology* **169**, 5046–53.

196   *Arun K. Chatterjee* et al.

Watanabe, T., Hayashi, S., and Wu, H.C. (1988). Synthesis and export of outer membrane lipoprotein in *Escherichia coli* mutants defective in generalized protein export. *Journal of Bacteriology* **170**, 4001–7.

Willis, J.W., Engwall, J.K., and Chatterjee, A.K. (1987). Cloning of genes for *Erwinia carotovora* subsp. *carotovora* pectolytic enzymes and further characterization of the polygalacturonases. *Phytopathology* **77**, 1199–205.

Zink, R.T., Engwall, J.K., McEvoy, J.L., and Chatterjee, A.K. (1985). *recA* is required in the induction of pectin lyase and carotovoricin in *Erwinia carotovora* subsp. *carotovora*. *Journal of Bacteriology* **164**, 390–6.

Addendum

Since the submission of this manuscript several significant advances have occurred in our investigations of the regulation and export of pectinases and other extracellular enzymes. Some of the salient findings are listed below.

The *E. carotovora* subsp. *carotovora* gene activating extracellular enzyme production (see the section on regulation), designated as *aep*, has been localized on a 1.0 kb DNA stretch. The *aep* product acts in *trans* to activate specifically the production of extracellular Pel, Cel, Peh, and Prt but not the periplasmic or cytoplasmic enzymes. Our recent data, although somewhat preliminary, suggest that the *aep* gene expression is stimulated by substances known to induce the production of the extracellular enzymes, in particular the Pels.

There is now substantial evidence indicating specific activation of pectolytic enzyme production by plant components. For example, *E. rhapontici* strain ER1 produces at least three basic Pel species exclusively during growth in plant tissue and not in artificial media supplemented with pectate or pectin. In EC16 derivatives deficient in genes for the four endo-Pel activities, new Pel species are produced following incubation in the presence of plant cell walls. These data provoke the hypothesis that Pel production in *Erwinia* spp. could occur via different regulatory circuits in response to diverse signals.

An *E. chrysanthemi* (EC16) Out⁺ gene cluster required for enzyme export (see the section on enzyme export) has been cloned. This gene cluster allows *E. coli* to export to the cell exterior Echr Pel and Peh proteins but not an Ecc71 Pel protein or periplasmic proteins such as β-lactamase. Thus, the export (Out) system reconstituted in *E. coli* appears to have the same recognition specificity as the Echr strain EC16. Additionally, sequence data for some of the *out* genes suggest that their products are extracytoplasmic.

With regard to the regulation of pectin lyase production by DNA-damaging agents (see the section on regulation), we now have cloned a gene, the product of which activates the transcription of the pectin lyase structural gene, pnlA. The activator gene function also appears necessary for the production of carotovoricin and the induction of cellular lysis. In light of this generalized effect the gene (previously designated as *pnlR*) has now been redesignated *digR* for the regulator of damage-inducible genes. We have shown that at least in *E. coli*, *digR* expression is regulated by the RecA–LexA pathway. The data suggest that *pnlA* expression is subject to a cascade type of regulation wherein LexA represses the gene encoding the transcriptional activator of *pnlA* and not *pnlA* itself. The RecA effect may therefore manifest at the level of *digR* expression and not DigR processing. This hypothesis is currently being tested.

12 The early events in *Agrobacterium* infection

CHARLES H. SHAW, GARY J. LOAKE,
ADRIAN P. BROWN, and CHRISTINE S.
GARRETT

Department of Biological Sciences, University of Durham, Science Laboratories, South Road, Durham DH1 3LE, UK

Introduction

There have been many articles in recent years in which the process of crown gall tumour formation has been reviewed (see for example Melchers and Hooykaas 1987; Zambryski *et al.* 1989). In this review, I shall summarize briefly the salient points of this process, and then concentrate on the very earliest events in the interaction. In my laboratory we are particularly interested in the role of chemotaxis in plant–microbe interactions, a topic which has received scant attention until very recently. Although most treatments of this topic commence with bacterial attachment to the plant cell, it seems clear that the bacteria do not originate there, but must arrive from somewhere else. Our studies have demonstrated the existence of a highly evolved method of identifying wounded cells and a system for guiding *A. tumefaciens* towards them. This involves a multifunctional system capable of triggering either chemotaxis, or gene-induction depending upon the ligand concentration. As a similar system seems to be present in *Rhizobium*, it may represent a fundamental, common feature in plant–microbe interactions.

Crown gall tumour

Crown gall tumour is a neoplastic overgrowth induced upon dicotyledonous plants by Ti-plasmid-harbouring *Agrobacterium tumefaciens*. *A. tumefaciens*, a Gram-negative soil bacterium, is prevalent in the rhizosphere (Kerr 1969; 1974) where often less than 1 per cent of the isolates are virulent (Kerr 1969). Wounding is a prerequisite for infection, a requirement explained by observations that wound-exuded phenolics, such as acetosyringone and sinapinic acid induce the Ti-plasmid operons for virulence (*vir*) *via* an interaction with the *virA & G* products (Okker *et al.* 1984;

Table 12.1 Functions of *Vir*-genes

Locus	Genes	Functions
A	1	Membrane-receptor for acetosyringone induction and chemotaxis
B	11	Transmembrane pore?
C	2	Overdrive binding?
D	4	Site-specific endonuclease
E	2	ss-DNA binding protein
G	1	Transcriptional activator for acetosyringone induction and chemotaxis

Stachel *et al.* 1985; 1986a; Stachel and Zambryski 1986; Winans *et al.* 1986; Melchers *et al.* 1986; Leroux *et al.* 1987).

The *vir* gene products are involved in transfer of the T-DNA to the plant cell (Table 12.1). The consensus view is that the *virD2* endonuclease cleaves within the 25bp repeats which flank the T-DNA (Stachel *et al.* 1986b; 1987; Wang *et al.* 1987; Albright *et al.* 1987) and a single stranded T-DNA intermediate is produced with the *virD2* protein covalently attached to the 5' end (Young and Nester 1988). This process may be enhanced by binding of *vir*C to overdrive (E.W. Nester, personal communication) a sequence that is adjacent to the T-DNA border. The single strand is stabilized by the *vir*E product (Gietl *et al.* 1987; Christie *et al.* 1988; Citovsky *et al.* 1988; Das 1988). The exact mechanism of T-DNA transfer, including the site of second strand synthesis and subsequent integration into the plant chromosome, is unknown. However, it has been proposed that the *vir*B products may function as a transmembrane pore, through which the T-DNA could pass (Engstrom *et al.* 1987; Ward *et al.* 1988; Thompson *et al.* 1988). Genes expressed from the integrated T-DNA effect an increased phytohormone production (Akiyoshi *et al.* 1984; Barry *et al.* 1984; Schröder *et al.* 1984; Thomashow *et al.* 1984; 1987; Buchmann *et al.* 1985; Kemper *et al.* 1985) and opine biosynthesis.

In addition to these functions determined by the Ti-plasmid a number of chromosomal genes are directly involved in tumour formation: *chvA* and *B*, required for β-1, 2-glucan biosynthesis (Douglas *et al.* 1985; Zorreguita *et al.* 1988); *chvD* & *ros*, involved in *vir* gene regulation (Close *et al.* 1985; 1987; Winans *et al.* 1988); *cel*, involved in cellulose fibril synthesis (Matthysse 1983); *pscA* which affects cyclic glucan and acidic succinoglycan biosynthesis (Cangelosi *et al.* 1987; Marks *et al.* 1987); and *att* which affects cell surface proteins (Matthysse 1987). Many of these loci are involved in the early interactions between *Agrobacterium* and the plant cell, particularly in the attachment process. However, the earliest event in the interaction is the initial sensing of chemoattractant chemicals by the bacterium in the rhizosphere.

Motility and chemotaxis in *Agrobacterium tumefaciens*

A. tumefaciens C58C[1] (Van Larebeke *et al*. 1974) is vigorously motile, with speeds up to 60 μm sec^{-1}, over runs in excess of 500 μm (Loake, *et al*. 1988). In comparison other strains, such as A136 (Watson *et al*. 1975) or LBA 4301 (Klapwijk *et al*. 1979) are very poor swimmers (Loake *et al*. 1988). Motility in unstimulated conditions is characterized by long, straight, or curvilinear, high speed runs with few tumbles. Moreover, there is a marked bias towards run curvature when the bacteria are close to glass: runs curve in a clockwise (CW) direction when the bacteria are close up against the coverslip, and counterclockwise (CCW) at the bottom of an indented slide. This pattern is distinct from that of *E. coli*, but reminiscent of *Rhizobium* (Götz and Schmitt 1987; Loake *et al*. 1988) and suggests that runs in these organisms are due to CW flagellar rotation and not CCW rotation as in *E. coli*. Further support for this proposal is provided by the fact that tethered *A. tumefaciens* cells rotate CW, and stop, but never CCW (Shaw unpublished). This suggests that *Agrobacterium*, like *Rhizobium*, employs undirectional flagellar rotation (Götz and Schmitt 1987; Loake *et al*. 1988).

In common with other bacteria *A. tumefaciens* when exposed to a chemical gradient, will migrate in a favourable direction. This behaviour can be monitored using either capillary (Alder 1973) or blindwell (Armitage *et al*. 1977) assays. A range of sugars and amino acids (Table 12.2) that are found in plant extracts and exudates (Kandler and Hopf 1980) are chemoattractants for *A. tumefaciens* C58C[1] (Loake *et al*. 1988). These fall into a number of classes which may be differentiated according to the molarity required for the peak response (Table 12.2). The most powerful and

Table 12.2 Groups of chemoattractants

Monosac -charide	Oligosac -charide	Amino acids	Optimum concn(M)
Glucose Fructose	Sucrose		10^{-6}
Galactose	Lactulose Maltose		10^{-5}
Arabinose	Raffinose Stachyose		10^{-4}
		Arginine Valine	10^{-3}
Xylose	Palatinose Lactose Cellobiose	Glycine Alanine Cysteine Methionine	Non-attrac- tant

sensitive response is that towards sucrose, which evokes a chemotactic maximum at 10^{-6}M, with a threshold at $<10^{-7}$M. These responses are up to 1000 times more sensitive than the equivalent attraction in *E. coli* (Adler 1973). Less easily reconciled is the differential attraction of *A. tumefaciens* towards Na$^+$ but not K$^+$ ions (Ashby *et al.* 1988).

In the *Enterobacteriaceae*, chemotactic responses are effected by a class of inner membrane proteins known as methyl-accepting chemotaxis proteins (MCPs; Springer *et al.* 1979). During the course of tactic responses, methyl groups are transferred from S-adenosyl methionine to specific sites in the highly conserved C-terminal domains of the MCPs (Kehry and Dahlquist 1982; Kehry *et al.* 1983; Terwilliger *et al.* 1986). The extent of methylation of the MCPs is determined by the balanced activities of the *cheR* methyltransferase, and the *cheB* methylesterase (Springer and Koshland 1977; Stock and Koshland 1978).

A number of pieces of evidence indicate that a similar system operates in *A. tumefaciens* (Loake *et al.* 1991) Methionine auxotrophs of *A. tumefaciens*, although fully motile, are non-chemotactic unless supplemented with methionine; the degree of restoration of chemotaxis by methionine analogues is proportional to their affinity for S-adenosyl-methionine synthetase; in wild type cells, transfer of methyl groups from methionine to MCP-like proteins can be detected; oligonucleotide probes complementary to the conserved domain of MCPs from *E. coli* hybridise to specific fragments in the *Agrobacterium* chromosome.

To elucidate the mechanism of chemotaxis in *Agrobacterium*, we have created using Tn5 mutogenesis a range of behavioural mutants (Loake and Shaw 1991). These mutants are deficient in flagellum production (*fla*), or motor function (*mot*) or generalized chemotaxis (*che*). Using CHEF pulsed-field gel electrophoresis, we are both establishing a chromosomal restriction map of *A. tumefaciens*, and mapping the Tn5 insertions. The Tn5 and flanking sequences have been isolated from six of the mutants, and they have been used as probes to clone 3 of the corresponding genes. Early indications are of extensive sequence similarity between *Agrobacterium* and *Rhizobium fla*, *mot* and *che* genes.

One of the mutants (*che-2*) is particularly interesting. It is non-chemotactic, motile, but tumbles excessively, and thus is prevented from moving in a favourable direction. In *E. coli* such mutants usually have lesions in either *cheZ* or *cheB*, and also show elevated levels of methylation of MCP. However, *che-2* exhibits reduced methylation of MCP (Loake *et al.* 1991). Thus it will be interesting to isolate the gene corresponding to this mutation.

The role of chemotaxis in *Agrobacterium*: plant interactions

The attraction of *Agrobacterium* towards sugars and amino acids in plant exudates may in part explain the prevalence of the bacterium in the rhizosphere. Accumulation of *A. tumefaciens* in the vicinity of plant roots was noted some time ago (Schroth *et al.* 1971). Excised root cap cells act as good attractants, and mutants have been isolated which are deficient in chemotaxis towards root cap cells or exudates (Hawes *et al.* 1988). Soluble factors from homogenates of the roots and shoots of tobacco, *Kalanchoë*, and wheat evoke strong chemotactic responses from *A. tumefaciens* (Ashby, *et al.* 1988). Furthermore, the presence of a Ti-plasmid enhances chemotaxis of *A. tumefaciens* towards soluble wheat factors. These results lend credence to an involvement of chemotaxis in the early interactions between *Agrobacterium* and plants. Moreover, the enhancement of chemotaxis by the Ti-plasmid suggests that Ti-plasmid functions provide in addition to their other roles, functions for chemotaxis. Finally, the results with wheat suggest that attraction towards the plant is not the failure in interactions of *Agrobacterium* with monocots.

Wounded plant tissues exude a mixture of aromatic and aliphatic chemicals (Stachel *et al.* 1985). Octopine and nopaline Ti-plasmids confer upon *A. tumefaciens* an ability to respond chemotactically to a range of *vir*-inducing phenolics including acetosyringone (Ashby *et al.* 1987; 1988; Shaw *et al.* 1988). Phenolics that are characteristic of plant wound exudates fall into three distinct groups (Table 12.3): strong *vir*-inducers that require the presence of Ti-plasmid for chemotaxis; weak or non-inducers that are attractants for cured and Ti-plasmid harbouring *A. tumefaciens*; and non *vir*-inducing, non-attracting compounds. Thus, *vir*-inducers are chemoattractants only in the presence of the Ti-plasmid, the responses they evoke being of a significantly greater sensitivity than those evoked by non-inducers. For strong *vir*-inducers, such as acetosyringone, sinapinic acid, and syringic acid, the response threshold is $<10^{-8}$M, with the maximum response at 10^{-7}M, some 100 fold lower than the concentration for maximum *vir*-induction (Stachel *et al.* 1985). Furthermore, for maximal chemotaxis and *vir*-induction, 4' hydroxyl (Bolton *et al.* 1986), 3' and 5' O-methyl, and 1' polar side chains are essential (Ashby *et al.* 1988). In addition to the highly sensitive Ti-plasmid-encoded phenolic receptor, there appears to be a separate, less sensitive, and chromosomally-encoded phenolic receptor (Ashby *et al.* 1987; 1988; Parke *et al.* 1987). It is possible that the lower sensitivity results from a poor fit of the agonist in a receptor that has evolved for some other attractant, e.g. tyrosine.

VirA and *G* are the Ti-plasmid loci required for chemotaxis towards acetosyringone (Shaw *et al.* 1988) suggesting a multifunctional role for the

Table 12.3 Chemotaxis and *vir*-induction by phenolics

Plant phenolic	Chemotaxis –concentration for maximum response (M)[a]	Ti-plasmid required[a]	*Vir*-inducer
1. *Vir*-inducers, requiring Ti-plasmid for chemotaxis			
Acetosyringone	10^{-7}	+	+[b]
Sinapinic acid	10^{-7}	+	+[b]
Syringic acid	10^{-7}	+	+[b]
Vanillin	10^{-4}	+	+[c]
Ferulic acid	10^{-4}	+	+[a]
3, 4-dihydroxy-benzoic acid	10^{-2}	+	+[c]
2. Weak/non-*vir*-inducers, chemoattractant for cured strain			
Catechol	10^{-2}	−	+/−[c, d]
p-hydroxy-benzoic acid	10^{-3}	−	+/−[c, d]
Vanillyl Alcohol	10^{-2}	−	−[a]
3, 4-dihydroxy-benzaldehyde	10^{-2}	−	−[a]
3. Non-*vir*-inducing, non-chemoattractants			
Vanillic acid	−	−	−[c]
Isovanillic acid	−	−	−[a]

Data taken from (a) Ashby *et al.* (1988); (b) Stachel *et al.* (1985); (c) Bolton *et al.* (1986); (d) Melchers and Hooykaas (1987).

virA/G system: at low concentrations of *vir*-inducer chemotaxis is triggered; at high concentrations the inducer effects *vir*-induction. *Vir*-induction does not appear to be required for chemotaxis; the low level of *virA/G* expressed constitutively (Rogowsky *et al.* 1987) would appear to be sufficient to effect chemotaxis. Moreover, under conditions optimal for chemotaxis, *vir*-induction is unlikely to occur. Indeed, *vir*-induction at high acetosyringone concentrations may even suppress chemotaxis or motility (Shaw *et al.* 1988) and thus prevent exit from the wound site.

Because the *virA/G* system is similar to other two-component chemoreceptor-regulator systems, such as *envZ/ompR* and *ntrB/C* (Winans *et al.* 1986; Melchers *et al.* 1986; Leroux *et al.* 1987), a model for *vir*-induction has been proposed in which *virA* is an inner membrane chemoreceptor that transduces a signal to *virG* functioning in the cytoplasm as a transcriptional regulator. *VirA* resembles MCPs structurally (Leroux *et al.* 1987) and has sequence conservation with *cheA* (Stock *et al.* 1988). *VirG* also possesses similarity to *cheB* & Y (Winans *et al.* 1986; Melchers *et al.* 1986). Perhaps *vir*A functions as a binding protein or MCP during chemotaxis, and *vir*G transmits a signal to the flagellar apparatus. NtrB is a protein kinase, involved in a phosphorylation cascade that regulates expression of glutamine synthetase (Ninfa and Magasanik 1986; Ninfa *et al.* 1987), and *che*A appears to transmit the excitatory signal to the flagellar motor by phosphorylation/dephosphorylation events involving *che*B & Y (Hess *et al.* 1988; Oosawa *et al.* 1988). Thus, *vir*A/G may be part of a phosphorylation cascade, with a branch point allowing chemotaxis or *vir*-induction.

As most of the chemotaxis and motility functions are chromosomally encoded, *vir*A and G must interact with this system to mediate chemotaxis towards *vir*-inducers. Acetosyringone chemotaxis can be conferred upon *E. coli*, but only *vir*A is required (Shaw *et al.* 1989). Presumably an *E. coli* function complements for *vir*G. Introduction of cosmid clones from the *vir*-region into defined *E. coli* mutants has demonstrated that acetosyringone chemotaxis requires at least *mot*A and B, *che*A, B, Y, and Z (Shaw *et al.* 1991). This suggests that some features of the chemotaxis systems of *E. coli* and *A. tumefaciens* are functionally conserved.

Chemotaxis in *Rhizobium*

Recently, it has been discovered that certain flavones, found in legume root exudates are chemoattractants for *Rhizobium* (Caetano-Anolles *et al.* 1988; Aguilar *et al.* 1988; Armitage *et al.* 1988). These same flavones are also inducers of the *nod* genes on the Sym plasmids in *Rhizobium*, that are concerned with nodulation (Rossen *et al.* 1987). Both *nod*-induction and chemotaxis, in response to flavones, require *nod*D (Caetano-Anolles *et al.* 1988). Different *Rhizobium* strains seem to recognize a different spectrum of flavones as attractants, consequently chemotaxis may be one factor in determination of host-range in *Rhizobium* legume interactions. The commonality of gene regulation and chemotaxis in *Agrobacterium* and *Rhizobium* in response to plant exudates, suggests that this type of response may be a common factor in bacterial–plant interactions.

Agrobacterium in the rhizosphere

A scenario describing the behaviour of *A. tumefaciens* in the rhizosphere can now be postulated (Shaw *et al.* 1986). The prevalence of *A. tumefaciens* in this habitat is partly explained by an attraction to plant saccharides. This highly sensitive response is shown by cured and Ti-plasmid containing strains alike, allowing both to accumulate in the vicinity of the plant. However, Ti-plasmid harbouring strains possess an additional receptor system, permitting them to respond to *vir*-inducing phenolic wound exudates. This attraction is probably only significant on the micrometre scale, both in the

204 Charles H. Shaw et al.

rhizosphere or on the rhizoplane. The bacteria will migrate up the concentration gradient, towards the wounded plant cells where the inducer concentrations will be highest, thus effecting *vir*-induction. Therefore, *A. tumefaciens* has evolved an extremely sensitive mechanism through which it will be attracted to and guided towards plant wounds. Acetosyringone release is an active process (M. Van Montagu and A. Darvill, personal communication) and it can be surmised that this mechanism has evolved to filter out signals from leaf litter, a potential source of phenolic lignin-degradation products. Thus *A. tumefaciens* would be drawn towards susceptible and not dying cells in a wound site.

References

Adler J. (1973). A method for measuring chemotaxis and use of the method to determine optimum conditions for chemotaxis by *Escherichia coli*. *Journal of General Microbiology* **74**, 77–91.

Aguilar, J.M.M., Ashby, A.M., Richards, A.J.M., Loake, G.J., Watson, M.D., and Shaw, C.H. (1988). Chemotaxis of *Rhizobium leguminosarum biovar phaseoli* towards flavonoid inducers of the symbiotic nodulation genes. *Journal of General Microbiology* **134**, 2741–6.

Akiyoshi, D.E., Klee, H., Amasino, R.M., Nester, E.W., and Gordon, M.P. (1984). The T-DNA of *Agrobacterium tumefaciens* encodes an enzyme of cytokinin biosynthesis. *Proceedings of the National Academy of Sciences USA* **81**, 5994–8.

Albright, L.M., Yanofsky, M.F., Leroux, B., Ma, D., and Nester, E.W. (1987). Processing of the T–DNA of *Agrobacterium tumefaciens* generates border nicks and linear single-stranded T-DNA. *Journal of Bacteriology* **169**, 1046–55.

Armitage, J.P., Josey, D.P., and Smith, D.G. (1977). A simple quantitative method for measuring chemotaxis and motility in bacteria. *Journal of General Microbiology* **102**, 199–202.

Armitage, J.P., Gallagher, A., and Johnston, A.W.B. (1988). Comparison of the chemotactic behaviour of *Rhizobium leguminosarum* with and without the nodulation plasmid. *Molecular Microbiology* **2**, 743–9.

Ashby, A.M., Watson, M.D., and Shaw, C.H. (1987). A Ti-plasmid determined function is responsible for chemotaxis towards the plant wound product acetosyringone. *FEMS Microbiology Letters* **41**, 189–92.

Ashby, A.M., Watson, M.D., Loake, G.J., and Shaw, C.H. (1988). Ti-plasmid specified chemotaxis of *Agrobacterium tumefaciens* C58C[1] to *vir*-inducing phenolics and soluble factors from monocotyledonous and dicotyledonous plants. *Journal of Bacteriology* **170**, 4181–7.

Barry, G.F., Rogers, D.A., Fraley, R.T., and Brand, L. (1984). Identification of a cloned cytokinin biosynthetic gene. *Proceedings of the National Academy of Sciences USA* **81**, 4776–80.

Bolton, G.W., Nester, E.W., and Gordon, M.P. (1986) Plant phenolic compounds induce expression of the *Agrobacterium tumefaciens* loci needed for virulence. *Science* **232**, 983–5.

Buchmann, I., Marner, F.J., Schröder, G., Waffenschmidt, S., and Schröder, J. (1985) Tumour genes in plants: T-DNA encoded cytokinin biosynthesis. *EMBO Journal* **4**, 853–9.

Caetano-Anolles, G., Crist-Estes, D.K., and Bauer, D. (1988). Chemotaxis of *Rhizobium meliloti* to the plant flavone luteolin requires functional nodulation genes. *Journal of Bacteriology* **170**, 3164–9.

Cangelosi, G.A., Hung, L., Puvanesarajah, V., Stacey, G., Ozga, D.A., Leigh, J.A., and Nester, E.W. (1987). Common loci for *Agrobacterium tumefaciens* and *Rhizobium meliloti* exopolysaccharide synthesis and their roles in plant interaction. *Journal of Bacteriology* **169**, 2086–91.

Christie, P.J., Ward, J.E., Winans, S., and Nester, E.W. (1988). The *Agrobacterium tumefaciens virE2* gene product is a single stranded DNA binding protein that associates with T-strands. *Journal of Bacteriology* **170**, 2659–67.

Citovsky, V., DeVos, G., and Zambryski, P. (1988). Single stranded DNA binding protein encoded by the *virE* locus of *Agrobacterium tumefaciens*. *Science* **240**, 501–4.

Close, T.J., Tait, R.C., and Kado, C.I. (1985). Regulation of Ti-plasmid virulence genes by a chromosomal locus of *Agrobacterium tumefaciens*. *Journal of Bacteriology* **164**, 774–81.

Close, T.J., Rogowsky, P.M., Kado, C.I., Winans, S.C., Yanofsky, M.F., and Nester, E.W. (1987). Dual control of *Agrobacterium tumefaciens* Ti-plasmid virulence genes. *Journal of Bacteriology* **169**, 5113–18.

Das, A. (1988). The *A. tumefaciens virE* encodes a single stranded DNA binding protein. *Proceedings of the National Academy of Sciences USA* **85**, 2909–13.

Douglas, C., Staneloni, R.J., Rubin, R.A., and Nester, E.W. (1985) Identification and genetic analysis of an *Agrobacterium tumefaciens* chromosomal virulence region. *Journal of Bacteriology* **161**, 850–60.

Engstrom, P., Zambryski, P., Van Montagu, M., and Stachel, S.E. (1987). Characterisation of *Agrobacterium tumefaciens* virulence proteins induced by the plant factor acetosyringone. *Journal of Molecular Biology* **197**, 635–45.

Gietl, C., Koukolikova-Nicola, Z., and Hohn, B. (1987). Mobilization of T-DNA from *Agrobacterium* to plant cells involves a protein that binds single stranded DNA. *Proceedings of the National Academy of Sciences USA* **84**, 9006–10.

Götz, R. and Schmitt, R. (1987). *Rhizobium meliloti* swims by unidirectional, intermittent rotation of right-handed flagellar helices. *Journal of Bacteriology* **169**, 3146–50.

Hawes, M.C., Smith, L.Y., and Howarth, A.J. (1988). *Agrobacterium tumefaciens* mutants deficient in chemotaxis to root exudates. *Molecular Plant–Microbe Interactions* **1**, 182–6.

Hess, J.F., Oosawa, K., Kaplan, N., and Simon, M.I. (1988). Phosphorylation of three proteins in the signalling pathway of bacterial chemotaxis. *Cell* **53**, 79–87.

Kandler, O. and Hopf, H. (1980). Occurrence, metabolism and function of oligosaccharides. In *The biochemistry of plants*, Vol. 3, pp.221–70. Academic Press, New York.

Kehry, M.R. and Dahlquist, F.W. (1982). The methyl-accepting chemotaxis proteins of *Escherichia coli*: identification of the multiple methylation sites on methyl-accepting chemotaxis protein I. *Journal of Biological Chemistry* **257**, 10378–86.

Kehry, M.R., Engstrom, P., Dahlquist, F.W., and Hazelbauer, G.L. (1983). Multiple covalent modifications of Trg, a sensory transducer of *Escherichia coli*. *Journal of Biological Chemistry* **258**, 5050–5.

Kemper, E., Waffenschmidt, S., Weiler, E.W., Rausch, T., and Schröder, J. (1985) T-DNA encoded auxin formation in crown gall cells. *Planta* **163**, 257–62.

Kerr, A. (1969). Crown gall of stone fruit. I. Isolation of *Agrobacterium tumefaciens* and related species. *Austrian Journal of Biological Sciences* **22**, 111–16.

Kerr, A. (1974). Soil microbiological studies on *Agrobacterium radiobacter* and biological control of crown gall. *Soil Science* **118**, 168–72.

Kersters, K. and De Ley, J. (1984) Genus III *Agrobacterium* Conn 1942. In *Bergeys manual of systematic bacteriology*, Vol. I. pp.244–54. The Williams and Wilkins Co., Baltimore.

Klapwijk, P.M., van Beelen, P., and Schilperoort, R.A. (1979). Isolation of a recombination deficient *Agrobacterium tumefaciens* mutant. *Molecular and General Genetics* **173**, 171–5.

Leroux, B., Yanofsky, M.F., Winans, S.C., Ward, J.E., Ziegler, S.F., and Nester, E.W. (1987). Characterisation of the *virA* locus of *Agrobacterium tumefaciens*: a transcriptional regulator and host range determinant. *EMBO Journal* **6**, 849–56.

Loake, G.J. and Shaw, C.H. (1991). Isolation and characterisation of chemotaxis and motility mutants of *Agrobacterium tumefaciens* (in preparation).

Loake, G.J., Ashby, A.M., and Shaw, C.H. (1988). Attraction of *Agrobacterium tumefaciens* C58C[1] towards sugars involves a highly sensitive chemotaxis system. *Journal of General Microbiology* **134**, 1427–32.

Loake, G.J., King, T.J., and Shaw, C.H. (1991). Involvement of a methionine-dependent MCP system in chemotactic responses of *Agrobacterium tumefaciens* (in preparation).

Marks, J.R., Lynch, T.J., Karlinsey, J.E., and Thomashow, M.F. (1987). *A. tumefaciens* virulence locus *pscA* is related to the *Rhizobium meliloti exoC* locus. *Journal of Bacteriology* **169**, 5835–7.

Matthysse, A.G. (1983). Role of bacterial cellulose fibrils in *Agrobacterium tumefaciens* infection. *Journal of Bacteriology* **154**, 906–15.

Matthysse, A.G. (1987). Characterisation of nonattaching mutants of *Agrobacterium tumefaciens*. *Journal of Bacteriology* **169**, 313–23.

Melchers, L.S. and Hooykaas, P.J.J. (1987). Virulence of *Agrobacterium*. *Oxford Surveys of Plant Molecular and Cell Biology* **4**, 167–220.

Melchers, L.S., Thompson, D.V., Idler, K.B., Schilperoort, R.A., and Hooykaas, P.J.J. (1986). Nucleotide sequence of the virulence gene *virG* of the *Agrobacterium tumefaciens* octopine Ti-plasmid: significant homology between *virG* and the regulatory genes *omp*R, *pho*B and *dye* of *Escherichia coli*. *Nucleic Acids Research* **14**, 9933–42.

Ninfa, A.J. and Magasanik, B. (1986). Covalent modification of the *glnG* product, NRI, by the *glnL* product, NRII, regulates the transcription of the *glnALG* operon in *Escherichia coli*. *Proceedings of the National Academy of Sciences USA* **83**, 5909–13.

Ninfa, A.J., Reitzer, L.J., and Magasanik, B. (1987). Initiation of transcription at the bacterial *gln*Ap2 promoter by purified *E. coli* components is facilitated by enhancers. *Cell* **50**, 1039–46.

Okker, R.J.H., Spaink, H., Hille, J., van Brussel, T.A.N., Lugtenberg, B., and Schilperoort, R.A. (1984). Plant inducible promoter of the *Agrobacterium tumefaciens* Ti-plasmid. *Nature* (London) **312**, 564–6.

Oosawa, K., Hess, J.F., and Simon, M.I. (1988). *che*A mutants defective in chemotaxis show modified protein phosphorylation. *Cell* **53**, 89–96.

Parke, D., Ornston, L.N., and Nester, E.W. (1987). Chemotaxis to plant phenolic inducers of virulence genes is constitutively expressed in the absence of the Ti-plasmids in *Agrobacterium tumefaciens*. *Journal of Bacteriology* **169**, 5336–8.

Rogowsky, P.M., Close, T.J., Chimera, J.A., Shaw, J.J., and Kado, C.I. (1987). Regulation of the *vir* genes of *Agrobacterium tumefaciens* plasmid pTiC58. *Journal of Bacteriology* **169**, 5101–12.

Rossen, L., Davis, E.O., and Johnston, A.W.B. (1987). Plant induced expression of *Rhizobium* genes involved in host specificity and early stages of nodulation. *Trends in Biochemical Sciences* **12**, 430–3.

Schröder, G., Waffenschmidt, S., Weiler, E.W., and Schröder, J. (1984). The T-region of Ti-plasmid codes for an enzyme synthesising indole-3-acetic acid. *European Journal of Biochemistry* **138**, 387–91.

Scroth, M.N., Weinhold, A.R., McCain, A.H., Hildebrand, D.C., and Ross, N. (1971). Biology and control of *Agrobacterium tumefaciens*. *Hilgardia* **40**, 536–52.

Shaw, C.H., Ashby, A.M., and Watson, M.D. (1986). Plant tumour induction. *Nature* **324**, 415.

Shaw, C.H., Ashby, A.M., Brown, A., Royal, C., Loake, G.J., and Shaw, C.H. (1988). *Vir*A and G are the Ti-plasmid functions required for chemotaxis of *Agrobacterium tumefaciens* towards acetosyringone. *Molecular Microbiology* **2**, 413–18.

Shaw, C.H., King, T.J., Duffell, A., and Loake, G.J. (1991). Acetosyringone chemotaxis in *E. coli* requires *virA, CheA, B, Y & Z, motA & B*, but not *virG* (in preparation).

Springer, W.R. and Koshland, D.E. (1977). Identification of a protein methyl-transferase as the *che*R gene product in the bacterial sensing system. *Proceedings of the National Academy of Sciences USA* **74**, 533–7.

Springer, M.S., Goy, M.F., and Adler, J. (1979). Protein methylation in behavioural control mechanisms and in signal transduction. *Nature* **280**, 279–84.

Stachel, S.E. and Zambryski, P. (1986). *vir*A and *vir*G control the plant-induced activation of the T-DNA transfer process of *A. tumefaciens*. *Cell* **46**, 325–33.

Stachel, S.E., Messens, E., Van Montagu, M., and Zambryski, P. (1985). Identification of the signal molecules produced by wounded plant cells that activate T-DNA transfer in *Agrobacterium tumefaciens*. *Nature* **318**, 624–9.

Stachel, S.E., Nester, E.W., and Zambryski, P.C. (1986a) A plant cell factor induces *Agrobacterium tumefaciens vir* gene expression. *Proceedings of the National Academy of Sciences USA* **83**, 379–83.

Stachel, S.E., Timmerman, B., and Zambryski, P. (1986b). Generation of single stranded T-DNA molecules during the initial stages of T-DNA transfer from *Agrobacterium tumefaciens* to plant cells. *Nature* **322**, 706–12.

Stachel, S.E., Timmerman, B., and Zambryski, P. (1987). Activation of the *Agrobacterium tumefaciens vir* gene expression generates multiple single stranded T-strand molecules from the pTiA6 T-region: requirement for 5' *virD* products. *EMBO Journal* **6**, 857–63.

Stock, J.B. and Koshland, D.E. (1978). A protein methylesterase involved in bacterial chemosensing. *Proceedings of the National Academy of Sciences USA* **75**, 3659–63.

Stock, A., Chen, T., Welsh, D., and Stock, J. (1988). *CheA* protein, a central regulator of bacterial chemotaxis, belongs to a family of proteins that control gene expression in response to changing environmental conditions. *Proceedings of the National Academy of Sciences USA* **85**, 1403–7.

Terwilliger, T.C., Wang, J.Y., and Koshland, D.E. (1986). Surface structure recognised for covalent modification of the aspartate receptor in chemotaxis. *Proceedings of the National Academy of Sciences USA* **83**, 6707–10.

Thomashow, L.S., Reeves, S., and Thomashow, M.F. (1984). Crown gall oncogenesis: evidence that a T-DNA gene from the *Agrobacterium* Ti-plasmid pTiA$ encodes an enzyme that catalyses synthesis of indoleacetic acid. *Proceedings of the National Academy of Sciences USA* **81**, 5071–5.

Thomashow, M.F., Karlinsey, J.E., Marks, J.R., and Hurlbert, R.E. (1987). Identification of a new virulence locus in *A. tumefaciens* that affects polysaccharide composition and plant cell attachment. *Journal of Bacteriology* **169**, 3209–16.

Thompson, D.V., Melchers, L.S., Idler, K.B., Schilperoort, R.A., and Hooykaas, P.J.J. (1988). Analysis of the complete nucleotide sequence of the *Agrobacterium tumefaciens virB* operon. *Nucleic Acids Research* **16**, 4621–36.

Van Larebeke, N., Engler, G., Holsters, M., Van den Elsacker, S., Zaenen, I., Schilperoort, R.A., and Schell, J. (1974). Large plasmid in *Agrobacterium tumefaciens* essential for crown gall inducing ability. *Nature* **252**, 169–70.

Wang, K., Stachel, S.E., Timmerman, B., Van Montagu, M., and Zambryski, P. (1987). Single strand nick in the T-DNA border as a result of *Agrobacterium vir* gene expression. *Science* **235**, 587–91.

Ward, J.E., Akiyoshi, D., Regier, D., Datta, A., Gordon, M.P., and Nester, E.W. (1988). Characterisation of the *virB* operon from an *Agrobacterium tumefaciens* Ti-plasmid. *Journal of Biological Chemistry* **263**, 5804–14.

Watson, B., Currier, T.C., Gordon, M.P., Chilton, M-D., and Nester, E.W. (1975). Plasmid required for virulence of *Agrobacterium tumefaciens*. *Journal of Bacteriology* **123**, 255–64.

Winans, S.C., Ebert, P.R., Stachel, S.E., Gordon, M.P., and Nester, E.W. (1986). A gene essential for *Agrobacterium* virulence is homologous to a family of positive regulatory loci. *Proceedings of the National Academy of Sciences USA* **83**, 8278–82.

Winans, S.C., Kerstetter, R.A., and Nester, E.W. (1988) Transcriptional regulation of the *virA* & *G* genes of *Agrobacterium tumefaciens*. *Journal of Bacteriology* **170**, 4047–54.

Young, C. and Nester, E.W. (1988) Association of the *virD2* protein with the 5′ ends of T-strands in *Agrobacterium tumefaciens*. *Journal of Bacteriology* **170**, 3367–74.

Zambryski, P., Tempe, J., and Schell, J. (1989). Transfer and function of T-DNA genes from *Agrobacterium* Ti and Ri plasmids in plants. *Cell* **56**, 193–201.

Zorreguita, A., Geremia, R.A., Cavaigne, S., Cangelosi, G.A., Nester, E.W., and Ugalde, R.A. (1988). Identification of the product of an *Agrobacterium tumefaciens* chromosomal virulence gene. *Molecular Plant–Microbe Interactions* **1**, 121–7.

13 Recognition events associated with specific interactions between plants and pathogenic fungi

J.A. BAILEY

Department of Agricultural Sciences, University of Bristol, AFRC Institute of Arable Crops Research, Long Ashton Research Station, Bristol BS18 9AF, UK

Introduction

In the present symposium, emphasis has been rightly placed on the molecular biology and genetics of interactions between plants and bacteria or viruses. This is consistent with present interest in and knowledge gained from studies with these organisms. In contrast, molecular work on pathogenic fungi is still in its infancy. There is an urgent need to redress this balance, because on a world scale, fungi are the most important and damaging group of plant pathogens. In human disease also, there is increasing awareness of the importance of fungi, notably in diseases associated with the use of immunosuppressive drugs in transplant surgery and with the spread of auto-immune diseases.

The reason for the relatively small amount of activity with plant pathogenic fungi compared with that with the other organisms is mainly the ease with which bacteria and viruses can be analysed by molecular genetics. A comparison of genome size of various organisms highlights the numerical advantage of molecular studies on these micro-organisms (Table 13.1,

Table 13.1 Comparison of genome size

Species	No. base pairs
Wheat (*Triticum aestivum*)	1.7×10^{10}
Tobacco (*Nicotiana tabacum*)	4.8×10^{9}
Parsley (*Pteroselinum sativum*)	1.9×10^{9}
Arabidopsis thaliana	7.0×10^{7}
Man (*Homo sapiens*)	3.3×10^{9}
Fruit Fly (*Drosophila melanogaster*)	1.4×10^{8}
Yeast (*Saccharomyces cerevisiae*)	1.5×10^{7}
Escherichia coli	4.2×10^{3}

Boulnois 1987). It is also very significant that, like *Escherichia coli*, the bacterium on which molecular genetics has been developed, many plant pathogenic bacteria are Gram-negative. Furthermore the genus Erwinia, which contains important plant pathogens, is from the same family, the Enterobacteriaceae. As a result, advances made with *E. coli* are often applied quickly to plant pathogenic bacteria. These characteristics have combined to facilitate isolation of genes the products of which were known, for example degradative enzymes, and also genes, for example avirulence genes, where the products were not known. This has been poss- ible because of the development of successful gene complementation tech- niques and because an entire bacterial genome, approximately 10^3 base- pairs, can be represented in less than 1000 transformants.

In contrast, the genome size of most crop plants is extremely high, often exceeding 10^9 base-pairs, and thus cloning genes by complementation is virtually impossible. This means that isolation of genes that determine phenomena such as specificity will not be possible without considerable biochemical knowledge. However, the importance of plants as major foods and hence the enormous interest shown by commercial enterprises, principally as a result of diversification from pesticide development to plant breeding, has led to a concentration on the molecular biology of plants. Many plant genes whose functions are known have already been isolated.

To date, plant pathogenic fungi, although responsible for the greatest crop losses, have not attracted much attention from industrial concerns. Efficient methods for gene cloning are not widely available. Even today good cloning systems are available for only a few fungi, for example, *Usti- lago maydis*. As a result molecular biology has yet to have a significant influence on our understanding of mechanisms of fungal pathogenicity and specificity. The more important reports are those that describe the cloning and expression of pisatin demethylase (Schafer *et al*. 1989), cutinase (Dick- man *et al*. 1989), cellulase, and mating type genes (Kronstad and Leong 1989).

Bacterial plant pathogens are limited to very few genera and so research has also benefited from the concentration of work on the relatively few related bacterial species that cause diseases. As described elsewhere in this book, Erwinia, Pseudomonas, and Xanthomonas are the genera which have been studied in most detail. Unfortunately pathogenic fungi show extreme diversity: there are many different families, each with many differ- ent genera. They also show different modes of infection, pathogenesis and nutrition. Some fungi invade through wounds, others invade through sto- mata or lenticels, whilst others invade directly by penetrating the cuticle and/or epidermal cell walls (Heath 1986). Once in the plant they have many ways of obtaining nutrients: some are biotrophic, that is they feed off living cells, while others are necrotrophic, that is they kill cells prior to feeding from the resulting released nutrients.

Furthermore, many pathogenic fungi, for example rusts and mildews which are considered to be the most important commercial targets, are obligate pathogens: these can only be grown on their host plants. The inability to grow and manipulate these pathogens in simple culture media remains a major barrier to biochemical and molecular studies. There is, however, a group of pathogens, the hemi-biotrophic facultative fungi, which grow readily in culture, yet have an absolute requirement for an initial biotrophic infection to permit initial invasion and colonization of plant tissues. These fungi include species of *Fulvi* (*Cladosporium*), *Colletotrichum*, *Magnaporthe* (*Pyricularia*), and *Phytophthora*. Thus, although there is no accepted well-researched model for plant–fungus interactions, the facultative biotrophs have already been used extensively for biochemical studies (Hahlbrock and Scheel 1989) and now present attractive systems for future molecular work. Facultative hemibiotrophs combine the best of many worlds; they are biotrophic, yet grow readily in culture, and they exhibit specific resistance and susceptibility on different plants.

Despite their individuality, pathogenic fungi do have properties in common. The present paper aims to highlight the more important aspects of fungal pathogenicity, with emphasis placed on identifying phenomena that exist as unifying features in the expression of successful pathogenesis and specificity by many different plant pathogenic fungi.

Physiological and genetic aspects of cultivar and species specificity

When considering interactions between different plants and pathogens, two types of specificity leading to either resistance or susceptibility are generally recognized. The first, sometimes misleadingly referred to as 'non-host resistance', is the specificity shown by different plant and pathogen species; I will refer to this as 'species–species specificity' (see also Valent and Chumley 1987). The second is 'race–cultivar specificity', which refers to specificity shown by different cultivars or forms of the same crop or weed species to different races or forms of the same pathogenic species. Examples of both are illustrated in Table 13.2, where interactions of *Colletotrichum* spp. with various legumes are outlined. Two additional points need to be made. First, some species of pathogenic fungi may not show such precise specificity. Thus, for example, *C. gloeosporioides* appears to attack a range of different tropical crops. Secondly, the examples of species–species specificity illustrated in Table 13.2 refer to interactions between species within the same genus. It is probable that any comments and conclusions on the nature of specificity

Table 13.2 Specificity exhibited by the genus Colletotrichum

a) Race–cultivar specificity of *C. lindemuthianum*

Race	Cultivars of *Phaseolus vulgaris*			
	La victoire	Kievitsboon Koekoek	Dark Red Kidney	Cornell
Beta	S	R	S	R
Gamma	S	S	R	R
Lambda	S	S	S	S

b) Species–species specificity of *Colletotrichum* species

Fungal spp.	Species of plant			
	French bean	Cucumber	Linseed	Banana
C. lindemuthianum (all races)	S	R	R	R
C. orbiculare	R	S	R	R
C. linum	R	R	S	R
C. musae	R	R	R	S

R indicates resistance; S indicates susceptibility

between such related species will not be relevant to interactions between unrelated species, for example the presumed resistance shown by oak to *Colletotrichum lindemuthianum*.

Although generally accepted, the distinction between the specificity shown by related species and that shown by races and cultivars may not be valid. A comparison of different species and different cultivars indicates that the actual expression of resistance can be similar. For example, resistance of some legumes to various *Colletotrichum* species and of cultivars of French bean to different races of *Colletotrichum lindemuthianum* appear very similar. In many interactions resistance is expressed as a severe hypersensitive response causing the rapid death of only the initially infected epidermal cell (Bailey *et al*. 1990; Bailey and O'Connell 1989).

The genetic basis of race–cultivar specificity has been studied extensively, though most data come from work with the host rather than the pathogen. There are many examples where plant resistance is controlled by a single 'resistance' gene. There is less evidence from fungi, but that available indicates that pathogens have a corresponding gene for 'avirulence'. Thus the simplest model for specificity is one in which resistance is controlled by a single plant gene, the product of which interacts with the product of an avirulence gene. This is usually referred to as a gene-for-gene relationship. There is now clear evidence from molecular work with several different bacteria that avirulence can be controlled by single genes: when

DNA was transferred from an avirulent strain to a virulent strain, among the resulting transformants were some that induced resistance (see Minsavage *et al.* 1990). So far none of the products of bacterial avirulence genes have been characterized. In fungi, however, although no avirulence genes have been isolated, the most significant work is the biochemical studies from de Wit's laboratory where a specific toxin (equivalent to a specific elicitor), probably the product of an avirulence gene, has been isolated from *Cladosporium fulvum*. This toxin only affects the appropriate specifically-resistant cultivars (de Wit *et al.* 1989).

There has been very little work concerning the genetic basis of species–species specificity, though there have been important results reported recently. These arise from studies with the rice blast pathogen *Magnaporthe oryzae*. Isolates were identified that specifically attacked different species of grass, finger millet (*Eleusine coracana*), goose grass (*Eleusine indica*), weeping lovegrass (*Eragrostis curvula*), and rice (*Oryza sativa*) (Valent and Chumley 1987). Genetic analysis of the progeny resulting from matings of these isolates indicates that a single avirulence gene determines the different resistance and susceptibility of these grasses. This is the first indication that species–species specificity may be under simple genetic control.

Thus on the basis of the limited biological and genetic data available, it appears that the mechanisms which regulate race–cultivar and species–species specificity are similar. It should also be noted that during the breeding process the genes introduced into resistant cultivars have frequently been obtained from closely related, but distinct plant species. For example, the modern tomato plant has important agronomic and defensive properties that have come from other species of Lycopersicon: pest and disease resistance has been obtained from *L. hirsutum*, *L. pimpinellifolium*, *L. peruvianum* and *L. demissum* (Hoyt 1988).

Furthermore, those avirulence genes from bacteria that determine cultivar specificity also effect species specificity (Minsavage *et al.* 1990). So why are race–cultivar and species–species specificity usually considered to be distinct? The distinction seems to arise from the different durability of resistance shown by cultivars to that shown by species: the resistance of cultivars is often short-lived as a result of genetic changes in the pathogen, while resistance of species has persisted unaltered for millennia. If similar genes are involved in both cultivar and species resistance, it is clearly important to understand the mechanisms that regulate durability. Perhaps the answer lies in the multitude of genes that together may effect species resistance, compared with the single genes that often are used to produce resistance cultivars. The recent evidence that single genes determine the pathogenesis of *Magnaporthe* spp. on different plant species would, however, cast doubt on this conclusion. Alternatively, durability may arise because of the different genetic backgrounds, or more simply because different, though related, plant species are not exposed to the extremely high selection pressures that fall on a new cultivar. For example, species of *Colletotrichum* that attack *Vigna unguiculata* (a plant of African origins) would not have encountered *Phaseolus vulgaris* (a plant originating from South America) until relatively recently. In addition, even when grown in the same area it would be rare to find them growing as close together as different cultivars of the same species.

These are important considerations, not only when attempting to order priorities towards understanding the basis of plant–pathogen interactions, but also when considering future prospects for genetic engineering of resistant plants. It is often suggested that, because the resistance of a species is more durable than that of cultivars, the genes determining resistance of a species are more likely to produce durably resistant plants than the genes that regulate cultivar resistance. Existing evidence from traditional breeding programmes provides little to support this view. When such genetically-engineered crops become available they will require the same careful monitoring that is needed for traditionally-produced cultivars.

Recognition events during expression of resistance and susceptibility

A simple outline of the major areas of interaction between plants and pathogens is illustrated in Fig. 13.1. This summary is based on studies with *Colletotrichum*, but is generally relevant to many interactions between plants and biotrophic fungi. It highlights several areas; arrival and subsequent development of a propagule on the surface of a plant; penetration of the plant surface; and then interactions between the infecting hyphae and host cells. This often appears to be an interaction of a hypha with the plasmalemma of the plant, and leads to the subsequent development of the pathogen in tissues and production of symptoms, or alternatively, to the subsequent expression by the host of a range of defence responses. It can be seen from Fig. 13.1 that the ability of plant cells to survive infection is presented as the key to subsequent colonization of tissues by fungi, while early host cell death determines the restriction of pathogen development.

Differentiation on plant surfaces

After arrival of propagules, usually spores, on the plant, the first event in the process of pathogenesis will be adhesion of the propagule to the surface. As discussed below, there is much evidence that appressoria (structures produced by many pathogens following germination of spores)

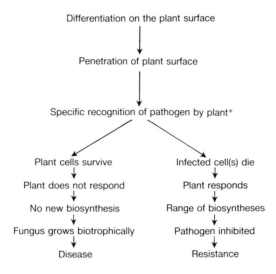

Differentiation on the plant surface

Penetration of plant surface

Specific recognition of pathogen by plant*

Plant cells survive	Infected cell(s) die
Plant does not respond	Plant responds
No new biosynthesis	Range of biosyntheses
Fungus grows biotrophically	Pathogen inhibited
Disease	Resistance

*Recognition that determines specificity is assumed to take place on the plasmalemma

Fig. 13.1.   Recognition events during interactions between plants and pathogenic fungi.

attach firmly to surfaces, but recently, by describing a mechanism for the immediate and persistent attachment of spores of *Magnaporthe grisea* to various surfaces, Hamer *et al.* (1988) have highlighted the contribution of spores to the adhesion process. Microscopy of hydrating conidia showed that mucilage was expelled from one end of the conidia before the germ tube emerged. The mucilage was shown to effect attachment not only to the host plant, but also to a synthetic Teflon surface.

The next essential requirement for development of infection is germination and differentiation of the spore to produce structures which facilitate penetration of the surface. The most common of these structures are appressoria, structures which are produced by a wide range of pathogenic fungi. Two aspects of appressoria production deserve attention: the mechanisms by which appressoria attach to plant surfaces and those that are involved in differentiation of such complex structures.

In ways similar to those described for spores of *M. grisea* there are much data indicating that mucilage, secreted by appressoria, or more likely, by the germ tubes during appressoria differentiation, is responsible for sticking appressoria very firmly to plant surfaces. However, the nature of the adhesives and the mechanisms by which they interact with the waxes on plant surfaces are still not understood and they remain important topics for future research. Any method of preventing pathogenic fungi from attaching to plants would have great potential for controlling disease and could be applicable to a range of different fungi.

The importance of differentiation of appressoria on the surface cannot be understated: without appressoria most fungi would not be pathogenic. The mechanisms underlying the differentiation process are now being investigated. It has been shown that production of appressoria, which usually require a hard surface, involves expression of many new genes. In the case of some rust fungi appressoria are induced not by any hard surface, but only by surfaces that have precise structural features. The best example of this is shown by *Uromyces appendiculatus* on leaves of its host *Phaseolus vulgaris*. Hoch *et al.* (1987) established that ridges on the leaves acted as topographic signals. Studies with artificial materials revealed that similar inductive processes occurred if scratches were present on the surface. Further work showed that appressoria were produced only if ridges were present and only if these were 0.25–1.0 μm high and 0.5–6.7μm apart.

While differentiation on plant surfaces is an essential feature of pathogenesis there is little evidence to suggest that the behaviour of pathogenic fungi on plant surfaces is involved in determining either specific resistance or susceptibility of cultivars or species. Spores of most pathogenic fungi, even the obligate biotrophs, will germinate in the presence of water or simple nutrients. Adhesion appears to be non-specific: spores of *M. grisea* stuck to plant and synthetic surfaces, while studies with various *Colletotrichum* species have shown that conidia will produce appressoria on a range of hard surfaces, for instance, glass, plastic and cellophane, as well as on the leaves and stems of a range of resistant and susceptible cultivars and species (Bailey *et al.* 1990; O'Connell, unpublished data).

Penetration of plant surfaces

The ultrastructure of the initial penetration of plant surfaces has been studied extensively (Cooper 1981). The notable feature of penetration by biotrophic hyphae is that passage through the cuticle and the epidermal cell wall is achieved in a very localized manner, there is no dissolution of the wall away from the penetrating hyphae. In addition, the holes produced by biotrophic hyphae are extremely narrow, much narrower than the hyphae that emerge. This situation contrasts greatly with the much larger holes and extended wall dissolution that are produced by necrotrophic hyphae.

In studies of race–cultivar specificity, detailed descriptions of the initial infection process have nearly always been restricted to the susceptible interaction. The difficulties encountered when searching for, and sectioning through, very small infection hyphae have precluded detailed studies of many resistant interactions. Nevertheless, it does appear that initial penetration of resistant and susceptible cultivars occurs equally readily, and

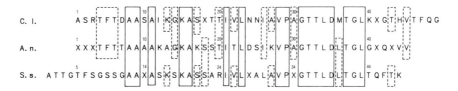

Fig. 13.2. Comparison of N-terminal amino acid sequences of polygalacturonases from plant pathogenic fungi. C.l., *Colletotrichum lindemuthianum*; A.n., *Aspergillus niger*; S.s. *Sclerotinia sclerotiorum*.

thus it is unlikely that differential penetration of plant surfaces can explain specific resistance.

The mechanisms of cuticle and wall penetration have not been fully established, though several enzymes capable of dissolving cuticles or cell walls have been isolated and described. Studies of cutinases have not yet extended to a comparison of enzymes from different pathogenic fungi, or of their effects on different cultivars or species. However, a cutinase gene has been isolated from *Fusarium solani* f. sp. *pisi*. When this gene was transferred to a species of *Mycosphaerella* that does not normally penetrate the cuticle of papaya, this non-pathogen of papaya was able to penetrate and rot the papaya tissues (Dickman *et al*. 1989). Although these findings show that cutinase can degrade cuticles, and that the pathogenicity of a fungus can be altered by gene transfer, they do not prove that such enzymes are involved in the penetration of papaya by natural pathogens. There are considerable amounts of data to show that appressoria can generate sufficient force to penetrate cuticles purely by mechanical means. With the advent of molecular techniques there are great opportunities for further research in this area.

Two classes of cell wall degrading enzyme have been studied extensively, the endo-polygalacturonases (PGs) and the endo-pectin lyases. They are produced by most pathogenic fungi and are believed to be responsible for degradation of plant walls during necrotrophic fungal growth (Cooper 1981). Their roles in biotrophic infections and in determining specificity are less clear. An analysis of PGs from two races of *Colletotrichum lindemuthianum* has failed to reveal any biochemical or structural differences (Keon *et al*. 1990), although differences were found between *C. lindemuthianum* and other fungi (Fig. 13.2). The physical properties of the enzymes from *C. linde-muthianum*, including their iso-electric points, their amino acid and sugar compositions, and most significantly their N-terminal amino acid sequences, were identical. Their abilities to macerate hypocotyls of different cultivars of bean were also indistinguishable. These findings suggest that neither the properties of the PG enzymes from different races, nor the properties of the walls of the different cultivars can explain specific resistance. Other studies of the composition of walls of *Phaseolus vulgaris* indicate the presence of a proteinaceous inhibitor of PG (Cervone *et al.* 1985). Whether this protein is involved in specificity or whether the PGs from different races release specific elicitors remains a question for further study.

Consequences of specific recognition

There is considerable evidence from studies with various facultative bio-trophic fungi, including *Colletotrichum, Cladosporium,* and *Phytophthora,* and also from work with rusts and mildews, that the recognition event(s) that determine resistance and susceptibility occur soon after initial penetration (Callow *et al.* 1989). Except for the work from de Wit's laboratory there is a paucity of significant biochemical and molecular data.

It has been argued previously that the immediate consequence of specific recognition are the survival or death of the cells initially infected (Bailey 1983). Studies with *C. lindemuthianum* showed that the first detectable difference between the responses of resistant and susceptible cultivars is dysfunction of the plasmalemma of the first infected cell. Infected cells of resistant cultivars lost the ability to plasmolyse when the infection hyphae were extremely small (Bailey and O'Connell 1989). In contrast, in susceptible cultivars, cells survived infection and an extensive biotrophic infection became established. This scheme is based on studies with *C. lindemuthianum*, but is consistent with information available from other species of *Colletotrichum* and from other pathogens. Thus although the hyphae produced by *C. lindemuthianum* in cowpea (Bailey *et al.* 1990) and by *C. truncatum* in pea tissues (O'Connell unpublished data) differed from those of *C. lindemuthianum* in bean, the initial growth in each plant was biotrophic. Resistance to these fungi was expressed as a typical hypersensitive reaction.

Proposed gene-for-gene relationships are readily explained in terms of a yes–no recognition event leading to survival or death of infected cells. On this basis these genes should be referred to as 'recognition genes'. The term 'resistance gene' should be used to describe those genes that are activated as a consequence of recognition, that is those genes that are activated by the death of infected cells.

Importance of host cell injury/death in resistance

Early death of infected cells often leads to restriction of the pathogen within the dead cells or tissues, a hypersensitive reaction. What is the significance of this rapid cell injury and death? For obligate pathogens, the death of an infected cell *per se* would be sufficient to explain the cessation of pathogen growth. These fungi could not continue to grow as they cannot obtain nutrients from dead cells. However, death of cells *per se* cannot explain why the growth of facultative fungi becomes restricted. For facultative fungi, growth inhibition appears to be due to one of several host responses. These include production of phytoalexins, hydroxyproline-rich glycoproteins (HRGPs), chitinases, and other pathenogenesis-related proteins (Bailey 1987). It is now clear that the enzyme synthesis required for production of all these defence responses involves gene transcription. The latest information on this topic is discussed by Dixon, Chapter 17 (see also Hahn *et al*. 1985).

It has been proposed that death of cells in the hypersensitive reaction may arise from dysfunction of the plasmalemma of infected cells (Bailey 1983). On this basis an attractive hypothesis is that recognition of the avirulent pathogen causes disruption of the membrane and that this leads to new gene transcription during the early stages of death, especially in the adjacent cells which remain alive: products of the defence responses accumulate soon afterwards. Experimental support for this proposal comes from the demonstration that the specific elicitor isolated from the facultative biotroph *Cladosporium fulvum* is a toxin, that kills cells of resistant cultivars but does not affect cells of susceptible cultivars. Other less well defined elicitors are also toxic (for example Doares *et al*. 1989).

There are also data to show that the biochemical basis of hypersensitivity is very localized. This is well illustrated by immunocytochemical and *in situ* hybridization studies (Schmelzer *et al*. 1989). In the hypersensitive reaction of *P. vulgaris* to *C. lindemuthianum* enhanced production of phenylalanine ammonia lyase was located in the cytoplasm of cells that had died, and also in a few living cells immediately around the dead cells (Bailey and O'Connell 1989), whilst HRGPs accumulated in the walls of the adjacent living cells (O'Connell *et al*. 1990).

Importance of host cell survival in pathogenesis and specificity

Survival of plant cells infected by micro-organisms is a wide spread phenomenon. Most successful plant–microbe interactions involve the existence and growth of the microbe in living plant cells (Fig. 13.3). This phenomenon is not restricted to pathogenic microorganisms. It is also widespread in symbiotic interactions (Gianinazzi-Pearson 1986) and is essential for infection of plants by parasitic nematodes (see Bowles Chapter 14) and by parasitic plants, e.g. *Striga* and *Orabanche*. In the context of pathogenesis, survival of infected cells has two important consequences: (1) it provides the environment necessary for obligate pathogens to obtain nutrients; and (2) the defence responses associated with cell death are not instigated and thus the required enzyme syntheses are not

Fig. 13.3 Importance of survival of infected cells for pathogenesis and symbiosis

Life style	Organisms	Requirement for host cell survival
Obligate	Viruses Pathogenic fungi	Essential for replication
	Parasitic plants Mycorrhizal fungi	Essential for infection, growth, and reproduction
Facultative	Pathogenic bacteria Symbiotic bacteria Pathogenic fungi Cyst nematodes Parasitic plants	Essential for initial infection and colonization of tissues

induced. As a result both obligate and facultative organisms can grow and prosper. Continued growth and eventual reproduction of obligate fungi depends on the cells of the infected tissues remaining alive. Eventually, however, facultative fungi switch their behaviour to become necrotrophic, at which time the rate and extent of tissue death must be sufficient to prevent the significant production of defence responses (Bailey and O'Connell 1989).

The nature of the biotrophic relationships established by obligate and facultative fungi appear to be very similar. Detailed ultrastructural analysis, however, has revealed significant differences. As an example of a facultative pathogen, O'Connell (1987) studied the plant–pathogen interface between *C. lindemuthianum* and *P. vulgaris*. Initially, all infected cells were alive. Plasmolysis caused the plasmalemma to withdraw from the plant cell wall and from the infection hyphae (vesicle). Similar work with obligate pathogens has shown that the host plasmalemma does not withdraw from the pathogen because there is a connection between the host plasmalemma and the infection hypha (haustorium). In addition, the plant plasmalemmae which surround the haustoria of obligate pathogens show several modifications, for example increased ATPase activity and modified surface structure. The plasma membranes which surround the vesicles of *Colletotrichum lindemuthianum* were identical to the rest of the plasmalemma. This suggests that because a fungus is obligate, it has nutritional requirements that can only be provided by modifying the plant–fungus interface. For the facultative organism, which grows readily in a range of nutrients, no modifications are needed. Facultative organisms can readily scavenge any nutrients present in the apoplast, they simply feed on the available nutrients. Thus although the feeding characteristics of obligate and facultative organisms are different, successful pathogenicity has an essential requirement—survival of infected cells.

Conclusion

A survey of the interactions that determine successful fungal pathogenesis has identified several common features that deserve attention in the future. These include phenomena associated with the differentiation of pathogens on plant surfaces, the mechanisms by which pathogens penetrate plant cell walls and, of greatest importance, the mechanisms by which plant cells survive infection. An understanding of the mechanisms that determine and regulate race–cultivar and species–species specificity will depend on increasing our knowledge of cell survival and death.

References

Bailey, J.A. (1983). Biological perspectives of host–pathogen interactions. In *The dynamics of host defence*, (ed. J.A. Bailey and B.J. Deverall), pp. 2–32. Academic Press, Australia.

Bailey, J.A. (1987). Phytoalexins: a genetic view of their significance. In *Genetics and plant pathogenesis*, (ed. P.R. Day and G.J. Jellis), pp. 233–43. Blackwell Scientific Publications, Oxford.

Bailey, J.A. and O'Connell, R.J. (1989). Plant cell death: a determinant of disease resistance and susceptibility. In *Phytotoxins and plant pathogenesis*, (ed. A. Graniti, R.D. Durbin, and A. Ballio), pp. 275–84. Springer-Verlag, Berlin.

Bailey, J.A., Nash, C., O'Connell, R.J., and Skipp, R.A. (1990). Infection process and host specificity of a *Colletotrichum* species causing anthracnose of cowpea, *Vigna unguiculata*. *Mycological Research* (in press).

Boulnois, G.J. (1987). *Gene cloning and analysis: a laboratory guide*. Blackwell Scientific Publications, Oxford.

Callow, J.A., Ray, T., Estrada-Garcia, T.M., and Green, J.R. (1989). Molecular signals in plant cell recognition. In *Cell to cell signals in plant, animal and microbial symbiosis* (ed. S. Scannerini, D. Smith, P. Bonfonte-Fasolo, and V. Gianinazzi-Pearson), pp. 167–82. Springer-Verlag, Berlin.

Cervone, F., De Lorenzo, G., Degra, L., Salvi, G., and Bergami, M. (1985). Purification and characterization of a polygalacturonase-inhibiting protein from *Phaseolus vulgaris* L. *Plant Physiology* **85**, 631–7.

Cooper, R.M. (1981). Pathogen-induced changes in host ultrastructure. In *Plant disease control: resistance and susceptibility* (ed. R.C. Staples and G.H. Toenniessen), pp. 105–42. John Wiley, New York.

De Wit, P.J.G.M., Van Den Ackerveken, G.F.J.M., Joosten, M.H.A.J., and Van Kau, J.A.L. (1989). Apoplastic proteins involved in communication between tomato and the fungal pathogen *Cladosporium fulvum*. In *Signal molecules in plants and plant microbe interactions* (ed. B.J.J. Lugtenberg), pp. 273–80. Springer-Verlag, Berlin.

Dickman, M.B., Podila, G.K., and Kolattukudy, P.E. (1989). Insertion of cutinase gene into a wound pathogen enables it to infect intact host. *Nature* **342**, 446–8.

Doares, S.H., Bucheli, P., Albersheim, P., and Darvill, A.G. (1989). Host-pathogen interactions XXXIV. A heat-labile activity secreted by a fungal phytopathogen releases fragments of plant cell walls that kill plant cells. *Molecular Plant–Microbe Interactions* **2**, 346–53.

Gianinazzi-Pearson, V. (1986). Cellular modification during host–fungus interactions in endomycorrhizae. In *Biology and molecular biology of plant–pathogen interactions*, (ed. J.A. Bailey), pp. 29–37. Springer-Verlag, Berlin.

Hahlbrock, K. and Scheel, D. (1989). Physiology and molecular biology of phenylpropanoid metabolism. *Annual Review of Plant Physiology and Plant Molecular Biology* **40**, 347–69.

Hahn, M.G., Bonhoff, A., and Grisebach, H. (1985). Quantitative localization of the phytoalexin glyceollin I in relation to fungal hyphae in soybean roots infected with *Phytophthora megasperma* f. sp. *glycinea*. *Plant Physiology* **77**, 591–601.

Hamer, J.E., Howard, R.J., Chumley, F.G., and Valent, B. (1988). A mechanism for surface attachment in spores of a plant pathogenic fungus. *Science* **239**, 288–90.

Heath, M.C. (1986). Fundamental questions related to plant–fungal interactions: can recombinant DNA technology provide the answers? In *Biology and molecular biology of plant–pathogen interactions*, (ed. J.A. Bailey), pp. 15–27. Springer-Verlag, Berlin.

Hoch, H.C., Staples, R.C., Whitehead, B., Comeau, J., and Wolf, E. D. (1987). Signaling for growth orientation and cell differentiation by surface topography in *Uromyces*. *Science* **235**, 1659–62.

Hoyt, E. (1988). *Conserving wild relatives of crops*. International Board For Plant Genetic Resources, Rome.

Keon, J.P.R., Waksman, G., and Bailey, J.A. (1990). A comparison of the biochemical and physiological properties of a polygalacturonase from two races of *Colletotrichum lindemuthianum*. *Physiological and Molecular Plant Pathology* (in press).

Kronstad, J.W. and Leong, S.A. (1989). Isolation of two alleles of the b locus of *Ustilago maydis*. *Proceedings of the Natinal Academy of Sciences USA* **86**, 978–82.

Minsavage, G.V., Dahlbeck, D., Whalen, M.C., Kearney, B., Bonas, U., Staskawicz, B.J., and Stall, R.E. (1990). Gene-for-gene relationships specifying disease resistance in *Xanthomonas campestris* pv. *vesicatoria* – pepper interactions. *Molecular Plant–Microbe Interactions* **3**, 41–7.

O'Connell, R.J. (1987). Absence of a specialized interface between intracellular hyphae of *Colletotrichum lindemuthianum* and cells of *Phaseolus vulgaris*. *New Phytologist* **107**, 725–34.

O'Connell, R.J., Brown, I.R., Mansfield, J.W., Bailey, J.A., Mazau, D., Rumeau, D., and Esquerre-Tugaye, M.T. (1990). Immunocytochemical localization of hydroxyproline-rich glycoproteins accumulating in melon and bean at sites of resistance to bacteria and fungi. *Molecular Plant–Microbe Interactions* **3**, 33–40.

Schafer, W., Straney, D., Ciuffetti, L., Van Etten, H.D., and Yoder, O.C. (1989).

One enzyme makes a fungal pathogen, but not a saprophyte, virulent on a new host plant. *Science* **246**, 247–9.

Schmelzer, E., Johnson, W., and Hahlbrock, K. (1989). *In situ* localization of light-induced chalcone synthase mRNA, chalcone synthase and flavonoid end products in epidermal cells of parsley leaves. *Proceedings of the National Academy of Sciences USA* **85**, 2989–93.

Valent, B. and Chumley, F.G. (1987). Genetic analysis of host species specificity in *Magnaporthe grisea*. In *Molecular strategies for crop protection* (ed. N.T. Keen and C.J. Lamb), pp. 83–93. Alan R. Liss Inc., USA.

14 Local and systemic changes in plant gene expression following root infection by cyst nematodes

DIANNA J. BOWLES, SARAH J. GURR, CLAIRE SCOLLAN, and HOWARD J. ATKINSON

Centre for Plant Biochemistry and Biotechnology, University of Leeds, Leeds LS2 9JT, UK

KIM E. HAMMOND-KOSACK

The Sainsbury Laboratory, John Innes Institute, Colney Lane, Norwich NR4

Introduction

Nematodes are serious pests of many economically important crop plants. In the UK alone, losses to the potato harvest caused by *Globodera rostochiensis* and *Globodera pallida*, the two sibling species of potato cyst nematode (PCN) are estimated to cost agriculture around £56M per year (Anon 1986). Current control of PCN infection is achieved by crop rotation, which is expensive and restrictive to specialist growers, and by the use of toxic nematicides with their attendant environmental hazards (Hague and Gowen 1987). The wide-spread planting of resistant cultivars, such as the cv. Maris Piper, provides a more attractive means of containing PCN infection. Unfortunately this cultivar is resistant to only two pathotypes of *G. rostochiensis* (Ro1 & Ro4), and is susceptible to the remaining three forms of this species (Ro2, Ro3, & Ro5) and to each form of *G. pallida* (Pal−3). All species are as described by the European Pathotyping Scheme (Kort *et al.* 1977). The extensive use of the cv. Maris Piper has led to a widespread build up of *G. pallida* populations in Britain, and there is now an urgent need for the breeding or engineering of cultivars resistant to all forms of PCN.

The time-course of invasion and establishment of the nematode in roots has been outlined for PCN (Hammond-Kosack *et al.* 1989a), and described in greater detail for soybean cyst nematode (Atkinson and Harris 1989). The infective juvenile uses a mouth stylet to cut through plant cell walls.

Subsequently a syncytium is formed that progresses as a column of responding plant cells from the head of the sedentary animal to the central vascular tissues (Jones 1981). The syncytium resembles a transfer cell system and its complete formation is essential for the development of adult females. Expression of the single dominant gene for resistance (H_1) in the cv. Maris Piper leads to the degeneration of the syncytium before the structure associates with the vascular tissue, and is accompanied by necrosis in adjacent unmodified plant cells (Turner and Stone 1984; Rice *et al.* 1985).

Other resistance mechanisms to PCN exist (Sidhu and Webster 1981). Several are oligogenic or polygenic in nature and come into play post-invasion, during nematode migration, or following the initiation of the syncytium (Rice *et al.* 1985; Robinson *et al.* 1988).

Our work involves the cv. Maris Piper and *G. rostochiensis* pathotypes Ro1 and Ro2. This allows gene expression in both susceptible and resistant responses to be studied by selecting the appropriate pathotype for plant infection (Ro1 incompatible; Ro2 compatible). The cell biology of the system is well described; the resistant response is defined and synchrony of infection can be maintained in the interaction throughout the early determinative stages of invasion. This enables us to study stage-specific changes in gene expression in both the nematode and the host plant, which in turn provides the basis for the design of novel resistance strategies using transgenic plants.

This article will summarize our findings from the use of three approaches. Firstly, stage-specific changes in the abundance of translatable mRNAs will be described. Thus, root tissue infected with Ro1, Ro2, or sham-inoculated, was harvested at four time-points post-invasion, RNA was isolated and translated *in vitro*. In addition, aerial tissue from the infected plants was analysed to provide a comparison of the local and systemic effects induced by the invading pathogen. In total, this approach gives an insight into the broad spectrum of changes induced during the infection process. To gain information on the nature of defined gene products two alternative approaches have also been taken. In these, we have focused on two classes of known defence-related products: pathogenesis-related proteins (PR proteins) and proteinase inhibitors (PI proteins). The former are known to be targeted to the apoplast, and in consequence we have prepared intercellular washing fluid from infected plants as a means to identify and isolate novel proteins induced to accumulate by the nematode pathotypes. Changes in PI proteins have been investigated both at the level of PI activity and, using cDNA probes to PI genes, at the level of mRNA abundance in Northern analyses.

Thus, for each stage in the invasion of cv. Maris Piper by either pathotype of *G. rostochiensis* we have been able to build a picture of the effects of the nematode on the plant.

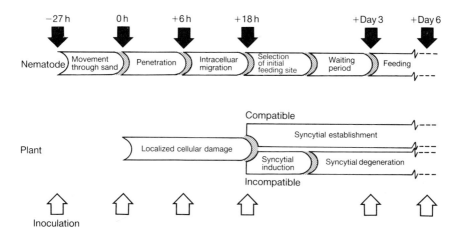

Fig. 14.1. Summary of events during compatible and incompatible (H$_1$) PCN-interactions.

Changes in translatable mRNA populations

Changes in gene expression have been investigated both local to the site of invasion within the root, and elsewhere, in young leaves (Hammond-Kosack *et al.* 1989a). The time-course of nematode invasion for both the susceptible and the resistant plant response is summarized in Fig. 14.1. It should be noted that there is an 18 h migration period prior to the juvenile selecting the initial feeding cell and inducing the syncytium. Messenger RNA was prepared from roots and leaves of both healthy and infected plants, translated *in vitro* in the presence of [^{35}S]-methionine, and the translation products were analysed by two-dimensional gel electrophoresis and fluorography.

Local responses

These involve changes in the roots following invasion. Since nematodes remain within the root tissue they were necessarily homogenized with the infected plant roots and therefore contribute to the patterns of *in vitro* translation products. Several novel products in the range 16–24 kDa were common to Ro1 and Ro2 infection, but they were induced transiently at either early or late phases of infection. A sub-set of these could also be induced also by mechanical wounding of uninfected roots and were therefore clearly host-encoded. A few products were detected in response to just one of the two nematode pathotypes. A novel product of 18.8 kDa and pI 7.4 was transiently abundant at 21 h and 3 days post-invasion in the incompatible interaction. This product was not detected on mechanical wounding and its presence suggests that by this time a differential host response to the two pathotypes had been established.

Systemic responses

These involve changes in the leaves following nematode infection of the roots. Since the nematodes remain in the root tissue throughout the inter-action all changes observed were host-encoded. Surprisingly, many sys-temic changes were observed even within 6 h of root infection. Novel gene products accumulated in leaves at both early and later stages of the infec-tion sequence. Several changes detectable from 3 days post-invasion onwards were specific to just one of the two interactions. The detection of these pathotype-specific events is both unexpected and exciting as they have not previously been described for other plant–pathogen interactions (reviewed, Bowles 1990).

A general effect was the disappearance of many translation products normally found in healthy leaves. At later time-points following the invasion process, however, the abundance of several of the mRNA subsets increased, returning to levels similar to those found in control plants.

Conclusions

In other experimental systems, cellular damage has been reported to lead to a simplification in the pattern of gene expression at both local and dis-tant sites (Theillet *et al.* 1982; Davies *et al.* 1986). Overall the cellular damage caused by nematodes is confined to cells close to the animal which typically migrates about 1 mm in potato roots (Robinson *et al.* 1988). Wounding is principally limited to the cortex and is most severe during the migratory phase of root invasion. The lack of gross disturbance to root gene expression may reflect the biotrophic nature of the interaction which must be maintained until normal maturity of the female at about 40 days. In addition the immature animal must benefit nutritionally from minimiz-ing disturbance of normal metabolic processes in the leaves as well as roots. Conversely, the plant must limit the overall disruption caused by the parasite and this may partly explain why such gross changes in leaf gene expression are transient. Further work is required to define which of these responses correlates with the expression of the H_1 resistance response, and which are cultivar specific, that is, specific to cv. Maris Piper.

Changes in proteins targeted to the apoplast

Because of the importance of the apoplast in the defence strategies of the plant, regulation of its composition is likely to be influenced by a diverse array of internal and external signals (reviewed, Bowles 1990). Apoplastic contents can be extracted by vacuum infiltration of water into the tissue and intercellular fluids recovered by low speed centrifugation (DeWit and Spikeman 1982).

Pathogenesis-related proteins

This family of defence-related proteins has been defined on the basis of their physical properties (solubility at low pH, resistance to proteinases, small molecular mass), extracellular location, and local and systemic accumulation in response to viruses and other pathogens (Van Loon 1985). The products are known also to occur at certain stages during healthy plant development, and to accumulate locally in response to various biotic and abiotic elicitors, including aspirin and ethylene (Fraser 1981; Parent and Asselin 1984; Lotan *et al.* 1989). First discovered in tobacco plants, sero- logically-related proteins have now been found in numerous plant species including potato (Nassuth and Sanger 1986; White *et al.* 1987; Parent *et al.* 1988). Currently their precise role in plant defence is unknown, although some have recently been identified as chitinases and 1,3-β-glucanases (Kauffmann *et al.* 1987; Legrand *et al.* 1987; Kombrink *et al.* 1988). A study of PR proteins in the PCN-potato interaction was undertaken to investigate whether systemic signals originating from a root-localized path- ogen could lead to PR accumulation in the leaves (Hammon-Kosack *et al.* 1989b).

Analysis of intercellular fluids (IF) recovered from leaves of healthy and nematode-infected plants has revealed that root invasion induces a marked increase in the total extractable protein, and major changes in the spec- trum of polypeptides present. Several major polypeptides found in healthy leaves are no longer present, but numerous novel gene products are observed. For example, on infection there is a drastic decrease in abun- dance of doublet polypeptides of 52.4 and 61.6 kDa. In parallel to these changes the nematode induced the accumulation of eight novel polypep- tides in the range 14.3–34.7 kDa, and with pI values in the range 4.2–9.3. The novel PR proteins are detectable in leaves 6 days after root invasion. Two of these novel molecular species, of M_r 25.8 and 34.7 kDa, are multi- ple polypeptides (4 and 2 respectively). The existence of multiple isomeric forms of a PR-protein has been described previously for tomato p14 (DeWit and Van de Meer 1986). Similar changes are observed in all the interactions examined involving *G. rostochiensis* pathotypes Ro1–5 and *G. pallida* pathotypes Pa1, 2, and 3 with the cvs. Maris Piper and Pentland Crown (universally susceptible).

The M_r and isoelectric points of several of the novel polypeptides induced by nematodes are similar to those of leaf PR proteins reported to accumulate in potato leaves following either aspirin treatment or leaf infection with citrus exocortis viroid, tobacco mosaic virus, potato virus X *Phytophthora infestans* (Conjero *et al.* 1979; Parent and Asselin 1987; White *et al.* 1987; Kombrink *et al.* 1988). In particular polypeptides of 32, 32.7 and 34.7 kDa correspond in size to three chitinases previously identified by Kombrink *et al.* (1988).

Conclusions

The data provide the first demonstration that the composition of the leaf apoplast can be regulated by systemic signals arising from pathogen invasion of root tissue. However, the pattern of changes observed was not specific for potato cultivar, nematode pathoype or nematode species. This suggest that the systemic PR-protein response reflects a general stress response rather than one related to resistance or susceptibility.

Peroxidases

Changes in the patterns of peroxidase isoenzymes have been demonstrated both at the local site of pathogen invasion (Greppin *et al.* 1986; Lagrimini *et al.* 1987) and in systemically-responding tissue (Smith and Hammerschmidt 1988). Compared with unchallenged plants, the rate of induction is frequently observed to be faster in the healthy tissues of plants challenged on a previous occasion by a pathogen (Taylor 1987). At least 3 sub-groups of peroxidase isoenzymes are distinguishable in plants: the anionic (pI 3.5–4.0), the moderately anionic (pI 4.5–6.5), and cationic (pI 8.1–11). The former two groups are associated with the cell wall whilst the third is confined to the central vacuole. It is thought that the anionic peroxidases function in lignification and in cross-linking of cellulose, pectin, hydroxy-proline-rich glycoproteins, and lignin during secondary cell wall formation (Lamport 1986; Greppin *et al.* 1986). The moderately anionic peroxidase isoenzymes possess intermediate activity towards lignin precursors and may function in wound healing and suberisation (Espelie *et al.* 1986). The function of the cationic isoenzymes within the vacuole is unknown.

As a means of determining whether any of the PR proteins we identified were peroxidases, enzyme assays were carried out on intercellular fluids recovered from the leaves of healthy and nematode-infected plants. Marked changes occurred in the extractable peroxidase isoenzyme population following cyst-nematode invasion (Fig. 14.2). Two isozymes found in healthy leaves decrease in activity (2 and 8) while the activities of 5 others increase (1, 4, B, C, and D). Isozymes C and D with pI values of 7.0 and 5.3 respectively correspond to the novel PR with a M_r of 34.7 kDa described in the earlier section. The Rf values of isozymes A (0.07), C (0.39), and D (0.43) are similar to those reported for novel acidic PR peroxidases

Peroxidase activity (pH 6.0)

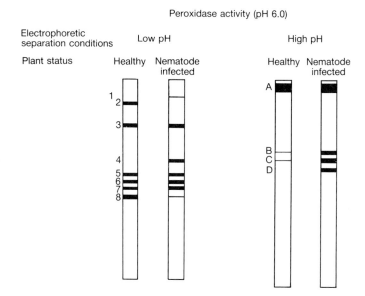

Fig. 14.2. Diagrammatic representation of peroxidase isoenzyme patterns of IF samples recovered from the leaves of healthy and nematode infected plants (Day 16). IF (60 µg protein) were separated by native PAGE (Davis, 1964, Reisfeld *et al.* 1962) and peroxidase activity was revealed with guaicol/H_2O_2 (Mäder *et al.* 1975).

detected in potato leaves after leaf infection by tobacco mosaic virus (Parent *et al.* 1988).

Conclusions

These results reflect changes in the pattern of biosynthesis of peroxidase isoenzymes, or changes in the structure of the cell wall such that the extractability of peroxidases into the IF alters. In either case, the data demonstrate that systemic changes in peroxidases do occur following root invasion by the nematode. Since race-specific changes were not observed, it is probable that the peroxidase response, like the PR-protein response, reflects a general stress rather than interactions associated with resistance or susceptibility. Earlier work (Zacheo 1987) has demonstrated increases in peroxidase activity within the root in the vicinity of the nematode, but the study did not determine whether the peroxidases were of animal or plant origin.

Changes in proteinase inhibitors

Wounding is known to induce the accumulation of a family of proteinase inhibitors (PI) in many monocotyledonous and dicotyledonous plant species (Ryan 1978). These potent inhibitors of serine proteases are themselves proteins, and are thought to be defence-related in that they reduce the nutritional value of the plant tissue and deter grazing pests (Green and Ryan 1972; Brown *et al.* 1985).

The proteinase inhibitor proteins I and II are well characterized (Ryan 1978) and their corresponding genes have been isolated in potato (Sanchez-Serrano *et al.* 1986; Cleveland *et al.* 1987) and tomato (Graham *et al.* 1985a,b; Cleveland *et al.* 1987). These two proteinase inhibitors (I & II) are abundantly expressed in potato tubers (Sanchez-Serrano *et al.* 1986) and, following wounding, at both the local wound site and systemically in non-wounded leaves (Pena-Cortes *et al.* 1988). Furthermore, Northern blot analysis of total leaf RNA has demonstrated accumulation of mRNA of the PI II gene following wounding of potato leaves or attached tubers (Pena-Cortes *et al.* 1988).

Two complementary analytical assays were used to investigate the effect of invasion by Ro1 and Ro2 in the plant (i) the detection of PI activity based on the inhibition of α-chymotrypsin-induced degradation of iodoinsulin substrate (Doherty *et al.* 1988), (ii) Northern blot analysis using a potato cDNA clone to PI II gene and a heterologous cDNA clone to tomato PI I gene (Hammond-Kosack *et al.* 1990).

Synchronized nematode root infections were found to induce accumulation of high abundance PI II (800bp) and PI I (600 bp) transcripts in leaves and upper stem sections, but no mRNA species were detected in roots or the basal portion of the stem. There was no quantitative difference in level of either transcript following infection with pathotypes Ro1 or Ro2 nor did the patterns of RNA accumulating in various tissues change. Mechanical wounding of roots, stems, or leaves gave rise to PI I and PI II transcript accumulation in all leaves and upper parts of the stem, but no induction in roots or the lower stem. This is consistent, and extends the observations of Pena-Cortes *et al.* (1988), who demonstrated PI II expression in upper and lower leaves and upper stem, but no induction in lower stem or in roots following wounding of potato leaves and tubers. There was no difference in the levels of the transcript encoding the small subunit of RUBISCO between Ro1, Ro2, and sham-inoculated potato plants. Rather, an accumulation of ssRUBISIO transcript was shown to be light and developmentally regulated, appearing only in photosynthetic tissue and most abundantly in young developing leaves. Total proteinase inhibitor activity levels were found to parallel the PI I and PI II transcript data except in the lower stem region where PI activity could be detected.

Conclusions

These results provide the first demonstration of the systemic induction of PIs in response to signals originating from root tissue following pathogen invasion or mechanical damage (Hammond-Kosack *et al.* 1990). The pattern of induction was identical for infection by both pathotypes of nematode, again suggesting that PI accumulation reflects a general stress response.

General Conclusions

The results described indicate that cyst-nematode infection induces many changes in plant gene expression, both at the local site of pathogen invasion and elsewhere in the infected potato plant. These changes encompass general stress and/or wound responses, and more specific responses to particular nematode pathotypes. As yet, the signalling events that determine these changes are unknown, but could in principle arise from the nematode (secretions, excretions, or surface components), and/or from the interaction of the pathogen with the plant cells. For example, it is clear that tissue damage is caused during the early stages of nematode migration through the root, and this could lead to the release of bioactive oligosaccharides (Ryan 1987) that may be responsible for systemic accumulation of PIs. Similarly, tissue damage is known also to affect ethylene levels and these in turn may be related causally to the induction of PR proteins. In addition to these general effects, however, there are highly specific events that occur rapidly and must reflect an equally high specificity in recognition between the plant and pathogen.

The data presented in this article provide a basis for analysing the signalling events involved in this interaction, as well as the means for understanding the spatial and temporal regulation of gene expression during a defence response.

Acknowledgements

The research has been supported by research grants to DJB/HJA (Perry Foundation; AFRC grant no. PG24/249; Enichem America) and a University of Leeds Pool Post Fellowship (KEH).

References

Anon (1986). Laws Agricultural Trust, Rothamsted.
Atkinson, H.J. and Harris, P.D. (1989). Changes in nematode antigens recognised by monoclonal antibodies during early infections of soya beans with the cyst nematode *Heterodera glycines*. *Parasitology* **98**, 479–87.
Bowles, D.J. (1990). Defence-related proteins in higher plants. *Annual Review of Biochemistry* **59**, 873–907.
Brown, W., Takio, K., Titani, K., and Ryan, C.A. (1985). Wound-induced trypsin

inhibitor in alfalfa leaves. Identity as a member of the Bowman-Birk inhibitor family. *Biochemisty* **24**, 2105–12.

Cleveland, T.E., Thornburg, R.W., and Ryan, C.A. (1987). Molecular characterisation of a wound-inducible inhibitor I gene from potato and the processing of its mRNA and protein. *Plant Molecular Biology* **8**, 199–207.

Conjero, V., Picazo, I, and Segado, P. (1979). Citrus exocortis viroid (CEV): protein alterations in different hosts following viroid infection. *Virology* **97**, 454–6.

Davies, E., Ramaiah, K.V.A., and Abe, S. (1986). Wounding inhibits protein synthesis yet stimulates polysome formation in aged, excised pea epicotyls. *Plant and Cell Physiology* **27**, 1377–86.

Davis, B.J. (1964). Disc electrophoresis II. Method and application to human serum proteins. *Annals of the New York Academy of Sciences* **121**, 404–27.

DeWit, P.J.G.M. and Spikeman, G. (1982). Evidence for the occurrence of race cultivar-specific elicitors of necrosis in intercellular fluids of compatible interaction of *Cladosporium fulvum* and tomato. *Physiological Plant Pathology* **21**, 1–11.

DeWit, P.J.G.M. and Van Der Meer, F.E. (1986). Accumulation of the pathogenesis related tomato leaf protein P14 as an early indicator of incompatibility in the interaction between *Cladosporium fulvum* (syn. *Fulvia fulva*) and tomato. *Physiological and Molecular Plant Pathology* **28**, 203–14.

Doherty, H.M., Selvendran, R.R., and Bowles, D.J. (1988). The wound response to tomato plants can be inhibited by aspirin and related hydroxy-benzoic acids. *Physiological and Molecular Plant Pathology* **33**, 377–84.

Espelie, K.E., Franceschi, V.R., and Kolattukudy, P.E. (1986). Immunocytochemical localisation and time course of appearance of an anionic peroxidase associated with suberization in wound-healing potato tuber tissue. *Plant Physiology* **81**, 487–92.

Fraser, R.S.S. (1981). Evidence for the occurrence of the 'pathogenesis-related' proteins in leaves of healthy tobacco plant during flowering. *Physiological Plant Pathology* **19**, 69–76.

Graham, J.J., Pearce, G., Merryweather, J., Titani, K., Ericsson, L., and Ryan, C.A. (1985a). Wound-induced proteinase inhibitors from tomato leaves. I. The cDNA deduced primary structure of pre-inhibitor 1 and its post-translational processing. *Journal of Biological Chemistry* **260**, 6555–60.

Graham, J.S., Pearce, G., Merryweather, J., Titani, K., Ericsson, L., and Ryan, C.A. (1985b). Wound-induced proteinase inhibitors from tomato leaves. II. The cDNA deduced primary structure of pre-inhibitor II. *Journal of Biological Chemistry* **260** 651–4.

Green, T.R. and Ryan, C.A. (1972). Wound-induced proteinase inhibitor in plant leaves: a possible defence mechanism against insects. *Science* **175**, 776–7.

Greppin, H., Penel, C., and Gaspar, Th. (1986). *Molecular and physiological aspects of plant peroxidases*. University of Geneva Press. Geneva.

Hague, N.G.M. and Gowen, S.R. (1987). Chemical control of nematodes. In *Principle and practice of nematode control in crops*, (ed. R.H. Brown and B.R. Kerry). Academic Press, Sydney. 447pp.

Hammond-Kosack, K.E., Atkinson, H.J., and Bowles, D.J. (1989a). Changes in abundance of translatable mRNA species in potato roots and leaves following root invasion by cyst-nematode *G. rostochiensis* pathotypes. *Physiological and Molecular Plant Pathology* (in press).

Hammond-Kosack, K.E., Atkinson, H.J., and Bowles, D.J. (1989b). Systemic accumulation of novel proteins in the apoplastic space of leaves following invasion of potato plant roots by the cyst nematode *Globodera rostochiensis Physiological and Molecular Plant Pathology* **35**, 495–506.

Hammond-Kosack, K.E., Gurr, S.J., Atkinson, H.J., and Bowles, D.J. (1990). Systemic induction of proteinase inhibitor gene expression in response to cyst nematode invasion of potato plant roots. *Physiological and Molecular Plant Pathology* (submitted).

Jones, M.G.K. (1981). In Plant Parasitic Nematodes Vol. III (ed. B.M. Zuckerman and R.A. Rohde) Academic Press, New York. p. 255.

Kauffmann, S., Legrand, M., Geoffroy, P., and Fritig, B. (1987). Biological function of 'pathogenesis-related' proteins: four PR proteins of tobacco have 1, 3-*β*-glucanase activity. *EMBO Journal* **6**, 3209–12.

Kombrink, E., Schroder, M., and Hahlbrock, K. (1988). Several 'pathogenesis-related' proteins in potato are 1, 3-*β*-glucanases and chitinases. *Proceedings of the National Academy of Sciences USA* **85**, 782–6.

Kort, J., Ross, H., Rumpenhorst, H.J., and Stone, A.R. (1977). An international scheme for identifying and classifying pathotypes of potato cyst-nematode *Globodera rostochiensis* and *G. pallida*. *Nematologica* **23**, 333–9.

Lagrimini, L.M. Burkhart, W., Moyer, M., and Rothstein, S. (1987). Molecular cloning of complementing DNA encoding the lignin-forming peroxidase from tobacco: Molecular analysis and tissue specific expression. *Proceedings of the National Academy of Sciences USA* **84**, 7542–6.

Lamport, D.T.A (1986). In *Molecular and physiological aspects of plant peroxidases* (eds. H. Greppin, C. Penel and T. Gaspar). University of Geneva Press, Geneva. p. 199.

Legrand, M., Kauffmann, S., Geoffroy, P., and Fritig, B. (1987). Biological function of 'pathogenesis-related' proteins: Four tobacco PR-proteins are chitinases. *Proceedings of the National Academy of Sciences USA.* **84**, 6750–4.

Lotan, T., Ori, N., and Fluhr, R. (1989). Pathogenesis-related proteins are developmentally regulated in tobacco flowers. *The Plant Cell* **1**, 881–7.

Mader, M., Munch, P., and Bopp, M. (1975). Regulation of peroxidase patterns during shoot differentiation in callus cultures of *Nicotiana tabacum* L. *Planta* **123**, 257–65.

Nassuth, A. and Sanger, H.L. (1986). Immunological relationship between 'pathogenesis-related' leaf proteins from tomato, tobacco and cowpea. *Virus Research* **4**, 229–42.

Parent, J.G. and Asselin, A. (1984). Detection of pathogenesis-related proteins and other proteins in the intercellular fluid of hypersensitive plants infected with tobacco mosaic virus. *Canadian Journal of Botany* **62**, 564–9.

Parent, J.G. and Asselin, A. (1987). Acidic and basic extracellular pathogenesis-related leaf proteins from fifteen potato cultivars. *Phytopathology* **77**, 1122–5.

Parent, J.G., Hogue, R., and Asselin, A. (1988). Serological relationships between pathogenesis-related proteins from four *Nicotiana* species, *Solanum tuberosum*, and *Chenopodium amaranticolor*. *Canadian Journal of Botany* **66**, 199–202.

Pena-Cortes, H., Sanchez-Serrano, J., Rocha-Sosa, M., and Willmitzer, L. (1988). Systemic induction of proteinase-inhibitor-II gene expression in potato plants by wounding. *Planta* **174**, 84–9.

Reisfeld, R.A., Lewis, U.J., and Williams, D.E. (1962). Disk electophoresis of basic proteins and peptides on polyacrylamide gels. *Nature* **195**, 281–3.

Rice, S.L., Leadbeater, B.S.C., and Stone, A.R. (1985). Changes in cell structure in roots of resistant potatoes parasitized by potato cyst-nematodes. I. Potatoes with resistance gene H_1 derived from Solanum *tuberosum* spp. *andigena*. *Physiological Plant Pathology* **27**, 219–34.

Robinson, M.P., Atkinson, H.J., and Perry, R.N. (1988). The association and partial characterisation of a fluorescent hypersensitive response of potato roots to the cyst nematodes, *Globodera rostochienis* and *G. pallida*. *Revue Nematology* **11**, 99–107.

Ryan, C.A. (1978). Proteinase inhibitors in plant leaves: a biochemical model for pest-induced natural plant protection. *Trends in Biochemical Science* **3**, 148–50.

Ryan, C.A. (1987). Oligosaccharide signalling in plants. *Annual Review of Cell Biology* **3**, 295–317.

Sanchez-Serrano, J., Schmidt, R., Schell, J., and Willmitzer, L. (1986). Nucleotide sequence of proteinase inhibitor II encoding cDNA of potato and its mode of expression. *Molecular and General Genetics* **203**, 15–20.

Sidhu, G.S. and Webster, J.M. (1981). The genetics of plant nematode parasitic systems. *The Botanical Review* **47**, 387–419.

Smith, J.A. and Hammerschmidt, R. (1988). Comparative study of acidic peroxidases associated with induced resistance in cucumber, musk lemon and watermelon. *Physiological and Molecular Plant Pathology* **33**, 255–61.

Taylor, P.N. (1987). Inducible systemic resistance to bacterial and fungal diseases in plants. *Outlook on Agriculture* **16**, 198–202.

Theillet, C., Delpeyroux, F., Fiszman, M., Reigner, P., and Esnault, R. (1982). Influence of the excision shock on the protein metabolism of *Vicia faba* L. meristematic root tissue. *Planta* **155**, 478–85.

Turner, S.J. and Stone, A.R. (1984). Development of potato cyst-nematodes in roots of resistant *Solanum tuberosum* ssp. *andigena* and *S. vernei* hybrids. *Nematologica* **30**, 324–32.

Van Loon, L.C. (1985). Pathogenesis-related proteins. *Plant Molecular Biology* **4**, 111–16.

White, R.F., Rybicki, E.P., Von Wechmar, M.B., Dekker, J.L. and Antoniw, J.F. (1987). Detection of PR 1–type proteins *Amaranthaceae, Chenopodiaceae, Graminae* and *Solanaceae* by immunoelectroblotting. *Journal of General Virology* **68**, 2043–8.

Zacheo, G. (1987). In *Cyst nematodes NATO ASI series*, Vol. 121, p. 163. (ed. F. Lamberti and C.E. Taylor). Plenum Press, New York.

15 Molecular responses of potato to infection by *Phytophthora infestans*

ERICH KOMBRINK, KLAUS HAHLBROCK, KARIN HINZE, and MARTIN SCHRÖDER

Max-Planck-Institut für Züchtungsforschung, Abteilung Biochemie, Carl-von-Linné-Weg 10, D-5000 Köln 30, FRG

Introduction

In their natural environment plants are exposed to a large number of potentially pathogenic micro-organisms. Heterotrophically growing organisms, such as the majority of bacteria and fungi, depend on organic material synthesized by autotrophic plants as their source of essential and valuable nutrients. However, despite their large number only very few micro-organisms have acquired the ability to penetrate and colonize living plants successfully, thereby evolving from saprophytic organisms to true biotrophic or hemibiotrophic (perthotrophic) pathogens (Lewis 1973; Bailey 1983).

The host-range of a given pathogen is usually limited to specific genotypes within one or a few plant species, and sometimes even to specific organs of a plant. All other plants are resistant to that pathogen and are therefore called non-hosts (Bailey 1983). Most host–pathogen combinations have evolved from cultured crop plants where extensive breeding has selected a large diversity of distinct cultivars and, as a consequence, has led to co-evolution of an equally large number of races of a given pathogen (pathotypes). Besides yield and nutritional value, disease resistance has been therefore a major goal in breeding programmes, and for decades breeders have identified, utilized and phenotypically described the effects of resistance genes (R genes), despite their unknown biochemical functions. If the presence of an R gene renders a cultivar resistant, plant and pathogen undergo an incompatible interaction and the pathotype is avirulent. By contrast, a susceptible cultivar undergoes a compatible interaction with a virulent race of the pathogen. For several pathosystems such type of race–cultivar specificity which show, operationally defined gene-for-gene relationships have been demonstrated in which formally the R genes of the host correspond to avirulence or virulence genes of the pathogen (Flor 1971; Bailey 1983; Ellingboe 1982).

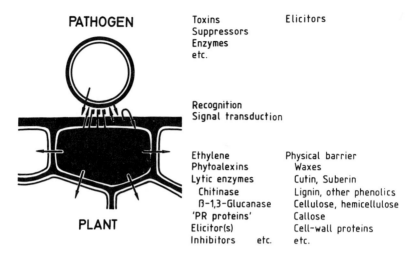

PATHOGEN

Toxins   Elicitors
Suppressors
Enzymes
etc.

Recognition
Signal transduction

Ethylene   Physical barrier
Phytoalexins   Waxes
Lytic enzymes   Cutin, Suberin
  Chitinase   Lignin, other phenolics
  β-1,3-Glucanase Cellulose, hemicellulose
'PR proteins'   Callose
Elicitor(s)   Cell-wall proteins
Inhibitors etc. etc.

PLANT

Fig. 15.1. Updated scheme adapted from Hahlbrock and Scheel (1987) summarizing major components of plant–pathogen interactions. It emphasizes the temporal and spatial sequence of events: recognition of a pathogen leads to hypersensitive cell death and the spread of signals activates defence responses in neighbouring cells.

Using potato (*Solanum tuberosum*) cultivars with phenotypically defined R genes controlling resistance to the late-blight fungus *Phytophthora infestans* (*Pi*) (Black *et al.* 1953), our research goal is to understand the molecular mechanisms leading to expression of resistance. Eventually this knowledge should help us to achieve better control of disease development in potato as well as in other crop plants.

General aspects of plant–pathogen interactions

Plants challenged with virulent or avirulent pathogens show a wide array of biochemical responses. Many of these reactions are considered to be directly or indirectly involved in the active defence of the plant and have been extensively reviewed elsewhere (Bell 1981; Sequeira 1983; Hahlbrock and Scheel 1987). A simple and generalizing scheme summarizing those biochemical reactions and components which have been observed most frequently to occur after inoculation of various host and non-host plants with various types of pathogen is presented in Fig. 15.1.

It is well established that a variety of mechanisms may be required for pathogenicity of fungi and bacteria for example enzymes capable of degrading plant cell walls, or toxins which inhibit reactions in the plant cell. Toxins may be identical with those molecules which have been postulated to suppress certain defence reactions of the plant, and thus convert incompatible interactions to compatible ones (Yoder 1980).

The defence mechanisms of the plant can formally be classified as preformed or induced. The rigid structure of the cell wall is one example of a preformed physical barrier. However, this barrier is often rapidly reinforced and thickened in infected tissue by the deposition of additional material, such as callose, cell-wall proteins, lignin and lignin-like, or other phenylpropanoid material. The classical example of an induced defence response is accumulation of phytoalexins, which are defined as low molecular weight substances with antimicrobial activity. Other defence-related reactions include formation of ethylene, synthesis of hydrolytic enzymes, such as 1,3-β-glucanase, and chitinase, and additional proteins with unknown functions, so-called 'pathogenesis-related' (PR) proteins.

Induction of phytoalexin synthesis, or of other typical defence reactions can be triggered by compounds termed elicitors, which can be molecules of very different types such as oligosaccharides, proteins, lipids, etc. Elicitors isolated from cell walls of both pathogen and plant, have been studied extensively (West 1981; Darvill and Albersheim 1984). Because of their putative function as signal molecules in plant–pathogen interactions, they have acquired great practical importance in phytopathology, since they trigger the same biochemical responses in plants and cultured cells as does infection (Kombrink *et al.* 1986; Fritzemeier *et al.* 1987; Hahlbrock and Scheel 1989).

Very little is known at present about the molecules mediating recognition and signal transduction between pathogen and plant cell, although the importance of these processes in plant–pathogen interactions has been recognized (Daly 1984; Ralton *et al.* 1987; Boller 1988b). In a recent review article Parker *et al.* (1989) discuss practical approaches to the examination of recognition and signalling events, with special emphasis on interactions involving *Phytophthora*.

We will concentrate in the following on a brief presentation of those three types of defence response which we have studied in some detail in potato tissue challenged with *P. infestans*: accumulation of phytoalexins, wall-bound phenolics, and PR proteins including the hydrolytic enzymes 1,3-β-glucanase and chitinase. The results presented emphasize our special interest in the elucidation of molecular mechanisms involved in expression of race-cultivar specific resistance, and the temporal and spatial dynamics of individual plant cell responses to the invading fungus.

Cytology of potato–*P. infestans* interactions

When the two potato cultivars Datura, carrying R gene 1, and Isola, carrying R gene 4, are inoculated separately with races 1 (*Pi*1) and 4 (*Pi*4) of

P. infestans, the double-reciprocal infection square typical for such race-cultivar specific interactions is obtained (Rohwer *et al.* 1987). The compatible interactions (Datura-*Pi*1 and Isola-*Pi*4) are characterized by the appearance of massive chlorosis and necrosis, and the formation of large water-soaked lesions. The fungus colonizes the whole leaf and usually sporulates on the whole leaf surface within 5 to 7 days post-inoculation.

By contrast, in the incompatible interactions (Datura-*Pi*4 and Isola-*Pi*1) the inoculated leaves remain green and healthy even after 7 days. The only visible sign of infection is the presence of small localized lesions, resulting from the death of a few cells around the infection sites. This typical hypersensitive response (HR) results in restriction of fungal growth within the necrotic areas. Depending on the plant cultivar and fungal pathotype used, the symptoms can vary considerably, from clearly visible to detectable only under the microscope (Cuypers and Hahlbrock 1988).

The earliest responses of plants to infection are detectable only at the cytological level. They include callose deposition in both compatible and incompatible interactions (Cuypers and Hahlbrock 1988). A yellow fluorescence, detectable only in incompatible interactions and which may possibly be linked to the expression of resistance, appears in plant cells directly surrounding those fungal hyphae that have successfully penetrated the plant tissue (N. Arabatzis, M. Schröder, unpublished). This fluorescence, visible under UV light and probably due to the accumulation of phenolic material, is detectable about 2–3 h postinoculation and is associated with hypersensitive death of the affected cells. In compatible interactions the fungus apparently escapes recognition by the plant, as indicated by the absence of both fluorescing cells and of the typical rapid browning reaction of the HR (Cuypers and Hahlbrock 1988; N. Arabatzis, M. Schröder, unpublished).

Phytoalexins

Phytoalexins represent a structurally diverse group of compounds (Bailey and Mansfield 1982; Darvill and Albersheim 1984; Kuć and Rush 1985). Depending on the plant species examined, the groups of compounds identified as phytoalexins include diterpenes, sesquiterpenes, furanocoumarins, isoflavonoids, polyacetylenes, and many more. However, they have in common both antimicrobial activity and accumulation in response to infection (Bailey and Mansfield 1982). It is interesting to note that the system of potato tubers inoculated with *P. infestans* has served as a basis for the phytoalexin concept proposed by Müller and Börger (1940). Since then the sesquiterpenoid phytoalexins, rishitin, lubimin, phytuberin, and solavetivone have been studied extensively as compounds accumulating in potato tubers in response to treatment with pathogens or elicitors (Kuć and Rush 1985).

However, none of these compounds could be detected under any conditions in either infected or elicitor-treated potato leaves (Rohwer *et al.* 1987), which are the primary sites for infection by *P. infestans* (Thurston and Schulz 1981; Rich 1983). Consequently, accumulation of these phytoalexins cannot be an essential part of the plant's defence response in this organ. This does not rule out the possibility that other types of phytoalexin, as yet unidentified, are synthesized in potato leaves.

Phenylpropanoid metabolism

As described above, rapid browning is associated with hypersensitive cell death. Although the biochemical reactions involved in this browning process are poorly understood, they probably include accumulation and oxidative linkage of phenylpropanoid compounds and their desposition in the cell wall (Tomiyama 1983). The first reactions of phenylpropanoid metabolism, the pathway converting phenylalanine to 4-coumaroyl-CoA is referred to as 'general phenylpropanoid pathway'. Its key role is to provide precursors for a variety of different soluble and insoluble secondary plant products, such as coumarins, flavonoids, stilbenoids, suberin, lignin, and other phenolic compounds (Hahlbrock and Scheel 1989). The importance of the general phenylpropanoid pathway in plant defence is indicated by the fact that it is activated in all plant–pathogen interactions analysed so far. The enzymes of this pathway are phenylalanine ammonia-lyase (PAL), cinnamate 4-hydroxylase (C4H), and 4-coumarate:CoA ligase (4CL). Of these PAL and 4CL have been studied extensively in potato at both the protein and gene levels (Fritzemeier *et al.* 1987; Hahlbrock and Scheel 1989).

Various soluble and insoluble phenylpropanoid compounds have been reported to accumulate in potato leaves and tubers in response to infection, elicitor treatment or wounding (for review, see Hahlbrock and Scheel 1989). Those wall-bound phenolics in which changes occur include lignin (Hammerschmidt 1984), lignin-like material (Friend 1981; Henderson and Friend 1979), as well as hydroxycinnamate derivatives, such as 4-coumaroyltyramine and feruloyltyramine (H. Keller and D. Scheel, unpublished).

In leaves, tubers, and cell cultures which have been infected or elicitor-treated, accumulation of phenolic compounds is generally preceded by co-ordinate increase in the activities of the biosynthetic enzymes PAL and 4CL (Henderson and Friend 1979; Fritzemeier *et al.* 1987). Accumulation of hydroxycinnamate amides also requires stimulation of tyrosine decarboxylase (TDC) (H. Keller and D. Scheel, unpublished). Induction of both PAL and 4CL was shown to be regulated at the transcriptional level (Fritzemeier *et al.* 1987). Nuclear run-off experiments demonstrated

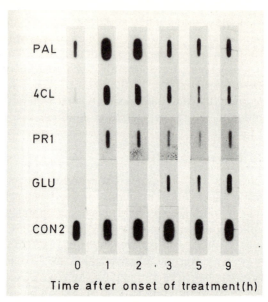

Fig. 15.2. Changes in transcription rates of various defence-related genes in nuclei isolated from potato leaves (cv. Datura) treated for the indicated times with P. *infestans*-derived elicitor. Equal amounts of ^{32}P-labelled run-off transcript were hybridized to 1 μg of each linerized plasmid immobilized on nitrocellulose filters. The amount of RNA bound to a specific cDNA clone was visualized by autoradiography. PAL, 4CL, and PR1 represent published potato cDNA clones (Fritzemeier *et al.*1987; Taylor *et al.* 1990), whereas the CON2 cDNA clone represents a constitutively expressed parsley gene of unknown function (Somssich *et al.* 1989). 1,3-β-glucanase (GLU) specific signal was detected using the tobacco 1,3-β-glucanase cDNA clone pGL43 (Mohnen *et al.* 1985).

that PAL and 4CL mRNAs belong to the most rapidly accumulating transcripts in elicitor-treated potato leaves (Fig. 15.2). In potato PAL is encoded by a large family of 30 to 40 genes (H.-J. Joos and G. Strittmatter, unpublished), whereas 4CL is encoded only by 2 genes per haploid genome (M. Becker-André and K. Hahlbrock, unpublished).

The importance of the phenylpropanoid metabolism in the resistance of potato was assessed by analysis of different race–cultivar combinations, with respect to the induction patterns of the enzyme activity and amount of mRNA for both PAL and 4CL. No appreciable differences were observed between compatible and incompatible interactions at early infection stages. In contrast at 20–30 h postinoculation higher levels of both PAL and 4CL mRNA accumulated in compatible interactions (Fritzemeier *et al.* 1987; Taylor *et al.* 1990; H.-J. Joos and G. Strittmatter, unpublished). In view of these results it appears unlikely that PAL and 4CL play a decisive role in race–cultivar specific resistance of potato to P. *infestans*. However, the experimental approach, i.e., extraction of RNA from whole leaves, is obviously insufficient to resolve localized differences in gene expression at the cellular level, as they are expected to occur at very early stages of the plant–pathogen interaction. To circumvent such problems, additional histological methods were applied (see below).

Pathogenesis-related proteins

Accumulation of PR proteins has been described in many plant species infected with various types of pathogen, such as fungi, bacteria, viruses, and viroids (van Loon 1985). Most PR proteins share some characteristic properties, such as low molecular weight (M_r range $\sim$ 10 000–40 000), selective solubility at low pH, high resistance to proteolytic degradation, and predominant accumulation in the intercellular space of the leaf (van Loon 1985). PR proteins were first identified in tobacco (*Nicotiana tabacum*) infected with tobacco mosaic virus (van Loon and van Kammen 1970), but have since been studied in numerous systems.

We have investigated the proteins accumulating in the intercellular space of potato leaves following inoculation with *P. infestans*. As with other defence responses, elicitor treatment of leaves induced accumulation of the same proteins as did infection with the pathogen, except that the effects with elicitor were more rapid and more marked (Kombrink *et al.* 1988). At least nine major proteins, in the M_r range of 10 000 to 40 000, accumulated in large quantities and by this and other criteria were classified as PR proteins. However, some of these proteins were also found to accumulate within the cell itself.

Several proteins were purified to apparent homogeneity from the intercellular washing fluid (IWF), and a predominant band with M_r 36 000 was shown to consist of two isoenzymes of 1,3-β-glucanase (Kombrink *et al.* 1988). Chitinase activity present in IWF and cellular extracts could be resolved into a total of six isoenzymes with M_r ranging from 33 000 to 38 000 (Kombrink *et al.* 1988). The combined activities of the separated isoforms accounted for most of the respective enzyme activity extractable from leaves. Oligomers of 2 to 5 units, occurring in variable amounts, were the main products of all six chitinases, whereas the monomer (N-acetyglucoseamine) was not detected. This is clear evidence that all six isoenzymes have endo-chitinase activity, a property shared with all other plant chitinases analysed to date (Boller 1988a). Analogous results enabled us to classify the 1,3-β-glucanases also as endo-acting enzymes. The assignment of 1,3-β-glucanase and chitinase activities to PR proteins is not restricted to potato (Hahlbrock *et al.* 1987), but has been reported also for tobacco (Kauffmann *et al.* 1987; Legrand *et al.* 1987), cucumber (Métraux *et al.* 1988), tomato (Joosten and de Wit 1989), and maize

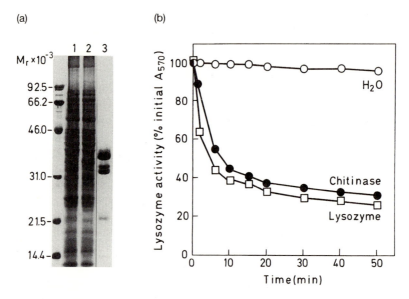

Fig. 15.3. Lysozyme activity of affinity purified potato chitinases. Total protein was extracted from potato leaves (cv. Datura) treated for 24 h with elicitor, passed through an affinity column of regenerated chitin, and bound chitinase was eluted with 20 mM acetic acid. For details, see Kombrink *et al.* (1988).

(A) Protein patterns of fractions obtained during the course of purification were resolved on SDS-PAGE: lane 1, total protein applied to the column (50 μg); lane 2, unbound fraction (50 μg); lane 3, bound fraction of purified chitinases (5 μg). Proteins were separated in 12.5 per cent acrylamide gels and stained with Coomassie brilliant blue. The major bands with M_r 33 000–38 000 represent six different chitinase isoenzymes.

(B) Lysozyme activity assay: 1 μg each of affinity purified potato chitinases (●) and of commercial chicken egg-white lysozyme (□) were incubated with 0.2 mg *Micrococcus luteus* cells suspended in 0.5 ml of 20 mM sodium acetate buffer (pH 5.5). The decrease in absorbance at 570 nm was determined at the times indicated. Heat-inactivated proteins and water controls (○) showed no appreciable decrease in absorbance.

(Nasser *et al.* 1988). Multiple forms of both enzymes have been purified and characterized from numerous plants (Boller 1988a)

An interesting feature of plant chitinases is their lysozyme activity (Boller *et al.* 1983; Boller 1985), again a feature shared by the potato chitinases. The protocol for purification of chitinases included an affinity-chromatography column by which almost all the chitinolytic activity present in potato leaf extracts could be isolated in a single step, at fairly high purity (Fig. 15.3A). This mixture of chitinases, when assayed for lysozyme

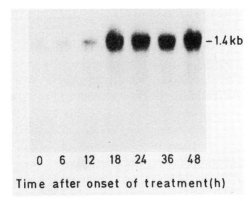

Fig. 15.4. Time course of 1,3-β-glucanase mRNA accumulation in potato leaves (cv. Datura) treated with elicitor. Total extractable RNA (20 μg/lane) was separated on a 1.2 per cent agarose-formaldehyde gel, blotted onto nitrocellulose and hybridized to the ^{32}P-labelled potato 1,3-β-glucanase cDNA clone pcG11–1, which was previously isolated using the tobacco homolog pGL43 [Hinze, K. (1987). Diploma thesis, Universität zu Köln]. Autoradiography revealed a single hybridizing band of about 1.4 kilobases (kb) in length. No appreciable increase in 1,3-β-glucanase mRNA was detectable with RNA extracted from water-treated leaves at the time points indicated (data not shown).

activity, was as active as commercially obtained chicken egg-white lysozyme tested at the same protein concentration (Fig. 15.3B).

In agreement with the timing of 1,3-β-glucanase and chitinase accumulation (Kombrink *et al.* 1988), the level of mRNA (Fig. 15.4) and the *in vitro* transcription rate for 1,3-β-glucanase (Fig. 15.2) increased strongly in elicitor-treated leaves, though more slowly than for PAL, 4CL, and PR1 (Fig. 15.2). Heterologous cDNA probes corresponding to tobacco 1,3-β-glucanase and chitinase (Mohnen *et al.* 1985; Shinshi *et al.* 1988; Shinshi *et al.* 1987) were used initially for these studies. Recent analysis of homologous cDNA and genomic clones has revealed a high degree (80–90 per cent) of sequence similarity to the tobacco clones, and that both proteins are encoded by gene families in potato (L. Beerhues, K. Hinze and E. Kombrink, unpublished).

As a consequence of many fungi containing 1,3-β-glucans, chitin, and other potential substrates in their cell walls (Bartnicki-Garcia 1968; Wessels and Seitsma 1981), it has long been proposed that 1,3-β-glucanase and chitinase participate in the active defence response of plants to pathogens (Abeles *et al.* 1971, Boller *et al.* 1983). The capability of both enzymes to degrade isolated fungal cell walls has been established (Young and Pegg 1982; Boller *et al.* 1983; Mauch *et al.* 1988) and it has recently been demonstrated that chitinase, either alone or in combination with 1,3-β-glucanase, can effectively restrict growth of several fungi (Schlumbaum *et al.* 1986; Mauch *et al.* 1988). Whether they also act on phytopathogenic bacteria, as the lysozyme activity suggests, is presently not known. Nevertheless 1,3-β-glucanase and chitinase appear to be part of the inducible defence response in all higher plants which have been investigated, even though definite proof of their participation in the mechanism of disease resistance is still lacking.

When comparing compatible and incompatible interactions of potato leaves with appropriate races of *P. infestans*, we were unable to detect appreciable differences in timing and total accumulating 1,3-β-glucanase and chitinase activities in leaf extracts. In the extracellular fraction, i.e., the IWF prepared from leaves, both enzymes behaved similarly with higher levels of enzyme activity accumulating in compatible interactions (unpublished results). The results are similar to those described previously for PAL (Fritzemeier *et al.* 1987). As with PAL, it is therefore difficult to deduce conclusively from these results whether or not 1,3-β-glucanase and chitinase play a decisive role in the mechanism of resistance to *P. infestans*. In tomato, *Verticillium* and *Fusarium* infection also induced accumulation of higher 1,3-β-glucanase and chitinase activities in compatible interactions (Pegg and Young 1981; Ferraris *et al.* 1987). In contrast the fungus *Cladosporium fulvum* growing extracellularly caused faster accumulation of both enzymes in the IWF prepared from leaves of incompatible interactions (Joosten and de Wit 1989).

Recently, the structure, genomic organization, and temporal pattern of expression of a gene encoding PR protein 1 (PR1) of potato have been described (Hahlbrock *et al.* 1989; Taylor *et al.* 1990). This protein differs from the PR proteins described above by the lack of any known function. It has a striking similarity, however, to the heatshock protein HSP 26 from soybean (Taylor *et al.* 1990). An interesting feature of PR1 is that activation of the gene and accumulation of its mRNA follow the same time courses as those previously observed for PAL and 4CL, and thus distinghishing PR1 from the 1,3-β-glucanase and chitinase accumulation kinetics (Fig. 15.2).

Histological analysis of defence responses in potato leaves

The lack of an unequivocal biochemical marker for incompatibility in the interaction of potato with *P. infestans* prompted us to analyse the temporal and spatial development of various defence responses in infected tissue by histochemical methods. Of particular interest was their occurrence in relation to penetration and growth of the fungus within the leaf. The methodology is restricted of course to those responses for which the appropriate probes, such as antisera and cDNAs, are available.

One prerequisite for the successful application of immunohistochemical methods is availability of monospecific antibodies. On protein blots the antiserum raised against potato 1,3-β-glucanase recognizes a single protein band in total leaf extracts and a double-band in the IWF (M_r 36 000), which migrates at the position of the two purified isoenzymes. In addition, a protein of M_r 38 000 is detected as a faint band, which has recently been identified as an additional 1,3-β-glucanase present in relatively minor amounts (unpublished results). In contrast, a complex pattern of cellular distribution and differential expression of isoenzymes has been observed for the chitinases of potato (B. Witte and E. Kombrink, unpublished).

When antiserum to 1,3-β-glucanase was applied to epoxy-embedded thin sections of chemically fixed leaves which had previously been inoculated with *P. infestans*, uniform labelling of the tissue sections revealed systemic accumulation of the antigen throughout the leaf (Hahlbrock *et al.* 1989). Again, no differential accumulation of 1,3-β-glucanase could be detected in sections derived from compatible and incompatible interactions (unpublished results). It should be mentioned, however, that this analysis was restricted to relatively late stages of infection (24–48 h postinoculation), when sufficient amounts of antigen had accumulated in the tissue. The partial extracellular localization of 1,3-β-glucanase as demonstrated by biochemical studies (see above), was confirmed by these investigations. Specific label was primarily associated with the cell wall, the intercellular space, and the cuticle, as well as with the cytoplasm. No label was detectable in non-infected tissue or with preimmune serum.

The apparent systemic accumulation of 1,3-β-glucanase was also observed at the level of mRNA. This was revealed by two types of experiment. Firstly, *in situ* hybridization using antisense RNA generated from tobacco 1,3-β-glucanase cDNA (Mohnen *et al.* 1985), indicated that 1,3-β-glucanase mRNA accumulation, though possibly initiated at infection sites, occurs throughout the infected leaf (Fig. 15.5). Secondly, accumulation of 1,3-β-glucanase mRNA and enzyme activity was not restricted solely to the infected areas of drop inoculated potato leaves, but occurred, with a short lag period, in the non-inoculated distal leaf areas as well, and was also found in adjacent, non-inoculated leaves (unpublished results). These results demonstrate the rapid, systemic spread of a signal that leads to a systemic gene activation in response to infection. Local as well as systemic induction of chitinase has been observed also in infected cucumber plants (Métraux and Boller 1986).

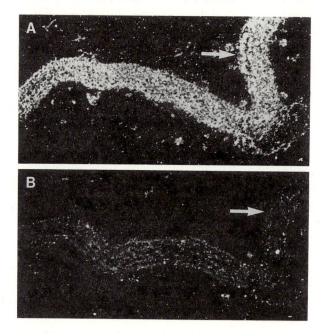

Fig. 15.5.   *In situ* localization of 1,3-β-glucanase transcripts in sections of potato leaves (cv. Datura) inoculated for 24 h with zoospores of *P. infestans* race 1 (compatible interaction).
(A) Autoradiography of a section after *in situ* hybridization to ^{35}S-labelled antisense transcript, and (B) of an adjacent section hybridized to sense transcript, both generated from the tobacco 1,3-β-glucanase cDNA clone pGL43 (Mohnen *et al.* 1985). Comparison of the silver grain distribution and density indicates the massive appearance of 1,3-β-glucanase mRNA throughout the leaf. The arrows point to the centre of an infection site.

The results obtained for induction of 1,3-β-glucanase are different from those reported for PAL and PR1 gene expression. In the last two cases, strong hybridization signals were observed locally around infection sites, whereas no significant changes above control values were found in distal tissue areas or non-infected leaf sections (Cuypers *et al.* 1988; Hahlbrock *et al.* 1989; Taylor *et al.* 1990), with the exception of cell-type specific constitutive expression of PAL in vascular bundles (unpublished results). A strictly localized accumulation of parsley PR1 and 4CL was observed also in infected parsley leaves (Hahlbrock *et al.* 1987; Somssich *et al.* 1988; Schmelzer *et al.* 1989). In agreement with the *in situ* hybridization data, we found that transient increases in PAL and mRNA and enzyme activity in drop inoculated leaves were also restricted to the vicinity of infection sites. No significant increases could be detected in distal non-infected leaf areas or beyond the inoculated leaf.

Conclusions

We have exploited the established race-cultivar specific interactions between potato and *P. infestans* to study various aspects of the multicomponent defence response in one system, and to correlate the biochemical responses with the cytological events. From these combined analyses a picture of the complex nature of plant–pathogen interactions starts to emerge. While participation of the classical sesquiterpenoid phytoalexins in the expression of resistance in potato can be ruled out, the role of the phenylpropanoid pathways or of the hydrolytic enzymes is still not clear.

In summary, the results obtained so far demonstrate that 1,3-β-glucanase and chitinase on the one hand, and PAL, 4CL, and PR1 on the other hand show different patterns of temporal as well as spatial expression in potato leaves upon infection by *P. infestans*. These differences in the mode of expression may reflect functional differences in the defence response. Rapid, localized gene activation (PAL, 4CL, PR1), probably following even more rapid reactions such as callose deposition and hypersensitive cell death, may be early lines of defence aimed at rapid localization and inhibition of growth of the invading pathogen. Once this has been achieved, a subsequent line of defence, involving genes which are activated more slowly and systemically (1,3-β-glucanase and chitinase) may serve as protection against additional later infections. The reason for this differential response could be related to cellular economy, thereby allowing an early full commitment to the essential, most immediate responses. Our cytological and histological studies do not exclude the possibility that the metabolic pathways represented by PAL and PR1 are involved in the expression of race-specific resistance in potato. However, it remains to be demonstrated whether differential patterns of expression occur at early stages of compatible and incompatible interactions.

In contrast it seems unlikely that 1,3-β-glucanase and chitinase determine race-specificity in potato, for the following reasons. *P. infestans* is an oomycete with a cell-wall structure of the cellulose-glucan type without chitin (Bartnicki-Garcia 1968). Consequently, chitinase cannot degrade the cell wall, unless an additional and as yet unknown enzymatic function is associated with the chitinase. It is interesting to note in this connection that, in compatible interactions between potato and *P. infestans* in which the leaves accumulate essentially the same amount of chitinase as in incompatible interactions, the fungus is not restricted in growth or prevented from sporulation.

Furthermore, crude plant extracts containing 1,3-β-glucanase and chitinase are apparently unable to inhibit growth of various *Phytophthora* species *in vitro* (Mauch *et al.* 1988; M. Schröder, B. Witte, unpublished results). However, an adverse effect of chitinase on other types of fungal or bacterial potato pathogens, and thereby a positive contribution to plant resistance, cannot of course be excluded.

In addition to a direct role in restricting fungal growth, at least in certain cases, 1,3-β-glucanase and chitinase have been discussed in relation to the recognition and signalling events between plant and pathogen, because both enzymes are capable of releasing elicitor molecules from fungal cell walls (Boller 1988b; Kurosaki *et al.* 1987). The extracellular localization of both enzymes, although not universally found in all plants (Boller 1988a; Mauch and Staehelin 1989), would be in agreement with such a function. It cannot be excluded, however, that both enzymes have additional roles in the metabolism or development of the plant which have not yet been identified.

Thus, many questions concerning disease resistance in plants and its biochemical basis remain, and more sophisticated techniques are required to unravel the complex molecular processes involved in plant-pathogen interactions. Although the plant may possess mechanisms to stop pathogen growth and development at any stage of the interaction, it is likely that the early events, recognition, signal transduction, and rapid gene activation, will determine the specificity and outcome of a given interaction.

References

Abeles, F.B., Bosshart, R.P., Forrence, L.E., and Habig, W.H. (1971). Preparation and purification of glucanase and chitinase from bean leaves. *Plant Physiology* **47**, 129–34.

Bailey, J.A. (1983). Biological perspectives of host–pathogen interactions. In *The dynamics of host defence* (ed. J.A. Bailey and B.J. Deverall), pp. 1–32. Academic Press, New York.

Bailey, J.A. and Mansfield, J.W. (1982). *Phytoalexins*. Blackie, Glasgow.

Bartnicki-Garcia, S. (1968). Cell wall chemistry, morphogenesis, and taxonomy of fungi. *Annual Review of Microbiology* **22**, 87–108.

Bell, A.A. (1981). Biochemical mechanisms of disease resistance. *Annual Review of Plant Physiology* **32**, 21–81.

Black, W., Mastenbroek, C., Mills, W.R., and Petersen, L.C. (1953). A proposal for an international nomenclature of races of *Phytophthora infestans* and of genes controlling immunity in *Solanum demissum* derivatives. *Euphytica* **2**, 173–8.

Boller, T. (1985). Induction of hydrolases as a defense reaction against pathogens. In *Cellular and molecular biology of plant stress* (ed. J.L. Key and T. Kosuge), pp. 247–62. Alan R. Liss, New York.

Boller, T. (1988a). Ethylene and the regulation of antifungal hydrolases in plants. In *Oxford surveys of plant molecular and cell biology* (ed. B.J. Miflin), Vol. 5, pp. 145–74. Oxford University Press, Oxford.

Boller, T. (1988b). Primary signals and second messengers in the reaction of plants to pathogens. In *Second messengers in plant growth and development* (ed. W.F. Boss and D.J. Marré), pp. 1–29. Alan R. Liss, New York.

Boller, T., Gehri, A., Mauch, F., and Vögeli U. (1983). Chitinase in bean leaves: induction by ethylene, purification, properties, and possible function. *Planta* **157**, 22–31.

Cuypers, B. and Hahlbrock, K. (1988). Immunohistochemical studies of compatible and incompatible interactions of potato leaves with *Phytophthora infestans* and of the nonhost response to *Phytophthora megasperma. Canadian Journal of Botany* **66**, 700–5.

Cuypers, B., Schmelzer, E., and Hahlbrock, K. (1988). *In situ* localization of rapidly accumulated phenylalanine ammonia-lyase mRNA around penetration sites of *Phytophthora infestans* in potato leaves. *Molecular Plant–Microbe Interactions* **1**, 157–60.

Daly, J.M. (1984). The role of recognition in plant disease. *Annual Review of Phytophology* **22**, 273–307.

Darvill, A.G. and Alversheim, P. (1984). Phytoalexins and their elicitors — a defence against microbial infection in plants. *Annual Review of Plant Physiology* **35**, 243–75.

Ellingboe, A.H. (1982). Genetical aspects of active defense. In *Active defense mechanisms in plants* (ed. R.K.S. Wood), pp. 179–92. Plenum Press, New York.

Ferraris, L., Abbattista Gentile, I., and Matta, A. (1987). Activation of glycosidases as a consequence of infection stress in *Fusarium* wilt of tomato. *Journal of Phytopathology* **118**, 317–25.

Flor, H.H. (1971). Current status of the gene-for-gene concept. *Annual Review of Phytopathology* **9**, 275–96.

Friend, J. (1981). Plant phenolics, lignification and plant disease. *Progress in Phytochemistry* **7**, 197–261.

Fritzemeier, K-H., Cretin, C., Kombrink, E., Rohwer, F., Taylor, J., Scheel, D., and Hahlbrock, K. (1987). Transient induction of phenylalanine ammonia-lyase and 4-coumarate:CoA ligase mRNAs in potato leaves infected with virulent or avirulent races of *Phytophthora infestans. Plant Physiology* **85**, 34–41.

Hahlbrock, K. and Scheel, D. (1987). Biochemical responses of plants to pathogens. In *Innovative approaches to plant disease control* (ed. I. Chet), pp. 229–54. John Wiley, New York.

Hahlbrock, K. and Scheel, D. (1989). Physiology and molecular biology of phenylpropanoid metabolism. *Annual Review of Plant Physiology and Plant Molecular Biology* **40**, 347–69.

Hahlbrock, K., Cretin, C., Cuypers, B., Fritzemeier, K-H., Hauffe, K-D., Jahnen, W., Kombrink, E., Rohwer, F., Scheel, D., Schmelzer, E., Schröder, M., and Taylor, J. (1987). Tissue specificity and dynamics of disease resistance responses in plants. In *Plant molecular biology*, (ed. D. von Wettstein and N.H. Chua), Nato ASI Series, Vol. A 140, pp. 399–406. Plenum Press, New York.

252 *Erich Kombrink* et al.

Hahlbrock, K., Arabatzis, N., Becker-André, M., Joos, H-J., Kombrink, E., Schröder, M., Strittmatter, G., and Taylor, J. (1989). Local and systemic gene activation in fungus-infected potato leaves. In *Signal molecules in plants and plant–microbe interactions* (ed. B.J.J. Lugtenberg), Nato ASI Series, Vol. H36, pp. 241–9, Springer Verlag, Berlin, Heidelberg.

Hammerschmidt, R. (1984). Rapid deposition of lignin in potato tuber tissue as a response to fungi non-pathogenic on potato. *Physiological Plant Pathology* **24**, 107–18.

Henderson, S.J. and Friend, J. (1979). Increase in PAL and lignin-like compounds as race-specific resistance response of potato tubers to *Phytophthora infestans. Phytophologische Zeitschrift* **94**, 323–34.

Joosten, M.H.A.J. and de Wit, P.J.G.M. (1989). Identification of several pathogenesis-related proteins in tomato leaves inoculated with *Cladosporium fulvum* (syn. *Fulvia fulva*) as 1,3-β-glucanases and chitinases. *Plant Physiology* **89**, 945–51.

Kauffmann, S., Legrand, M., Geoffroy, P., and Fritig, B. (1987). Biological function of 'pathogenesis-related' proteins: four PR proteins of tobacco have 1,3-β-glucanase activity. *EMBO Journal* **6**, 3209–12.

Kombrink, E., Bollmann, J., Hauffe, K.D., Knogge, W., Scheel, D., Schmelzer, E., Somssich, I., and Hahlbrock, K. (1986). Biochemical responses of non-host plant cells to fungi and fungal elicitors. In *Biology and molecular biology of plant-pathogen interactions* (ed. J.A. Bailey), Nato ASI Series, Vol. H 1, pp. 253–62, Springer-Verlag, Berlin, Heidelberg.

Kombrink, E., Schröder, M., and Hahlbrock, K. (1988). Several 'pathogenesis-related' proteins in potato are 1,3-β-glucanases and chitinases. *Proceedings of the National Academy of Sciences USA* **85**, 782–6.

Kuć, J. and Rush, J.S. (1985). Phytoalexins. *Archives of Biochemistry and Biophysics* **236**, 455–72.

Kurosaki, F., Tashiro, N., and Nishi, A. (1987). Induction, purification and possible function of chitinase in cultured carrot cells. *Physiological and Molecular Plant Pathology* **31**, 201–10.

Legrand, M., Kauffmann, S., Geoffroy, P., and Fritig, B. (1987). Biological function of pathogenesis-related proteins: four tobacco pathogenesis-related proteins are chitinases. *Proceedings of the National Academy of Sciences USA* **84**, 6750–4.

Lewis, D.H. (1973). Concepts in fungal nutrition and the origin of biotrophy. *Biological Reviews* **48**, 261–78.

Mauch, F. and Staehelin, L.A. (1989). Functional implications of the subcellular localization of ethylene-induced chitinase and β-1,3-glucanase in bean leaves. *The Plant Cell* **1**, 447–57.

Mauch, F., Mauch-Mani, B., and Boller, T. (1988). Antifungal hydrolases in pea tissue. II. Inhibition of fungal growth by combinations of chitinase and β-1,3-glucanase. *Plant Physiology* **88**, 936–42.

Métraux, J.P. and Boller, T. (1986). Local and systemic induction of chitinase in cucumber plants in response to viral, bacterial and fungal infections. *Physiological and Molecular Plant Pathology* **28**, 161–9.

Métraux, J.P., Streit, L., and Staub, T. (1988). A pathogenesis-related protein in cucumber is a chitinase. *Physiological and Molecular Pathology* **33**, 1–9.

Mohnen, D., Shinshi, H., Felix. G., and Meins, F., Jr. (1985). Hormonal regulation of β1,3-glucanase messenger RNA levels in cultured tobacco tissues. *EMBO Journal* **4**, 1631–5.

Müller, K.O. and Börger, H. (1940). Experimentelle Untersuchungen über die *Phytophthora*-Resistenz der Kartoffel. *Arb. Biol. Reichsanst. Land Forstwirtsch. Berlin-Dahlem* **23**, 189–231.

Nasser, W., de Tapia, M., Kauffmann, S., Montasser-Kouhsari, S., and Burkard, G. (1988). Identification and characterization of maize pathogenesis-related proteins. Four maize PR proteins are chitinases. *Plant Molecular Biology* **11**, 529–38.

Parker, J.E., Knogge, W., and Scheel, D. (1990). Molecular aspects of host-pathogen interactions in *Phytophthora* (in press).

Pegg, G.F. and Young D.H. (1981). Changes in glycosidase activity and their relationship to fungal colonization during infection of tomato by *Verticillium albo-atrum*. *Physiological Plant Pathology* **19**, 371–82.

Ralton, J.E., Smart, M.G., and Clarke, A.E. (1987). Recognition and infection processes in plant pathogen interactions. In *Plant–microbe interactions* Vol. 2, *Molecular and genetic perspectives* (ed. T. Kosuge and E.W. Nester), pp. 217–52. Macmillan, New York.

Rich, A.E. (1983). *Potato diseases*, Academic Press, New York.

Rohwer, F., Fritzemeier, K-H., Scheel, D., and Hahlbrock, K. (1987). Biochemical reactions of different tissues of potato (*Solanum tuberosum*) to zoospores or elicitors from *Phytophthora infestans*. *Planta* **170**, 556–61.

Schlumbaum, A., Mauch, F., Vögeli, U., and Boller, T. (1986). Plant chitinases are potent inhibitors of fungal growth. *Nature* **342**, 365–7.

Schmelzer, E., Krüger-Lebus, S., and Hahlbrock, K. (1989). Temporal and spatial patterns of gene expression around sites of attempted fungal infection in parsley leaves. *The Plant Cell* **1**, 993–1001.

Sequeira, L. (1983). Mechanisms of induced resistance in plants. *Annual Review of Microbiology* **37**, 51–79.

Shinshi, H., Mohnen, D., and Meins, F., Jr. (1987). Regulations of a plant pathogenesis-related enzyme: inhibition of chitinase and chitinase mRNA accumulation in cultured tobacco tissues by auxin and cytokinin. *Proceedings of the National Academy of Science USA* **84**, 89–93.

Shinshi, H., Wenzler, H., Neuhaus, J.-M., Felix, G., Hofsteenge, J., and Meins, F., Jr. (1988). Evidence for N- and C-terminal processing of a plant defense-related enzyme: primary structure of tobacco prepro-β-1,3-glucanase. *Proceedings of the National Academy of Science USA* **85**, 5541–5.

Somssich, I.E., Schmelzer, E., Kawalleck, P., and Hahlbrock, K. (1988). Gene structure and in situ transcript localization of pathogenesis-related protein 1 in parsley. *Molecular and General Genetics* **213**, 93–8.

Somssich, I.E., Bollmann, J., Hahlbrock, K., Kombrink, E., and Schulz, W. (1989). Differential early activation of defense-related genes in elicitor-treated parsley cells. *Plant Molecular Biology* **12**, 227–34.

Taylor, J.L., Fritzemeier, K.-H., Häuser, I., Kombrink, E., Rohwer, F., Schröder, M., Strittmatter, G., and Hahlbrock, K. (1989). Structural analysis and activation by fungal infection of a gene encoding a pathogenesis-related protein in potato. *Molecular Plant–Microbe Interactions* **3**, 72–7.

Thurston, H.D. and Schultz, O. (1981). Late blight. In *Compendium of potato diseases* (ed. W.J. Hooker), pp. 40–2. American Phytochemical Society, St. Paul.

Tomiyama, K. (1983). Research on the hypersensitive response. *Annual Review of Phytopathology* **21**, 1–12.

van Loon, L.C. (1985). Pathogenesis-related proteins. *Plant Molecular Biology* **4**, 111–16.

van Loon, L.C. and van Kammen, A. (1970). Polyacrylamid disc electrophoresis of the soluble leaf proteins from *Nicotiana tabacum* var. 'Samsun' and 'Samsun NN'. II. Changes in protein constitution after infection with tobacco mosaic virus. *Virology* **40**, 199–211.

Wessels, J.G.H. and Sietsma, J.H. (1981). Fungal cell walls: a survey. In *Encyclopedia of plant physiology*, New Series, 13 B, *Plant carbohydrates II* (ed. W. Tanner and F.A. Loewus), pp. 352–94. Springer-Verlag, Berlin.

West, C.A. (1981). Fungal elicitors of the phytoalexin response in higher plants. *Naturwissenschaften* **68**, 447–57.

Yoder, O.C. (1980). Toxins in pathogenesis. *Annual Review of Phytopathology* **18**, 103–29.

Young, D.H. and Pegg, G.F. (1982). The action of tomato and *Verticillium albo-atrum* glycosidases on the hyphal wall of *V. albo-atrum*. *Physiological Plant Pathology* **21**, 411–23.

16 Signal transduction involved in elicitation of phytoalexin synthesis in *Medicago sativa* L.

C.J. SMITH

Biochemistry Research Group, School of Biological Sciences, University College of Swansea, Singleton Park, Swansea SA2 8PP, UK

Introduction

Incompatibility between plants and fungal, bacterial, or viral pathogens is frequently expressed as a localized host-cell death, or hypersensitive response (HR), that occurs at the point of interaction of host and microorganism. It is an induced defence response that results in failure of the organism to infect. Restriction of the microorganism to the point of infection is not, however, a consequence solely of the death of host cells and a number of metabolic changes are known to occur in those cells surrounding the area of necrosis. Such changes include production of pathogenesis related proteins (van Loon 1985; Bol *et al.* Chapter 6) and inhibitors of proteases and polygalacturonases (Walker-Simmons and Ryan 1984; Degra *et al.* 1988); deposition into the cell wall of lignins (Pearce and Ride 1982) and hydroxyproline-rich glycoproteins (Roby *et al.* 1985); and production of a number of enzymes, some with hydrolytic activity against components of the microorganism, for example chitinases and glucanases (Kombrink *et al.* 1988, and Chapter 15 this volume), and others that are concerned with synthesis of various defence-related molecules, for example the enzymes of phytoalexin synthesis (Cramer *et al.* 1985a). Thus the potential exists for an array of metabolic changes to be induced in association with the HR and the subsequent restriction of growth of the microorganism. Not all the changes described above will occur in a single plant of course. The HR for a given plant–pathogen interaction appears to be more characterized by the metabolic capabilities of the plant and is less dependent upon the particular pathogen involved.

The relationship between the death of cells and the associated metabolic changes in the live cells is not entirely clear, though Bailey (1982) has proposed a mechanism in which induced synthesis of phytoalexins in the cells adjacent to the area of necrosis is causally related to cell death via release of endogenous oligosaccharide-elicitors from the dead cells. Fritig *et al.* (1987) have described a generalized model for the mechanism of the HR in which defence responses are induced in response to oligosaccharide sequences derived from the cell walls of pathogen and/or the plant. What is clear is the existence of an induction phase during which events that lead to the HR are initiated but where macroscopic changes are not evident (see Slusarenko Chapter 7), and that significant changes occur in the metabolic pattern of the cells during this phase preceding the collapse of the cell structure.

The large volume of literature published indicates the level of activity aimed at defining and understanding the mechanisms involved in the induction of such defence responses. Starting with identification and characterization of such responses in terms of the nature of the product, the kinetics and specificity of the response, etc., significant progress has been made towards identifying the molecular mechanisms underlying the response. Thus the elegant studies of the last few years have demonstrated that increases in gene activity are central to this class of defence response and have in turn raised questions relating to control of gene activity in plants. At the other end of the interaction, the host, studies have been performed to understand why some microorganisms induce a response and why others do not, what makes some cultivars susceptible and others resistant? A considerable effort has been made to identify the microbial elicitors of the inducible defence responses and now attempts are being made to identify the next link, the host receptor. As progress has been made at the two ends of the system consideration is being given to the system that links those two elements, the so-called signal transduction system.

Accumulation of phytoalexins as an inducible defence response

Accumulation of phytoalexins is often observed as a major characteristic of the HR, and the view has been frequently expressed that their rapid production at the site of infection has the potential to be the mechanism of resistance to potential pathogens, at least in some interactions (Mansfield 1982). Certainly phytoalexin synthesis is one of the most studied of the inducible defence responses expressed in plants, and a number of studies have been carried out in which the ability of the host to accumulate phytoalexins at the site of infection has been related to the degree of resistance to a particular phytopathogen (for an overview see Dixon 1986). Despite the substantial body of evidence to support induction of phytoalexin synthesis as a major determinant of resistance, however, in many incompatible interactions it is not the sole or major response, and resistance depends upon a combination of the types of mechanism described earlier. The precise combination and relative effectiveness of each mechanism will be specific to the plant tissue under consideration.

Nevertheless phytoalexin accumulation has a number of features in common with the other inducible defence responses referred to earlier, and so leaving aside arguments concerning its precise position as a determinant of resistance, it has proved to be a useful model for studying the mechanisms involved in the induction of defence responses in plants. It would be naïve to assume that the details of the mechanism that emerges for the induction of phytoalexin synthesis will necessarily be the same as those for the other induced responses. However, similarities can be recognized and induction of phytoalexin synthesis is in some respects more amenable to experimentation.

The most obvious of these similarities of course is that the products of the defence response are not normally present in healthy plant tissue, or not in the quantities expressed in the incompatible interaction. Thus the first event in the induction of the response will be perception of the potential pathogen by the plant tissues, and initiation of the subsequent train of events that leads to synthesis of lignin, proteins, etc. The synthesis induced by that perception requires increases in the activities of the responsible enzymes and such increases result from increases in the transcriptional activity of the genes that encode them. Generally the subsequent appearance of the defence-related molecule is rapid and, in addition to exposure to the pathogen, may result from treatment of plant cells with molecules that are derived from the pathogen and which are characteristic of it. These components, or elicitors, may be carbohydrates, glycoproteins, peptides (including enzymes), or fatty acids (Darvill and Albersheim 1984). A further common feature of such active defence responses is that the process involves an induction of host genes that is selective and does not extend to a general increase in gene activity of the plant. Thus only those genes related to the specific processes of the defence mechanism appear to be activated. Since many of the responses in question involve activation of several genes, and since for a given plant the same multigene response may arise from interaction of the plant with different pathogens or elicitors, then either the receptor for the elicitor molecules must have a broad specificity and be capable of activating specific families of genes, or the cell must possess a number of different receptors each of which interacts with a different elicitor but focuses on the same gene family.

Over the last several years studies have shown that activation of plant defence responses results in synthesis of proteins specific to the defence response, including enzymes of the phenylpropanoid pathway (Dixon *et al.* 1983; Hahlbrock and Scheel 1989), hydroxyproline-rich glycoproteins

(HRGP) of the cell wall (Roy *et al*. 1985), proteinase inhibitors (Walker-Simmons and Ryan 1984) and pathogenesis-related proteins (PR) (see Bol Chapter 6). The increases in enzyme or protein usually occur rapidly and result from transient increases in the concentration of the corresponding mRNAs. In bean, accumulation of mRNA for phenylalanine ammonia lyase (PAL), the first enzyme of the phenylpropanoid pathway, and mRNA for chalcose synthase (CHS), the first enzyme from the branch leading to isoflavonoid phytoalexins, occurs within 10 min of treatment of cells with an elicitor from *Colletotrichum lindemuthianum* (Edwards *et al*. 1985). A similar rise in concentration of mRNA for HRGP can be demonstrated though accumulation is less rapid but more prolonged, the concentration reaching maximum 24–26 h after elicitor treatment (Showalter *et al*. 1985). Such increases have been shown to result from increases in the transcriptional activity of PAL, CHS, and HRGP genes resulting in increased rates of synthesis of all three mRNAs (Cramer *et al*. 1985b; Lawton & Lamb 1987). Activation of the transcription rate of the PAL and CHS gene was demonstrated within 5 min of exposure of cells to elicitor, followed by *de novo* synthesis of the enzymes shortly after.

Signal transduction

Studies such as these indicate that activation of transcription is one of the later events in a chain that begins with perception of the pathogen or elicitor by the plant cell. That perception must involve a component that is characteristic of the microorganism, either because of the structural information it contains or because of its functional activity, for instance in releasing endogenous elicitors from the host. It must also involve binding of the elicitor molecule to a receptor in the host. Linking these two elements, elicitor binding and activation of gene transcription, must be a signal transduction system that is capable of carrying information specific to the defined response. Whatever that system is, the results concerning activation of transcription of the PAL and CHS genes indicate that it is a very rapid one.

Receptors

The location and nature of the receptor has consequences for the mechanism which must operate to link perception to activation of gene transcription. An assumption that has frequently been made, presumably because of the size of some of the elicitor molecules concerned, is that a receptor would be situated at the cell surface within the plasma membrane. Yoshikawa *et al*. (1983) used [^{3}H]mycolaminarin, a 1, 3-β-glucan that is a weak elicitor of phytoalexin synthesis, to determine the location of a receptor in soyabeans. Binding was detected in a membrane fraction enriched in plasma membrane, and could be prevented by pretreatment of the membranes with pronase. Their conclusion, that a proteinaceous receptor is located in the plasma membrane, was given support by the studies of Cosio *et al.* (1988). In that study a glucan fraction isolated from *Phytophthora megasperma f. sp. glycinea* and labelled with [^{125}I] was used to demonstrate binding of elicitor to the plasma membrane of protoplasts. Binding was found to be both saturable and reversible, and a high correlation was observed between binding of this elicitor and the ability to induce glyceollin accumulation.

In contrast to these results are the findings of Hadwiger *et al.* (1981) demonstrating that an externally applied chitosan elicitor, as well as wall components purified from the fungus *Fusarium solani*, can be detected in the cytoplasm of pea cells within 20 minutes of application. Such a demonstration of the transport of an elicitor across the plasma membrane raises the possibility of an internal receptor and a different form of signal transduction system (see below). It should be borne in mind, however, that few if any studies have been undertaken in which an elicitor molecule has been re-isolated from the cytoplasm of cells and been shown to be structurally and functionally identical to the elicitor that was applied originally, rather than being a degradation product of the original elicitor that has been internalized subsequent to its interaction with a receptor in the plasma membrane. Nor, unless it has been specifically demonstrated, can it be assumed that the structure of an exogenously applied elicitor remains unaltered prior to its interaction with the receptor site. Many elicitors used in such studies are carbohydrate or have a substantial carbohydrate content, and depend for their activity upon the carbohydrate moiety. Plant cells contain a number of hydrolases associated with the cell wall that are capable of degrading oligo- and polysaccharide structures (Nock and Smith 1987). Thus it may be that the component which is elicitor-active and which interacts with the receptor is a fragment derived from the elicitor originally applied.

Penetration of the elicitor-active component into the cell leads to the possibility of an internal receptor and Kendra *et al.* (1987) have demonstrated a direct interaction of the elicitor chitosan with DNA. In consequence, they have proposed that elicitation may occur *via* direct alterations in the DNA within chromatin. These authors have also suggested that such a direct interaction of elicitors with DNA may account for the observed activation of the same family of defence-related genes by a variety of different elicitors.

With the exception of a model in which the elicitor interacts directly with DNA, other models for gene activation must include a signal transduction system. In the case of effectors or signal molecules that bind with a receptor on the plasma membrane, for example the amine and peptide hormones of mammalian systems, binding leads to the generation of a second messenger that subsequently interacts with other elements of the cell. A slightly more subtle form of signal transduction operates in the case of the steroid hormones which cross the plasma membrane and interact with a receptor protein located in the cytoplasm. That receptor undergoes a change in conformation in response to binding of the signal molecule, and is transported to the nucleus where the complex interacts directly with the DNA to bring about an effect. In this case transduction of the signal occurs through the change in conformation and subsequent transport of the receptor.

Our own studies concerning signal transduction have been performed with tissues of *Medicago sativa* (lucerne) and the fungal pathogen *Verticillium albo-atrum* R & B, the causative agent of wilt disease. A degree of specificity exists in the interaction so that isolates of the fungus taken from lucerne can successfully infect a range of crops, but isolates from other crop plants are not pathogenic to lucerne (Isaac 1957). In the incompatible interaction the hypersensitive response is accompanied by synthesis of phytoalexins including medicarpin, a pterocarpan, and vestitol and sativan, both of which are isoflavanols (Kahn and Milton 1978). Normally neither medicarpin nor sativan can be detected in cell-suspension cultures of lucerne but their synthesis can be induced in response to an elicitor isolated from culture filtrates of *V. albo-atrum* (Smith *et al.* 1989a). Induction of phytoalexin synthesis is accompanied by, and dependent upon, a rise in the level of PAL, the enzyme catalysing the committed step on the phenylpropanoid pathway of phytoalexin synthesis. The interaction of the fungal elicitor with cell-suspension cultures of lucerne as determined by its effect on phytoalexin synthesis and PAL activity, has therefore been a useful system for studying the signal transduction system involved in phytoalexin accumulation.

A number of studies of plant–pathogen interactions have already been performed with the aim of establishing the role of second messengers in the induction process, and the results of some of them implied that the signal represented by the interaction of the elicitor with the receptor was transduced prior to activation of gene transcription. Several second messenger molecules have been identified and are therefore candidates for involvement in the present system. Amongst those molecules which frequently occur as second messengers in signal systems are adenosine 3',5'-cyclic monophosphate (cyclic AMP), Ca^{2+}, or diacylglycerols (DAG). The latter two are products of the same signal system (Fig. 16.1) and so may function in combination in a given response. In fact there is evidence that elements of the cyclic AMP system interact with those of the Ca^{2+}/DAG system and so features of all three may be found in a single signal system.

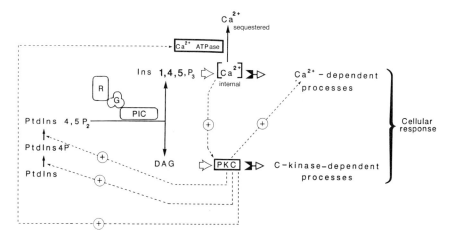

Fig. 16.1.   The phosphatidylinositide signal pathway. When an agonist binds to a receptor (R), it activates a G-protein (G) which simulates phosphoinositidase-C (PIC). Two intracellular signal molecules are generated from the hydrolysis of phosphatidylinositol-4, 5-bisphosphate (PtdIns 4, 5 P_2), diacylglycerol (DAG) and inositol 1, 4, 5-trisphosphate (Ins 1, 4, 5 P_3). Dashed lines represent the interaction of Ca^{2+} and protein kinase-C (PKC) with various elements of the system, + indicates a promotion. 'Ca^{2+} sequestered' refers to calcium ions which are removed to cellular compartments (vacuole, endoplasmic reticulum, mitochondria) and outside the cell *via* the activity of Ca^{2+} ATPases. See text for further details.

Cyclic AMP as an intracellular signal molecule

Despite prolonged debates concerning first its existence, then its possible function in plants, there is now unequivocal evidence that indicates the presence of a cyclic AMP system in plants (Newton and Brown 1987), including partial purification of an adenylate cyclase from lucerne (Carricarte *et al.* 1988), demonstration of phosphodiesterase activity specific to the 3',5'-cyclic isomer of adenosine monophosphate (Chiatante *et al.* 1987), identification of a cyclic AMP-dependent protein kinase (Janistyn 1988) and identification of cyclic AMP in a number of plant tissues including lucerne (Cooke *et al.* 1989).

Cyclic AMP has been implicated in the signal transduction process leading to phytoalexin synthesis by the demonstration that synthesis of terpenoid phytoalexins was induced in cells of sweet potato by cyclic AMP (Oguni *et al.* 1976). However, Hahn and Grisebach (1983), in their studies of the interaction between *Phytophthora megasperma* and soyabean could not detect cyclic AMP in cell-suspension cultures of the plant, and concluded therefore that it had no function in the induction process. In contrast to the findings from soyabean, our own studies with lucerne implicate cyclic AMP as a component of the signal mechanism and so support the results of Kurosaki *et al.* (1987a). These workers observed a rise in intracellular concentration of cyclic AMP that reached a maximum 30 minutes after addition of an elicitor prepared from cell walls of carrot to suspension-cultured carrot cells. Using a sensitive radioimmunoassay in our studies with lucerne suspension-cultured cells, cyclic AMP was detected at a concentration of 0.5 pmole g^{-1}F.wt. (Cooke *et al.* 1989). Identity of the cyclic AMP was confirmed by mass analysed ion kinetic energy mass spectroscopy. In response to treatment of cells with a crude elicitor preparation from culture filtrates of *V. albo-atrum* the intracellular cyclic AMP concentration increased 5-fold. The rise was both rapid, reaching a maximum within four minutes of the start of the treatment with elicitor, and transient, the concentration of cyclic AMP returning to the basal figure by 7 min after treatment began. The rapidity of the increase compares favourably to those responses in mammalian tissues induced by the fast-acting hormones and agonists and that feature cyclic AMP as second messenger. Such a response time would also be sufficient to account for the very early appearance of mRNA PAL transcripts observed in bean cells treated with elicitor (Lawton and Lamb 1987). The rather slower response observed in soyabean cells may result from a receptor that 'under the conditions of the experiment' is less accessible to the elicitor employed.

The responsiveness of both adenylate cyclase, the enzyme of synthesis, and phosphodiesterase, the enzyme of degradation, to challenge with elicitor, is consistent with a second-messenger role for cyclic AMP in the induction of phytoalexin accumulation in lucerne in response to the fungal elicitor. A rapid transient rise in adenylate cyclase activity was observed shortly after elicitor-treatment of lucerne cells, the kinetics of the response being similar to those of the changes in cyclic AMP concentration (Smith *et al.* unpublished). The activity of the phosphodiesterase was also increased in response to elicitor, though the rise was delayed relative to the increase in adenylate cyclase activity, beginning approximately 90 min after treatment was started and returning to basal activity only after several hours. In addition to these correlations between elicitor treatment and the effects upon the cyclic AMP system, it has been possible to demonstrate a direct effect of cyclic AMP upon both phytoalexin synthesis and PAL activity. Thus in cell-suspension cultures of lucerne in the absence of fungal elicitor, dibutyryl cyclic AMP, an analogue of cyclic AMP that can pass across membranes, induced a fivefold increase in PAL activity that was accompanied by phytoalexin accumulation (Cooke *et al.* 1989). In carrot cells treatment with either dibutyryl cyclic AMP or with cholera toxin, an agonist of adenylate cyclase which would be expected to lead to an increase in intracellular cyclic AMP concentration, induced synthesis of the phytoalexin 6-methoxymellein (Kurosaki *et al.* 1987a).

The role of Ca^{2+} in signal transduction

Many effectors bring about their cellular responses through the generation of elevated concentrations of intracellular Ca^{2+}, and in animal systems elevation of intracellular Ca^{2+} is a major second messenger operating as part of a signal transduction system (Berridge 1987). An essential feature of a system in which Ca^{2+} functions as a second messenger is a relatively low internal concentration in the unstimulated cell. In plants the Ca^{2+} concentration is normally maintained in the region of 100 nM primarily through the activity of Ca^{2+} ATPase pumps located in the plasma and tonoplast membranes, though significant ion pumping activity may also be found in the endoplasmic reticulum and mitochondria, and the pump from the endoplasmic reticulum has the highest affinity for Ca^{2+} (Hepler and Wayne 1985).

Whilst effects of Ca^{2+} upon the interaction of host and pathogen do not provide direct evidence of a second-messenger function for Ca^{2+}, resistance of barley to powdery mildew appears related to the intracellular Ca^{2+} concentration (Bayles and Aist 1987), and influx of Ca^{2+} in soyabean cells is necessary for the deposition of callose that occurs in response to treatment with a chitosan elicitor (Waldman *et al.* 1988). La^{3+} which competitively inhibits Ca^{2+} movement through plasma membrane channels, or the metal cation chelator EGTA can inhibit arachidonic acid-induced synthesis of the phytoalexin lubimin in potato, whilst increasing the exogenous Ca^{2+} concentration caused phytoalexin accumulation (Zook *et al.* 1987). Other studies that implicate an involvement of fluctuations in intracellular Ca^{2+} in the induction process include that of Kurosaki *et al.* (1987b) in which inhibition of the elicitor-induced synthesis of 6-methoxymellein was found to result from treatment of carrot cells with verapamil, an inhibitor of Ca^{2+} channels located in the plasma membranes. In contrast, treatment with the Ca^{2+} ionophore, A23187, stimulated phytoalexin accumulation in a dose-dependent manner, by facilitating the entry of exogenous Ca^{2+} into the cell. Stäb and Ebel (1987) demonstrated that antagonists of Ca^{2+} flux into cells inhibited the induction of both glyceollin synthesis and PAL activity normally observed in soyabean cells treated with a glucan elicitor from *Phytophthora megasperma*. However, amongst those results, which indicate a positive role for Ca^{2+} in mediation of elicitor-induced phytoalexin synthesis, it must be mentioned that Kendra and Hardwiger (1987) demonstrated that pisatin synthesis induced in pea tissue in response to treatment with either a chitosan elicitor or *Fusarium solani* is apparently independent of Ca^{2+}.

In lucerne cell cultures Ca^{2+} can also induce both an increase in PAL activity and phytoalexin accumulation, in a dose-dependent manner (Little 1989; Smith *et al.* 1989b). A seven to tenfold increase in PAL activity was observed at the optimal concentration of Ca^{2+} (3–4 mM), inhibition of the Ca^{2+} effect by the transcription inhibitor cordycepin indicating a requirement for synthesis of mRNA. A role for Ca^{2+} in mediating the synthesis of PAL induced by fungal elicitor is indicated by the fact that compounds antagonistic to the Ca^{2+} induced response are also antagonistic to the effect of fungal elicitor upon PAL activity and phytoalexin accumulation. Hence removal of Ca^{2+} by washing or EGTA treatment, treatment of cells with verapamil, or treatment with La^{3+} prevents induction brought about either by Ca^{2+} or fungal elicitor, and the increases in PAL activity or phytoalexin accumulation are not observed in cells treated thus. In contrast treatments which are expected to elevate intracellular Ca^{2+} enhance the response of lucerne cells to exogenous Ca^{2+} and to the fungal elicitor. For example, ionophore A23187 at 5 μM reduces the external concentration of Ca^{2+} required for the optimal increase in PAL activity to 1 mM or less, presumably by facilitating the Ca^{2+} flux across the plasma membrane. In the case of the fungal elicitor the concentration required to give the maximum increase in PAL activity in the presence of A23187 is only 10 per cent of that required in its absence.

Signal transduction through phosphatidylinositides

An involvement of Ca^{2+} flux in the induction process is indicated by such results. In animal systems Ca^{2+} mobilizing hormones and neurotransmitters generate messenger Ca^{2+} through the phosphatidylinositide signal transduction system outlined in Fig. 16.1. Receptor-mediated activation of phosphoinositidase C, an enzyme located in the plasma membrane, leads to hydrolysis of phosphatidylinsoitol-4,5-bisphosphate (PIP_2) and generation of two intracellular signal molecules, inositol-1,4,5-trisphosphate (IP_3) and 1,2-diacylglyccrol (DAG). IP_3 causes release of Ca^{2+} from internal sites (the endoplasmic reticulum) (Berridge 1987) but inositol-1,3,4,5-tetrakisphosphate, the phosphorylation product of IP_3 that has been identified in mammalian cells, has been reported to stimulate entry of Ca^{2+} across the plasma membrane (Irvine et al. 1988). DAG, the other product is capable of activating the membrane associated protein kinase C (Nishizuka 1986). Both Ca^{2+} and DAG function as signal transduction molecules through their subsequent effects upon cellular reactions. Ca^{2+} can trigger enzyme systems indirectly via calcium binding proteins such as calmodulin, or directly as it does in the case of calcium-dependent proteases (Marmé 1989; Suzuki 1987). A major route for its activity though is through its effect upon Ca^{2+}-dependent kinases that themselves affect metabolism by phosphorylation of enzymic and/or regulatory proteins (Marmé 1989). Two groups of such kinases are recognized, the Ca^{2+}-calmodulin dependent kinases (Blowers and Trewavas 1989), and the Ca^{2+} and phospholipid activated kinases (protein kinases C, Nishizuka 1986).

As described above, DAG has its effect by activating the protein kinases associated with the membrane, the protein kinases C. Since such kinases are also Ca^{2+} and phospholipid-activated in addition to the effect of DAG, they represent a convergence of the two elements of the signal pathway (Fig. 16.1). Once activated protein kinases C affect the activity of several enzymic systems.

There have been few studies directly related to the involvement of this signal transduction process in the activation of defence genes, though many elements of this system have been identified in plants. Thus PIP_2 and its precursors phosphatidylinositol (PI) and phosphatidylinositol-4-phosphate (PIP) have been identified in leaf and cell-suspension cultures (Boss 1989; Drøbak *et al.* 1988; Smith *et al.* unpublished); Ca^{2+}-dependent phosphoinositidase C systems are present in plants (McMurray and Irvine 1988), and Ca^{2+}-calmodulin dependent protein kinases and protein kinases similar to protein kinases C exist in plant tissues (Blowers and Trewavas 1989; Klucis and Polya 1988). In addition to the presence of these elements of the phosphoinositide signal transduction system IP_3 has been shown to cause rapid release of Ca^{2+} from microsomes in Zuccini (Drøbak and Ferguson 1985) and reduction of Ca^{2+} concentration in tonoplast vesicles from *Avena* (Schumaker and Sze 1987), indicating its potential for mediating Ca^{2+} mobilization in plants.

The studies that have been reported so far concerning involvement of the phosphoinositide transduction system and plant–pathogen interaction have largely been indirect. Kurosaki *et al.* (1987b), however, demonstrated that in carrot cells an elicitor of phytoalexin synthesis caused an increase in phospholipase C activity. At the same time increased breakdown of phosphatidylinositol lipids occurred and release of IP_3 was detected. In the same study 1-(5-isoquinolinesulphonyl)-2-methylpiperazine, a selective inhibitor of protein kinase C activity, was found to inhibit the ability of the elicitor to induce phytoalexin synthesis. Thus these studies implicate a protein kinase C-type activity mediating between the elicitor signal and synthesis of the phytoalexin. They further provide evidence that the elicitor can interact with a phosphoinositide transduction system to generate second messenger Ca^{2+}, via IP_3, and the second messenger DAG (the other product of the hydrolysis of PIP_2). Activation of the protein kinase C would occur through either or both of these second messengers. In addition Ca^{2+} could be acting on Ca^{2+}-calmodulin dependent kinase.

A number of observations that we have made appear to be consistent with a role for the phosphoinositide system in the induction of phytoalexin synthesis in lucerne tissues challenged with fungal elicitor. Hence 1-oleoyl-2-acetyl-glycerol, a synthetic diacylglycerol, at a concentration of 1 μM was found to stimulate both an increase in PAL activity and accumulation of phytoalexin to levels comparable with those achieved by the fungal elicitor alone (Little 1989). Such synthetic diacylglycerols which are generally accepted to mimic the effect of endogenous DAG arising from phosphoinositide hydrolysis, are known activators of protein kinase C (Nishizuka 1986). Similarly phorbol 12-myristate 13-acetate which is also a recognized activator of protein kinase C, was found to induce an increase in PAL activity, whilst compound 40/80, an antagonist of phospholipase C activity abolished the induction in PAL activity normally associated with treatment with fungal elicitor. Again the selective inhibitor of protein kinase C, ISMP completely inhibits the response of lucerne cells to elicitor.

In the operation of the phosphoinositide signal pathway protein kinase C is the focus of the two intracellular signals that are generated, Ca^{2+} and DAG (Fig. 16.1). The fact that compound 40/80 is an effective inhibitor of the elicitation of PAL by the fungal elicitor, combined with the fact that agonists and antagonists of protein kinase C are also agonists and antagonists of the elicitation by fungal elicitor, suggests that a signalling mechanism showing analogy to the agonist-induced phosphoinositide hydrolysis occurring in animal cells may be involved in elicitation of phytoalexin synthesis in lucerne.

Conclusion

Whilst studies of the intracellular signal transduction system operating in the induced synthesis of phytoalexins are still in their early stages, the results available give reason to continue examination of the potential role of Ca^{2+} and cyclic AMP, and the phosphatidylinositide system in the induction process. In lucerne a flux of Ca^{2+} into the cell is necessary for the induced increase in PAL activity and phytoalexin synthesis brought about by the fungal elicitor, but it is not known whether such flux is the source of the second messenger or whether it serves to refill internal pools of Ca^{2+} that are the true site of release of the second messenger Ca^{2+} (Taylor and Putney 1987). Internal fluxes of Ca^{2+} may be generated by IP_3 and fluxes across the plasma membrane by IP_4, both of which result from operation of the phosphatidylinositide signal transduction system. However, that sequence in which IP_3 is generated and releases Ca^{2+} from internal sites, in response to a pathogen or elicitor, has yet to be demonstrated unequivocally in plant tissues, though the effects of agonists and antagonists of protein kinase C upon the induction process in lucerne and carrot tissue, in addition to isolation of a C-type kinase, would indicate that particular element of the phosphatidylinositide signal system is present and plays a part. Significant progress is being made towards defining the metabolism of the phosphatidylinositides in plants (see for example Boss 1989; Drøbak *et al.* 1988), however, and our studies of the lucerne-*Vertillium* interaction have given indications that the system functions in signal transduction.

Many questions remain, not the least of which is the role played by cyclic AMP. Clearly there is evidence that implicates its involvement but its relationship both to the flux of Ca^{2+} that is required for induction and the actual mechanism of activation for gene transcription is a matter for future investigations. Interactions between the phosphatidylinositide and cyclic AMP signal transduction systems are recognized (Taylor and Putney 1987) so that down-regulation and 'cross-talk' between the two second messenger systems feature in a number of mammalian regulatory systems. This will certainly be a fruitful area for future research. Neither is it certain how the Ca^{2+} flux affects the rate of gene transcription. Whether it is by an effect upon the activity of protein kinases type C or through calmodulin directly as is the case in prolactin synthesis in cells of the pancreas (White and Bancroft 1987), or through the activity of Ca^{2+}-calmodulin dependent kinases must remain for the moment a matter of speculation.

References

Bailey, J.A. (1982). Physiological and biochemical events associated with the expression of resistance to disease. In *Active defence mechanisms in plants*, (ed. R. K. S. Wood), NATO ASI Series, pp. 39–65. Plenum Press, New York.

Bayles, C.J. and Aist, J.R. (1987). Apparent calcium mediation of resistance of an *ml-o* barley mutant to powdery mildew. *Physiological and Molecular Plant Pathology* **30**, 337–45.

Berridge, M.J. (1987). Inositol trisphosphate and diacylglycerol: two interacting second messengers. *Annual Review of Biochemistry* **56**, 159–93.

Blowers, D.P. and Trewavas, A.J. (1989). Second messengers: their existence and relationship to protein kinases. In *Second messengers in plant growth and development*, (ed. W. F. Boss and D. J. Morré), Plant Biology Vol. 6, pp. 1–28. Alan R. Liss, New York.

Boss, W. (1989). Phosphoinositide metabolism: its relationship to signal transduction in plants. In *Second messengers in plant growth and development* (ed. W. F. Boss and D. J. Morré), Plant Biology Vol. 6, pp. 29–56. Alan R. Liss, New York.

Carricarte, V.C., Bianchini, G.M., Muschietti, J.P., Tellez-Inon, M.T., Perticari, A., Torres, N., and Flawia, M.M. (1988). Adenylate cyclase activity in a higher plant, alfalfa (*Medicago sativa*). *Biochemical Journal* **249**, 807–11.

Chiatante, D., Newton, R.P., and Brown, E.G. (1987). Properties of a multifunctional 3',5'-cyclic nucleotide phosphodiesterase from *Lactuca* cotyledons: comparison with mammalian enzymes capable of hydrolysing pyrimidine cyclic nucleotides. *Phytochemistry* **26**, 1301–6.

Cooke, C.J., Newton, R.P., Smith, C.J., and Walton, T.J. (1989). Pathogenic elicitation of phytoalexin in lucerne tissue: Involvement of cyclic AMP in the intracellular mechanism. *Biochemical Society Transactions* **17**, 919–20.

Cosio, E.G., Popperl, H., Schmidt, W.D., and Ebel, J. (1988). High-affinity binding of fungal β-glucan fragments to soyabean (*Glycine max*. L) microsomal fractions and protoplasts. *European Journal of Biochemistry* **175**, 309–15.

Cramer, C.L., Bell, J.N., Ryder, T.B., Bailey, J.A., Schuch, W., Bolwell, G.P., Robbins, M.P., Dixon, R.A., and Lamb, C.J. (1985a). Co-ordinated synthesis of phytoalexin biosynthetic enzymes in biologically-stressed cells of bean (*Phaseolus vulgaris* L.). *EMBO Journal* **4**, 285–9.

Cramer, C.L., Ryder, T.B., Bell, J.N., and Lamb, C.J. (1985b). Rapid switching of plant gene expression by fungal elicitor. *Science* **227**, 1240–3.

Darvill, A.G. and Albersheim, P. (1984). Phytoalexins and their elicitors—a defence against microbial invasion in plants. *Annual Review of Plant Physiology* **35**, 243–75.

Degra, L., Salvi, G., Mariotti, D., DeLorenzo, D., and Cervone, F. (1988). A polygalacturonase-inhibiting protein in alfalfa callus cultures. *Journal of Plant Physiology* **133**, 364–6.

Dixon, R.A. (1986). The phytoalexin response: elicitation, signalling and control of host gene expression. *Biological Reviews* **61**, 239–91.

Dixon, R.A., Dey, P.M., and Lamb, C.J. (1983). Phytoalexins: enzymology and molecular biology. *Advances in Enzymology* **55**, 1–136.

Drøbak, B.K. and Ferguson, I.B. (1985). Release of Ca^{2+} from plant hypocotyl microsomes by inositol 1,4,5-trisphosphate. *Biochemical and Biophysical Research Communications* **130**, 1241–6.

Drøbak, B., Ferguson, I.B., Dawson, A., and Irvine, R.F. (1988). Inositol-containing lipids in suspension-cultured plant cells. An isotopic study. *Plant Physiology* **87**, 217–22.

Edwards, K., Cramer, C.L., Bolwell, G.P., Dixon, R.A., Schuch, W., and Lamb, C.J. (1985). Rapid transient induction of phenylalanine amonia-lyase mRNA in elicitor-treated bean cells. *Proceedings of the National Academy of Sciences USA* **82**, 6731–5.

Fritig, B., Kauffmann, S., Dumas, B., Geoffroy, P., Kopp, M., and Legrand, M. (1987). Mechanism of the hypersensitivity reaction of plants. In *Plant resistance to viruses*, Ciba Foundation Symposium 133, 92–108.

Hadwiger, L.A., Beckman, J.M., and Adams, M.J. (1981). Localization of fungal components in the pea-*Fusarium* interaction detected immunochemically with antichitosan and antifungal cell wall antisera. *Plant Physiology* **67**, 170–5.

Hahlbrock, K. and Scheel, D. (1989). Physiology and molecular biology of phenyl-propanoid metabolism. *Annual Review of Plant Physiology and Plant Molecular Biology* **40**, 346–69.

Hahn, M.G. and Grisebach, H. (1983). Cyclic AMP is not involved as a second messenger in the response of Soyabean to infection by *Phytophthora megasperma* f. sp. *glycinea*. *Zeitschrift fuer Naturforschung* **38c**, 578–82.

Hepler, P.K. and Wayne, R.O. (1985). Calcium and Plant development. *Annual Review of Plant Physiology* **36**, 297–349.

Irvine, R.F., Moor, R.M., Pollock, W.K., Smith, P.M., and Wregett, K.A. (1988). Inositol phosphates: proliferation, metabolism and function. *Philosophical Transactions of the Royal Society, London* **B320**, 281–98.

Isaac, I. (1957). Wilt of lucerne caused by species of *Verticillium*. *Annals of Applied Biology* **45**, 550–8.

Janistyn, B. (1988). Stimulation by manganese (II) sulphate of a cAMP-dependent protein kinase from *Zea mays* seedlings. *Phytochemistry* **27**, 2735–6.

Kendra, D.F. and Hadwiger, L.A. (1987). Calcium and calmodulin may not regulate the disease resistance and pisatin formation responses of *Pisum sativum* to chitosan or *Fusarium solani*. *Physiological and Molecular Plant Pathology* **31**, 337–48.

Kendra, D.F., Fritensky, B., Daniels, C.H., and Hadwiger, L.A. (1987). Disease resistance response genes in plants: expression and proposed mechanisms of induction. In *Molecular strategies for crop protection*, pp. 13–24. Alan R. Liss.

Khan, Z.F. and Milton, J.M. (1978). Phytoalexin production and the resistance of lucerne (*Medicago sativa* L.) to *Verticillium albo-atrum*. *Physiological Plant Pathology* **13**, 215–21.

Klucis, E. and Polya, G.M. (1988). Localization, solubilization and characterization of plant membrane-associated calcium-dependent protein kinases. *Plant Physiology* **88**, 164–71.

Kombrink, E., Schröder, M., and Hahlbrock, K. (1988). Several 'pathogenesis-related' proteins in potato are 1,3-β-glucanases and chitinases. *Proceedings of the National Academy of Sciences USA* **85**, 782–6.

Kurosaki, F., Tsurusawa, Y., and Nishi, A. (1987a). The elicitation of phytoalexins by Ca^{2+} and cyclic AMP in carrot cells. *Phytochemistry* **26**, 1919–23.

Kurosaki, F., Tsurusawa, Y., and Nishi, A. (1987b). Breakdown of phosphatidylinositol during the elicitation of phytoalexin production in cultured carrot cells. *Plant Physiology* **85**, 601–4.

Lawton, M.A. and Lamb, C.J. (1987). Transcriptional activation of plant defence genes by fungal elicitor, wounding and infection. *Molecular and Cellular Biology* **7**, 335–41.

Little, J.P. (1989). PhD thesis, University of Wales.

Mansfield, J. (1982). The role of phytoalexins in disease resistance. In *Phytoalexins* (ed. J. A. Bailey and J. W. Mansfield), p. 253–88. Blackie, Glasgow.

Marmé, D. (1989). The role of calcium and calmodulin in signal transduction. In *Second messengers in plant growth and development* (ed. W. F. Boss and D. J. Morré), Plant Biology Vol. 6, pp. 57–80. Alan R. Liss, New York.

McMurray, W.C. and Irvine, R.F. (1988). Phosphatidylinositol 4,5-bisphosphate phosphodiesterase in higher plants. *Biochemical Journal* **249**, 877–81.

Newton, R.P. and Brown, E.G. (1987). The biochemistry and physiology of cyclic AMP in higher plants. In *Hormones, Receptors and Cellular Interactions in Plants*, (ed. C. M. Chadwick and D. R. Garrod), p. 115–53. Cambridge University Press.

Nishizuka, Y. (1986). Studies and perspectives of protein kinase C. *Science* **233**, 305–12.

Nock, L.P. and Smith, C.J. (1987). Identification of polysaccharide hydrolases involved in autolytic degradation of *Zea* cell walls. *Plant Physiology* **84**, 1044–50.

Oguni, I., Suzuki, K., and Uritani, I. (1976). Terpenoid induction in Sweet Potato roots by Adenosine 3',5'-cyclic monophosphate. *Agricultural and Biological Chemistry* **40**, 1251–2.

Pearce, R.B. and Ride, J.P. (1982). Chitin and related compounds as elicitors of the lignification response in wounded wheat leaves. *Physiological Plant Pathology* **20**, 119–23.

Roby, D., Toppan, A., and Esquerre-Tugaye, M.T. (1985). Cell surfaces in plant-microorganism interaction. V. Elicitors of fungal and of plant origin trigger the synthesis of ethylene and of cell wall hydroxyproline-rich glycoprotein in plants. *Plant Physiology* **77**, 700–4.

Schumaker, K.S. and Sze, H. (1987). Inositol 1,4,5-trisphosphate releases Ca^{2+} from vacuolar membrane vesicles of oat roots. *Journal of Biological Chemistry* **262**, 3944–6.

Showalter, A.M., Bell, J.N., Cramer, C.L., Bailey, J.A., Varner, J.E., and Lamb, C.J. (1985). Accumulation of hydroxyproline-rich glycoprotein mRNAs in response to fungal elicitor and infection. *Proceedings of the National Academy of Sciences USA* **82**, 6551–5.

Smith, C.J., Little, J.P., and Milton, M. (1989a). Comparison of the effects of different isolates of *Verticillium* on phenylalanine ammonia lyase activity and phytoalexin biosynthesis in tissues of *Medicago sativa* L. *Biochemical Society Transactions* **17**, 199–200.

Smith, C.J., Newton, R.P., Mullins, C.J., and Walton, T.J. (1989b). Plant host-pathogen interaction: elicitation of phenylalanine ammonia lyase activity and its mediation by Ca^{2+}. *Biochemical Society Transactions* **16**, 1069–70.

Stäb, M.R. and Ebel, J. (1987). Effects of Ca^{2+} on phytoalexin induction by fungal elicitor in soyabean cells. *Archives of Biochemistry and Biophysics* **257**, 416–23.

Suzuki, K. (1987). Calcium activated neutral protease: domain structure and activity regulation. *Trends in Biochemical Sciences* **12**, 103–5.

Taylor, C.W. and Putney, J.W. (1987). Phosphoinositides and calcium signalling. In *Calcium and Cell Function*, Vol. VII, pp. 1–38. Academic Press, New York.

van Loon, L.C. (1985). Pathogenesis-related proteins. *Plant Molecular Biology* **4**, 111–16.

Waldman, T., Jeblisk, W., and Kauss, H. (1988). Induced net Ca^{2+} uptake and callose biosynthesis in suspension-cultured plant cells. *Planta* **173**, 88–95.

Walker-Simmons, M. and Ryan, C.A. (1984). Proteinase inhibitor synthesis in tomato leaves. Induction by chitosan oligomers and chemically modified chitosan and chitin. *Plant Physiology* **76**, 787–90.

White, B.A. and Bancroft, C. (1987). Regulation of gene expression by calcium. In *Calcium and Cell Function* Vol. VII, pp. 109–32. Academic Press, New York.

Yoshikawa, M., Keen, N.T. and Ming-Chang, W. (1983). A receptor on soyabean membranes for a fungal elicitor of phytoalexin accumulation. *Plant Physiology* **73**, 497–506.

Zook, M., Rush, J., and Kue, J. (1987). A role for Ca^{2+} in the elicitation of Rishitin and Lubimin accumulation in potato tuber tissue. *Plant Physiology* **84**, 520–5.

17 Transcription factors and defence gene activation

RICHARD A. DIXON, ARVIND D. CHOUDHARY, MARIA J. HARRISON, BRUCE A. STERMER, AND LLOYD YU

Plant Biology Division, The Samuel Roberts Noble Foundation, PO Box 2180, Ardmore, Oklahoma 73402, USA

SUSAN M. JENKINS, CHRISTOPHER J. LAMB, and MICHAEL A. LAWTON

Plant Biology Laboratory, The Salk Institute for Biological Studies, PO Box 85800, San Diego, California 92138, USA

Introduction

Plants possess both static and inducible defences against microbial pathogens. Examples of the latter type of defence include accumulation of low M_r antimicrobial compounds (phytoalexins), deposition of phenolic material and hydroxyproline-rich glycoproteins in the plant cell wall, and synthesis of hydrolytic enzymes such as chitinase and 1,3-β-D-glucanase. Genes encoding the proteins involved in these reponses, either directly or through their biosynthetic activity may be classified as defence response genes. Generally, synthesis of inducible antimicrobial barriers is rapidly activated in an incompatible host–pathogen interaction, and is believed to result from an interaction of elicitor molecules from the pathogen with specific receptors in the host. This response can be mimicked in suspension-cultured plant cells exposed to microbial elicitors. Much of this work has recently been reviewed (Dixon 1986; Dixon and Harrison 1989; Dixon *et al.* 1989; Lamb *et al.* 1989).

At the molecular level, expression of inducible plant resistance involves three main stages; elicitor binding, signal transduction and activation of defence-response gene. This article reviews recent work in our laboratories on factors which may be involved in the activation of phytoalexin biosynthesis at the gene level. An understanding of the DNA sequence requirements for activation of defence response gene promoters in response to elicitors or infection, and of the proteins that may bind to these sequences and bring about regulation, is important both fundamentally and for potential biotechnological applications. Transcriptional activation can be seen as the last stage in the signal transduction pathway that is initiated by interaction of elicitor with the host cell. Thus, a knowledge of the proteins which interact with gene regulatory sequences, and in particular, of how they interact in a stimulus-dependent manner, should be a valuable starting point to work back through the signal transduction pathway. In the applied context, inducible defence gene promoters potentially can be used for driving expression of novel gene constructs in attempts to engineer resistance in transgenic plants.

Molecular cloning of defence response genes in bean

Genomic and/or cDNA clones have been obtained for a number of inducible defence response genes from bean (*Phaseolus vulgaris* L.). These are summarized in Table 17.1. In the pathway leading to biosynthesis of isoflavonoid phytoalexins (Fig. 17.1), the key regulatory enzymes phenylalanine ammonia-lyase (PAL) and chalcone synthase (CHS) are encoded by multigene families of 3, and at least 6 members respectively (Cramer *et al.* 1989; Ryder *et al.* 1987). This arrangement results in the elicitor-induced appearance of a range of different subunit isoforms for the two enzymes (Bolwell *et al.* 1985; Ryder *et al.* 1987; Hamdan and Dixon 1987). In the case of PAL, these are represented by enzymically active tetramers with different Km values, elicitor preferentially inducing the lower Km forms (Bolwell *et al.* 1985). The function of the different forms of CHS is not yet clear; although the genes are activated differentially in response to elicitor, infection, wounding, or developmental cues, their coding sequences are much more homologous than those of the PAL family. It is possible that the presence of multiple CHS genes simply reflects requirements for extra enzymic capacity, or increased regulatory flexibility, that results from the role of this enzyme in defence (as well as in normal developmentally controlled flavonoid synthesis) in legumes. In contrast to this situation, the enzyme immediately following CHS, chalcone isomerase (CHI), is encoded by a single gene (Mehdy and Lamb 1987). The presence of a single elicitor-induced gene has also been shown for coniferyl alcohol dehydrogenase (CAD), which catalyzes the final step in the synthesis of the phenylpropanoid building blocks of lignin (Walter *et al.* 1988).

Details of the structure and organization of other defence response genes from bean and other sources have recently been reviewed (Dixon and Harrison 1989).

Table 17.1 Cloned defense response genes/cDNAs from bean

Gene encoding	Vector/Clone	Comments	Reference
Phenylalanine ammonia-lyase	pAT153/cDNA	One cDNA sequence from elicitor-induced cell suspension cultures.	Edwards *et al*. 1985
	λgt WES, λ1059/genomic	Family of 3 divergent genes; differential regulation.	Cramer *et al*. 1989
Chalcone synthase	pBR325/cDNA λ1059/genomic	From elicitor-induced cell suspension cultures. Family of 6-8 genes, some linked; differential regulation.	Ryder *et al*. 1984 Ryder *et al*. 1987
Chalcone isomerase	λgtll/cDNA	One cDNA sequence from elicitor-induced cell suspension cultures.	Mehdy and Lamb 1987
Cinnamyl alcohol dehydrogenase	λgtll/cDNA	Single cDNA from elicitor-induced cell suspension cultures; second gene involved in xylogenesis?	Walter *et al*. 1988
Hydroxyproline-rich glyco-protein(s)	pUC19/cDNA	Three different cDNAs corresponding to different length transcripts. 3 genes, differential induction.	Corbin *et al*. 1987
	pUC19/cDNA	Wound-induced cDNA, single gene (different from above).	Sauer *et al*. 1989
Chitinase	λgtll/cDNA λEMBL 4/ genomic	cDNAs from ethylene-treated leaves. Approx. 4 genes, at least 2 induced by ethylene.	Broglie *et al*. 1986
	λgtll/cDNA	cDNA for message from elicitor-induced cell suspension cultures.	Hendrick *et al*. 1988
1, 3-β-D-glucanase	pBR322/cDNA pUC19/cDNA	Putative cDNA for ethylene–induced message. cDNA for message from elicitor-induced cell suspension cultures.	Vogeli *et al*. 1988. B. Edington and R.A. Dixon, unpublished.

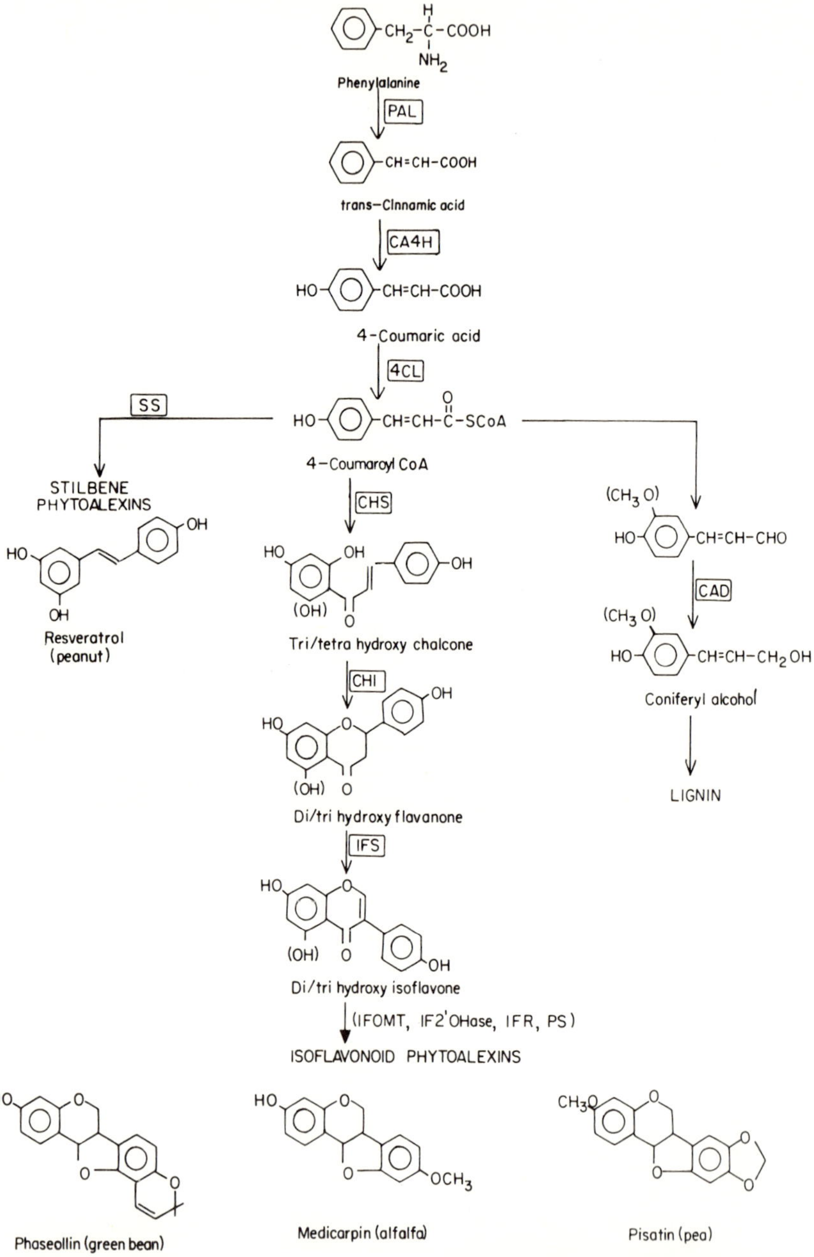

PAL = phenylanine ammonia-lyase; CA4H = cinnamic acid 4-hydroxylase; 4CL = 4-coumarate: CoA ligase; SS = stilbene synthase; CAD = coniferyl alcohol dehydrogenase; CHS = chalcone synthase; CHI = chalcone isomerase; IFS = isoflavone synthase; IFOMT = isoflavone O-methyl transferase; IF2'OHase = isoflavone 2'-hydroxylase; IFR = isoflavone reductase. PS = pterocarpan synthase.

Fig. 17.1.   The phenylpropanoid pathway leading to defence metabolites in bean.

Increased gene transcription underlies induced defence in bean cells

Elicitor-mediated induction of the bean PAL, CHS, CHI, CAD, chitinase, glucanase, and HRGP products is preceded by rapid but transient increases in the steady-state levels of their corresponding mRNAs. This situation results from increased rates of transcription (Lawton and Lamb 1987). Nuclear run-off transcript analysis with isolated nuclei has shown that transcriptional activation of PAL and CHS genes occurs within less than 5 min of exposure of cells to elicitor, thus revealing a rapid signal transduction chain. Measured at the steady state mRNA level, CAD induction is even more rapid than that of PAL and CHS (Walter *et al*. 1988). In contrast, there is a lag of around 2 h following elicitor addition before increased HRGP gene transcripts can be measured. This suggests that HRGP genes are in a different regulatory loop from the genes involved in phenolic biosynthesis. A functional comparison of the promoter sequences of the HRGP genes with those of genes such as PAL and CHS should lead to an understanding of these differences at the molecular level.

Functional analysis of the bean chalcone synthase promoter

Promoter regions for bean PAL (gPAL 2, Cramer *et al*. 1989) and CHS genes are currently being analysed. Figure 17.2 shows the sequence of the 5' upstream region of CHSλ15, a gene which is inducible by elicitor, UV light or infection. The CHSλ15 promoter has been fused to the *E. coli* β-glucuronidase (GUS) reporter gene, and introduced into tobacco by *Agrobacterium*-mediated transformation. The 336 bp upstream region shown in Fig. 17.2 is sufficient to render the GUS gene inducible by elicitors or UV light (Fig. 17.3). Furthermore, this sequence confers elicitor-inducibility on the bacterial chloramphenicol acetyltransferase (CAT) gene when introduced as a CHS 5'-CAT-NOS 3' fusion into tobacco, soybean or alfalfa protoplasts by electroporation (Dron *et al*. 1988; Choudhary and Dixon, unpublished). Analysis of the effects of deletions in the CHSλ15 promoter on its ability to drive CAT expression in electroporated soybean protoplasts has revealed the presence of a putative silencer region between positions −173 and −336, and a region absolutely required for elicitor inducibility between −130 and the TATA box. The effect of deleting the putative silencer region is to increase expression approximately twofold (Dron *et al*. 1988), and a similar increase is observed if the −173 to −336 region is co-electroporated *in trans* with the intact CHS promoter-CAT construct, presumably as a result of competition for negative regulatory proteins. In contrast, in alfalfa protoplasts, expression from the promoter

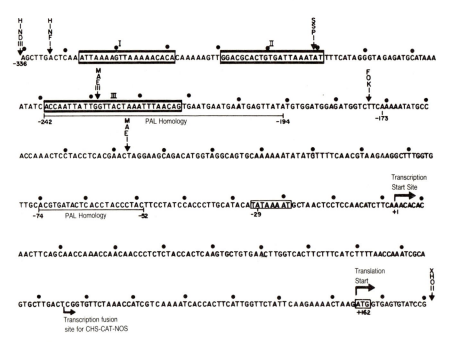

Fig. 17.2. Nucleic acid sequence of the bean chalcone synthase promoter. See text for explanation of boxes and homology regions.

deleted to −173 is approximately twofold lower than that from the −336 promoter, although the construct still retains elicitor inducibility. This suggests that the −173 to −336 region can act as an enhancer in alfalfa cells. Preliminary results of the expression of promoter deletions in transgenic tobacco are consistent with the idea that the sequences acting as a silencer in soybean protoplasts may in addition have cell-type-specific positive regulatory features.

The CHSλ15 promoter contains two regions of homology to the bean gPAL2 5' upstream region −242 to −194, and −74 to −52. These map within the putative silencer/enhancer and activator regions respectively. Furthermore, use of *in vivo* genomic footprinting has recently revealed a region in a parsley PAL promoter that appears to be a potential protein-binding site related to the elicitor response (Lois *et al.* 1989); the same sequences are found within the −74 to −52 region of the CHS 15 promoter. A sequence common to the promoters of a number of light-inducible genes (the so-called G-box, Giuliano *et al.* 1988) is also found overlapping the 5' end of the −74 to −52 region.

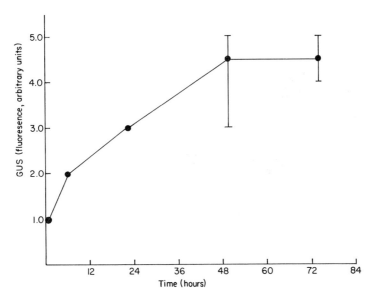

Fig. 17.3. Expression of the bean CHS λ 15 promoter (fused to the *E. Coli* β-glucuronidase gene) in leaves of transgenic tobacco plants after exposure to a 2 min pulse of UV (254 nm) irradiation.

In vitro binding studies confirm and define regulatory sites in the chalcone synthase promoter

A region of the CHS promoter, referred to as Mae 1–1, was obtained by cutting with enzymes Hind III (−332) and Mae 1 (−141). This fragment, which contains the above defined silencer region, was labelled with ^{32}P and was used as probe in gel-retardation experiments. Migration of Mae 1–1 in polyacrylamide gels was totally retarded if the probe was mixed with a protein extract from bean suspension-cell nuclei, bean leaf nuclei (Fig. 17. 4A) or alfalfa suspension-cell nuclei (data not shown). The exact size of the retarded complex depended upon the source of the nuclear extract. Binding activity was destroyed by pre-incubation of nuclear extracts with proteinase K, but not with DNase or RNase. Binding was reversible, could be competed by excess unlabelled Mae 1–1 or gPAL 2 promoter sequences (which contain homology to the −242 to −194 region included in Mae 1–1), but could not be competed by CHS cDNA coding sequences.

DNase 1 footprinting of the Mae 1–1 fragment *in vitro* revealed three sites at which potential binding of bean nuclear proteins could inhibit accessibility to the nuclease (M.A. Lawton, in preparation; Harrison *et al.* 1989). These sites (I–III) are shown as boxes in Fig. 17.2. Site III, which

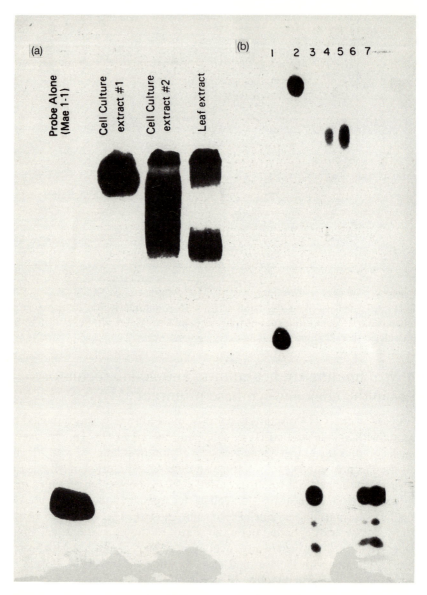

Fig. 17.4. Gel retardation of CHS promoter fragments by bean nuclear protein extracts. A. Retardation of Mae 1–1 by two extracts from bean cell cultures and an extract from leaves. B. Track 1, Mae 1–1 alone. Track 2, Mae 1–1 + cell suspension nuclear extract (NE). Track 3, end-filled site III oligo alone (see text). Tracks 4 and 5, end-filled site III oligo + NE. Tracks 6 + 7 end-filled site III oligo plus nuclear extract from bean leaves.

appeared to exhibit the strongest binding to bean nuclear proteins, lay within the −242 to −194 region with homology to the PAL promoter. Migration of a 33 bp double-stranded oligonucleotide complementary to site III was retarded by nuclear extracts from suspension-cultured bean or alfalfa cells, but not from bean leaves, in gel-retardation assays (Fig. 17. 4B). A ligated multimer of the site III oligonucleotide totally competed binding of bean nuclear extract to labelled Mae 1–1, and was itself retarded to a single position in gel-retardation assays. This suggests that site III is the major protein binding site within the functionally-defined silencer region.

The 33 bp site III oligonucleotide was modified to contain bromodeoxyuridine residues in place of thymine. The modified oligonucleotide was then mixed with nuclear extract from cultured bean cells and irradiated with UV light. Analysis of the UV-cross-linked mixture by SDS-PAGE revealed a protein–DNA complex of M_r approximately 90 000, which may therefore indicate the size of the subunit of the protein that binds to the site III region.

Gel-retardation assays have also revealed weak binding of bean nuclear proteins within the region of the bean CHSλ15 promoter that is believed to be necessary for elicitor activation. Attempts are currently being made to characterize these and other binding sites better by *in vivo* genomic footprinting (Saluz and Jost 1989) and more sensitive methods of *in vitro* footprinting (Chalepakis and Beato 1989). It is important in such studies to be able to relate putative protein binding sites with functionality *in vivo*, and for this reason more detailed sets of deletions are currently being examined both in transgenic plants and in transient assays utilizing electroporated protoplasts.

Strategies for the characterization and cloning of plant defence gene transcription factors

A number of DNA-binding transcription factors have been isolated and/or cloned from animal cells, either by classical biochemical purification methods, or by screening cDNA expression libraries with recognition site DNA sequences (Singh *et al.* 1989). These proteins are often modular, containing separate domains for DNA binding and functional activity. The main structural classes of DNA-binding protein are zinc finger, helix-turn helix, leucine zipper, and POU protein (Evans and Hollenberg 1988; Landschulz *et al.* 1988; Robertson 1988). To date, only a single example of the cloning of a plant DNA-binding protein has been reported. A factor from tobacco that binds to a functionally-defined regulatory region in the cauliflower mosaic virus 35S promoter was cloned by expression library screening (Lam *et al.* 1989); it appears to have similarity to the leucine zipper class of *trans*-acting factor (E. Lam, personal communication). Two main reasons probably underlie the apparent lack of progress in cloning or purifying plant *trans*-acting factors. Firstly, the lack of an *in vitro* transcription system for plants makes functional analysis of putative binding sites both time-consuming and subject to variability, especially if analysis relies on expression from deleted promoter constructs in transgenic plants. Secondly, levels of these proteins per gram tissue will probably be much lower than in animal cells, and large scale purifications from isolated nuclei will be required.

The CHSλ15 site III oligonucleotide has been immobilized as an affinity matrix to use in the purification of the bean protein(s) that bind to the silencer region. Binding activity to Mae 1–1 was retained when bean cell-culture nuclear extract was passed through heparin agarose, and could be selectively eluted from the site-specific affinity column. Problems of scale-up must now be addressed before the protein can be purified in sufficient quantity for raising of antibodies or N-terminal micro-sequencing.

A bean λgt11 cDNA expression library, constructed from mRNA from elicitor-induced cell cultures, has been screened with ligated site III oligonucleotide according to published procedures (Singh *et al.* 1989). One consistently positive clone (12–2) from the 2.5×10^5 recombinants that were screened was plaque purified. Lysogens were produced, and lysates were tested for their ability to bind specifically Mae 1–1 (Fig. 17.5) or site III in gel-retardation assays. Specific binding was inferred from data indicating lack of competition with high concentrations of poly dI:dC or CHS cDNA, lack of binding to CHS cDNA, efficient competition by unlabelled Mae 1–1 or site III, and total loss of binding to Mae 1–1 if the fragment was cut with Mae III near the centre of site III. However, nucleotide sequence analysis of the lac Z fusion protein contained in this clone revealed that the binding was artefactual, and some apparently non-specific competition was observed in gel-retardation assays (note the effect of plasmid pBR322 DNA in Fig. 17.5). The potential lack of reliability of gel-retardation assays is therefore a source of concern, especially as the binding shown in Fig. 17.5 appears so strong. This result highlights the need to define the exact contact nucleotides within a binding site in order to design mutant non-binding oligonucleotides as controls for library screening. It may also be necessary to determine optimal conditions for expression of binding protein synthesis and accumulation in cultured cells in order to produce enriched cDNA libraries.

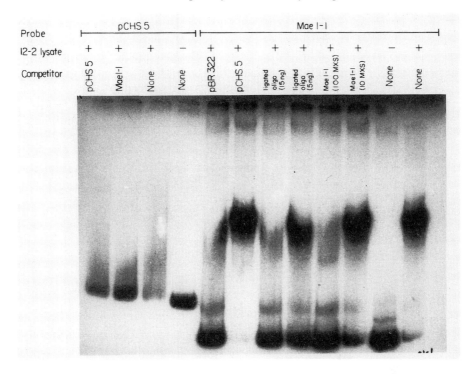

Fig. 17.5. Gel retardation of the CHS promoter Mae 1–1 fragment, or the pCHS5 cDNA sequence, by lysate from a gtll lysogen (12–2) selected by screening an expression library with ligated CHS promoter site III (ligated oligo, see text). Competition was by 100 molar excess (MXS) competitor DNA except where stated.

Implications for the co-ordinate regulation of gene expression during plant defence and development

In addition to its role in the phytoalexin defence response, transcriptional regulation of bean CHS genes is also associated with a number of developmentally regulated processes. These include flower pigmentation, production of protectants against UV radiation in young tissues, involvement in regulation of auxin transport, and synthesis of signal molecules for induction of the nodulation genes of *Rhizobium* (see Dixon and Lamb 1989 for references). In order to carry out these diverse functions, CHS genes must be under complex temporal and cell-type specific regulation. It is possible that the CHS promoter region that has been identified as a silencer in assays utilizing protoplasts isolated from apparently de-differentiated cell cultures may play an important role in the developmental regulation of this gene in the intact plant. This would then imply that potential strategies for manipulating the extent of defence response expression by engineering of chimaeric genes that contain modified enhancer functions, or that lack putative silencer regions, may also affect expression of the gene under non-stress conditions. It is hoped that further analyses of phenylpropanoid biosynthetic gene promoters in transgenic plants will help to unravel the complexity of the signal transduction loops that are presumed to interact in the programming of environmental and developmental responses. The next few years should see an increase in the number of defence gene promoters analysed, and hopefully a picture will emerge of the sequence elements necessary for co-ordinate induction of sets of biosynthetically related genes. Manipulation of these sequences may help in the orchestration of novel engineered defences.

Acknowledgement

We thank Scotty McGill for preparation of the manuscript.

References

Bolwell, G.P., Bell, J.N., Cramer, C.L., Schuch, W., Lamb, C.J., and Dixon, R.A. (1985). Phenylalanine ammonia-lyase from *Phaseolus vulgaris*: characterization and differential induction of multiple forms from elicitor-treated cell suspension cultures. *European Journal of Biochemistry* **149**, 411–19.

Broglie, K.E., Gaynor, J.J., and Broglie, R.M. (1986). Ethylene-regulated gene expression: molecular cloning of the genes encoding an endochitinase from *Phaseolus vulgaris*. *Proceedings of the National Academy of Sciences USA* **83**, 6820–4.

Chalepakis, G. and Beato, M. (1989). Hydroxyl radical interference: a new method for the study of protein-DNA interactions. *Nucleic Acids Research* **17**, 1783.

Corbin, D.R., Sauer, N., and Lamb, C.J. (1987). Differential regulation of a hydroxyproline-rich glycoprotein gene family in wounded and infected plants. *Molecular and Cellular Biology* **7**, 6337–44.

Cramer, C.L., Edwards, K., Dron, M., Liang, X., Dildine, S.L., Bolwell, G.P., Dixon, R.A., Lamb, C.J., and Schuch, W. (1989). Phenylalanine ammonia-lyase gene organization and structure. *Plant Molecular Biology* **12**, 367–83.

Dixon, R.A. (1986). The phytoalexin response: elicitation, signalling and the control of host gene expression. *Biological Reviews* **61**, 239–91.

Dixon, R.A. and Harrison, M.J. (1990). Activation, structure and organization of genes involved in microbial defence in plants. *Advances in Genetics* **28**, 165–234.

Dixon, R.A. and Lamb, C.J. (1990). Regulation of secondary metabolism at the biochemical and genetic levels. *Annual Proceedings of Phytochemical Society of Europe* **30**, 101–16.

Dixon, R.A., Blyden, E.R., and Ellis, J.A. (1990). Biochemistry and molecular genetics of plant–pathogen systems. In *Biochemical aspects of crop improvement*, (ed. K.R. Khanna). CRC Press, Boca Raton. In press.

Dron, M., Clouse, S.D., Lawton, M.A., Dixon, R.A., and Lamb, C.J. (1988). Glutathione and fungal elicitor regulation of a plant defense gene promoter in electroporated protoplasts. *Proceedings of the National Academy of Sciences USA* **85**, 6738–42.

Edwards, K., Cramer, C.L., Bolwell, G.P., Dixon, R.A., Schuch, W., and Lamb, C.J. (1985). Rapid transient induction of phenylalanine ammonia-lyase mRNA in elicitor-treated bean cells. *Proceedings of the National Academy of Sciences USA* **82**, 6731–5.

Evans, R.M. and Hollenberg, S.M. (1988). Zinc fingers: gilt by association. *Cell* **52**, 1–3.

Giuliano, G., Pichersky, E., Makik, V.S., Timko, M.P., Scilnik, P.A., and Cashmore, A.R. (1988). An evolutionarily conserved protein binding sequence upstream of a plant light-regulated gene. *Proceedings of the National Academy of Sciences USA* **85**, 7089–93.

Hamdan, M.A.M.S. and Dixon, R.A. (1987). Differential patterns of protein synthesis in bean cells exposed to elicitor fractions from *Colletotrichum lindemuthianum*. *Physiological and Molecular Plant Pathology* **31**, 105–21.

Harrison, M.J., Lawton, M.A., Lamb, C.J., and Dixon, R.A. (1989). Isolation and characterization of nuclear proteins binding to a silencer region in the promoter of a bean chalcone synthase gene. *Journal of Cellular Biochemistry* Suppl. **13D**, 236.

Hedrick, S.A., Bell, J.N., Boller, T., and Lamb, C.J. (1988). Chitinase cDNA cloning and mRNA induction by fungal elicitor, wounding and infection. *Plant Physiology* **86**, 182–6.

Lam, E., Benfey, P.N., Gilmartin, P.M., Katagiri, F., and Chua, N-H. (1989). *In vitro* and *in vivo* characterization of *trans*-acting factors which bind to the CaMV 35S promoter. *Journal of Cellular Biochemistry* Suppl. **13D**, 236.

Lamb, C.J., Lawton, M.A., Dron, M., and Dixon, R.A. (1989). Signals and transduction mechanisms for activation of plant defenses against microbial attack. *Cell* **56**, 215–24.

Landschulz, W.H., Johnson, P.F., and McKnight, S.L. (1988). The leucine zipper: a hypothetical structure common to a new class of DNA binding protein. *Science* **240**, 1759–64.

Lawton, M.A. and Lamb, C.J. (1987). Transcriptional activation of plant defense genes by fungal elicitor, wounding and infection. *Molecular and Cellular Biology* **7**, 335–41.

Lois, R., Dietrich, A., and Hahlbrock, K. (1989). A phenylalanine ammonia-lyase gene from parsley: structure, regulation and identification of elicitor and light responsive *cis*-acting elements. *EMBO Journal* **8**, 1641–8.

Mehdy, M. and Lamb, C.J. (1987). Chalcone isomerase cDNA cloning and mRNA induction by fungal elicitor, wounding and infection. *EMBO Journal* **6**, 1527–33.

Robertson, M. (1988). Homoeo boxes, POU proteins and the limits to promiscuity. *Nature* **336**, 522–4.

Ryder, T.B., Cramer, C.L., Bell, J.N., Robbins, M.P., Dixon, R.A., and Lamb, C.J. (1984). Elicitor rapidly induces chalcone synthase mRNA in *Phaseolus vulgaris* cells at the onset of the phytoalexin response. *Proceedings of the National Academy of Sciences USA* **81**, 5724–8.

Ryder, T.B., Hedrick, S.A., Bell, J.N., Liang, X., Clouse, S.D., and Lamb, C.J. (1987). Organization and differential activation of a gene family encoding the plant defense enzyme chalcone synthase in *Phaseolus vulgaris*. *Molecular and General Genetics* **210**, 219–33.

Saluz, H.P. and Jost, J.P. (1989). Genomic sequencing and *in vivo* footprinting. *Analytical Biochemistry* **176**, 201–8.

Sauer, N., Corbin, D.R., Keller, B., and Lamb, C.J. (1990). Cloning and characterization of a wound-specific hydroxyproline-rich glycoprotein in *Phaseolus vulgaris*. *Plant, Cell and Environment* **13**, 257–66.

Singh, H., Clerc, R.G., and Le Bowitz, J.H. (1989). Molecular cloning of sequence-specific DNA binding proteins using recognition site probes. *Biotechniques* **7**, 252–61.

Vogeli, U., Meins, Jr., F., and Boller, T. (1988). Co-ordinated regulation of chitinase and β-1,3-glucanase in bean leaves. *Planta* **174**, 364–72.

Walter, M.H., Grima-Pettenati, J., Grand, C., Boudet, A.M., and Lamb, C.J. (1988). Cinnamyl alcohol dehydrogenase; cDNA cloning and mRNA induction by fungal elicitor. *Proceedings of National Academy of Sciences USA* **85**, 5546–50.

Index